MATH
Connections®

A Secondary Mathematics Core Curriculum

William P. Berlinghoff
Clifford Sloyer
Robert W. Hayden
Eric F. Wood

Published by IT'S ABOUT TIME, Inc. © 2000 MATHconx, LLC

MATH
IT b5
2000
10
C9
V.2

Published in 2000 by
It's About Time, Inc.
84 Business Park Drive
Armonk, NY 10504
Phone (914)273-2233
Fax (914)273-2227
www.ITS-ABOUT-TIME.com
www.mathconnections.com

Publisher
Laurie Kreindler
Design
John Nordland
Production Manager
Barbara Zahm
Studio Manager
Leslie Jander

MATH *Connections*®: *A Secondary Mathematics Core Curriculum* was developed under the National Science Foundation Grant
No. ESI-9255251 awarded to the Connecticut Business and Industry Association.

ISBN 1-891629-85-9   MATH *Connections* Teacher Edition Book   2b   (Soft Cover)
ISBN 1-891629-89-1   MATH *Connections* Teacher Edition Books 2a, 2b Set   (Soft Cover)

ISBN 1-891629-21-2   MATH *Connections* Teacher Edition Book   2b   (3-Ring Binder)
ISBN 1-891629-79-4   MATH *Connections* Teacher Edition Books 2a, 2b Set   (3-Ring Binder)

2 3 4 5 D 02 01 00 99

This project was supported, in part,
by the
National Science Foundation
Opinions expressed are those of the authors
and not necessarily those of the Foundation.

Table of Contents

MATH *Connections* **Team** v

Welcome to MATH *Connections* vii

The History viii

The Vision and The Mission ix

Books in Brief x

Appendices xii

The Features xiii

Cooperative Learning xvii

Elements xviii

Pacing xx

Materials Included With Teacher Edition xxii

Ancillary Materials xxii

NCTM Standards Correlation xxiii

Planning Guide Book 2b xxiv

Chapter 4 Circles and Disks

Planning Guide T-2
Observations T-3
Profile Commentary T-4
4.1 Points of View T-7
      Problem Set T-25
4.2 Points on Circles T-35
      Problem Set T-51
4.3 Drawing Circles With a Graphing Calculator T-59
      Problem Set T-85
4.4 Area and Circumference T-91
      Problem Set T-115
4.5 Pieces of Circles T-123
      Problem Set T-143
4.6 Inscribed Angles T-149
      Problem Set T-161
4.7 Coming Around Full Circle T-165
      Problem Set T-173

Chapter 5 Shapes in Space

Planning Guide T-178
Observations T-179
Profile Commentary T-180
5.1 Another Dimension T-183
      Problem Set T-207
5.2 Three-Space in Layers T-215
      Problem Set T-239

Published by IT'S ABOUT TIME, Inc. © 2000 MATHconx, LLC

5.3 Finding Volumes by Stacking T-249
       Problem Set. T-263
5.4 Finding Volumes of Spheres, Cones and Pyramids. . . T-269
       Problem Set. T-289
5.5 Turning Around an Axis. T-297
       Problem Set. T-321
5.6 Space by the Numbers T-329
       Problem Set. T-343
5.7 Stacks of Lines T-349
       Problem Set. T-361
5.8 The Algebra of Three Dimensional Shapes T-365
       Problem Set. T-385
5.9 Time and Other Dimensions T-393
       Problem Set. T-411

Chapter 6 Linear Algebra and Matrices

Planning Guide T416
Observations. T-417
Profile Commentary T-418
6.1 Systems of Equations and Graphs. T-421
       Problem Set. T-441
6.2 Consistent and Inconsistent Systems. T-445
       Problem Set. T-455
6.3 Solving Systems of Equations T-459
       Problem Set. T-477
6.4 Solving Linear Systems Using Matrices T-481
       Problem Set. T-503
6.5 Matrix Operations T-507
       Problem Set. T-537

MATH *Connections®* Team

PRINCIPAL INVESTIGATORS

June G. Ellis, Director

Robert A. Rosenbaum
Wesleyan University

Robert J. Decker
University of Hartford

ADVISORY COUNCIL

James Aiello
Oakton Consulting Group, VA

Laurie Boswell
Profile High School, NH

Glenn Cassis
Connecticut Pre-Engineering Program

Daniel Dolan
Project to Improve Mastery in
Mathematics & Science

John Georges
Trinity College, CT

Renee Henry (retired)
Florida State Department of
Education

James Hogan Jr.
Connecticut Chapter of the National
Technology Association

Lauren Weisberg Kaufman
CBIA Education Foundation

James Landwehr
AT&T Bell Laboratories, NJ

Donald LaSalle
Talcott Mountain Science Center

Daniel Lawler (retired)
Hartford Public Schools

Steven Leinwand
Connecticut Department of Education

Valerie Lewis
Connecticut Department of Higher
Education

William Masalski
ex officio
University of Massachusetts

Gail Nordmoe
Cambridge Public Schools, MA

Thomas Romberg
Wisconsin Center for Educational
Research

Kenneth Sherrick
Berlin High School

Albert Shulte
Oakland Public Schools, MI

Irvin Vance
Michigan State University

Cecilia Welna
University of Hartford

SENIOR WRITERS

William P. Berlinghoff
Colby College, Maine

Clifford Sloyer
University of Delaware

Robert W. Hayden
Plymouth State College,
New Hampshire

Eric F. Wood
University of Western
Ontario, Canada

STAFF

Robert Gregorski
Associate Director

Lorna Rojan
Program Manager

Carolyn Mitchell
Administrative Assistant

CONTRIBUTORS

Don Hastings (retired)
Stratford Public Schools

Kathleen Bavelas
Manchester Community-
Technical College

George Parker
E. O. Smith High School, Storrs

Linda Raffles
Glastonbury High School

Joanna Shrader Panning
Middletown High School

Frank Corbo
Staples High School, Westport

Thomas Alena
Talcott Mountain Science Center,
Avon

William Casey
Bulkeley High School, Hartford

Sharon Heyman
Bulkeley High School, Hartford

Helen Knudson
Choate Rosemary Hall,
Wallingford

Mary Jo Lane (retired)
Granby Memorial High School

Lori White Moroso
Beth Chana Academy for Girls,
Orange

John Pellino
Talcott Mountain Science Center,
Avon

Pedro Vasquez, Jr.
Multicultural Magnet School,
Bridgeport

Thomas Willmitch
Talcott Mountain Science Center,
Avon

Leslie Paoletti
Greenwich Public Schools

Robert Fallon (retired)
Bristol Eastern High School

ASSESSMENT SPECIALIST

Don Hastings (retired)
Stratford Public Schools

Published by IT'S ABOUT TIME, Inc. © 2000 MATHconx, LLC

NSF ADVISORY BOARD

Margaret Cozzens
J. "Spud" Bradley
Eric Robinson
Emma Owens

FIELD-TEST SITES

Hill Career Magnet High School, *New Haven*:
John Crotty
Martin Hartog, SCSU
Harry Payne II
Angel Tangney

Cheshire High School:
Andrew Abate
Pauline Alim
Marcia Arneson
Diane Bergan
Christopher Fletcher
John Kosloski
Christina Lepi
Michael Lougee
Ann Marie Mahanna
Carol Marino
Nancy Massey
Denise Miller
John Redford
Cynthia Sarlo

Coginchaug Regional High School, *Durham*:
Theresa Balletto
Anne Coffey
John DeMeo
Jake Fowler
Ben Kupcho
Stephen Lecky
Philip Martel

Crosby High School, *Waterbury*:
Janice Farrelly
Rosalie Griffin

Danbury High School:
Adrienne Coppola
Lois Schwaller

Dodd Middle School, *Cheshire*:
William Grimm
Lara Kelly

Gay Sonn
Alexander Yawin

Manchester High School:
Pamela Brooks
Carl Bujaucius
Marilyn Cavanna
Paul DesRosiers
Mariamma Devassy-Schwob
Frank Kinel
Larry Olsen
Julie Thompson
Matthew Walsh

Montville High School:
Lynn Grills
Janice Hardink
Diane Hupfer
Ronald Moore
Mark Popeleski
Henry Kopij
Walter Sherwin
Shari Zagarenski

Oxford Hills Comprehensive High School, *South Paris, ME*:
Mary Bickford
Peter Bickford
Allen Gerry
Errol Libby
Bryan Morgan
Lisa Whitman

Parish Hill Regional High School:
Peter Andersen
Gary Hoyt
Vincent Sirignano
Deborah Whipple

Southington High School:
Eleanor Aleksinas
Susan Chandler
Helen Crowley
Nancy Garry
John Klopp
Elaine Mulhall
Stephen Victor
Bernadette Waite

Stonington High School:
Joyce Birtcher
Jill Hamel
Glenn Reid

TECHNOLOGY SUPPORT

Texas Instruments:
Graphing Calculators and
View Screens

Presto Press:
Desktop Publishing

Key Curriculum Press:
Geometer's Sketchpad

PROFILES

Barbara Zahm
David Bornstein
Alex Straus
Mimi Valiulis

EXTERNAL EVALUATORS

Donald Cichon Associates:
Donald Cichon
Nancy Johnson
Kevin Sullivan and
Sharon Soucy McCrone

Connecticut State Department of Education:
Michelle Staffaroni
Graduate Intern

EQUITY AND ACCESS REVIEWERS

Leo Edwards, Director
Mathematics and Science
Education Center
Fayetteville State University,
NC

Ray Shiflett
California Polytechnic
University, Pomona

READABILITY STUDY

Elaine Eadler Associates:
Elaine Eadler, University
of Maine at Farmington

Welcome to MATH Connections® A Secondary Mathematics Core Curriculum

MATH *Connections* is an exciting, challenging, three-year core curriculum for secondary mathematics! True to its name, this curriculum is built around connections of all sorts:

- between different mathematical areas;
- between mathematics and science;
- between mathematics and other subjects (history, literature, art, etc.);
- between mathematics and the real world of commerce, law, and people.

The following few pages tell you where this program came from, what it is designed to accomplish, and how it is constructed. We hope that reading this will help you, the teacher, become partners with us in making this curriculum come to life for your students.

The History

MATH *Connections* began in 1992 when a team of educators associated with the Connecticut Business and Industry Association (CBIA), Wesleyan University, and the University of Hartford collaborated on the development of a new secondary mathematics core curriculum based on the *NCTM Standards*. Funded by a $4.9 million grant from the National Science Foundation awarded to the CBIA Education Foundation, the principal investigators assembled a team of people, including:

1. Senior writers: university mathematics professors with considerable teaching and writing experience who developed and wrote drafts and final copy of all the books.

2. Contributing writers: sixteen high school and community college mathematics teachers and two science specialists who reviewed the chapter drafts and wrote additional problems, projects, and ancillary materials.

3. Advisory Council: twenty respected professionals from the fields of mathematics, science, education and business, who met annually to review the overall direction of the project and to provide advice about various aspects of the emerging materials.

4. Classroom teachers and students: more than 100 field-test teachers and approximately 2500 students from seventeen different inner city, urban, suburban and rural high schools who used preliminary versions of these materials and provided feedback from their classroom experiences.

A many-faceted external evaluation of the MATH *Connections* program was an integral part of the project. External evaluators provided comprehensive reports on student achievement on standardized tests and how well MATH *Connections* met its objectives. Consultants reviewed the materials for gender equity, multicultural equity, and readability.

The Vision and The Mission

The Vision: *All students can learn mathematics, be critical thinkers and be problem solvers.*

The Mission: *The conceptual understanding of the learner.*

The MATH *Connections* team designed the program to meet the following objectives:

MATH *Connections* is for *all* students. It is relevant to students as future citizens, parents, voters, consumers, researchers, employers and employees — people with a healthy curiosity about ideas. It serves well the needs and interests of all high school students, those who will go on to further math-intensive studies and science-related careers, as well as those who choose to pursue other fields of study. In fact, it enchances students' understanding of the interrelatedness of all fields and increases the attractiveness of mathematics and science as it is applied in the real world.

MATH *Connections* is flexible, allowing students with different learning styles equal opportunity to master the ideas and skills presented. This means providing both interactive group work and individual learning experiences, encouraging frequent student-teacher and student-student interaction, in finding answers and in discussing mathematics, relevant student experiences, opinions, and judgments.

The main technological tools of the world of work are an integral part of the material. Students learn to be comfortable with graphing calulators and computers, in order to cope with and profit from the opportunities of our increasingly technological world.

It is reality based. All the mathematical ideas in the curriculum are drawn from and connected to real world situations. Students can immediately see how the mathematics they are learning relates to their own lives and the world around them.

MATH *Connections* focuses on the conceptual aspects of mathematics — reasoning, pattern seeking, problem solving, questioning, and communicating with precision — because those are the features of a mathematical education that are important to lawyers, doctors, business leaders, teachers, politicians, social workers, military officers, entrepreneurs, artists, or writers as they are to scientists or engineers.

Books in Brief

Each of the three years of MATH *Connections* is built around a general theme which serves as a unifying thread for the topics covered. Each year is divided into two half-year books consisting of three or four large chapters. Every chapter has a unifying conceptual theme that connects to the general theme of the year.

Year 1 — Data, Numbers, and Patterns

Book 1a begins and ends with data analysis. It starts with hands-on data gathering, presentation, and analysis, then poses questions about correlating two sets of data. This establishes the goal of the term — that students be able to use the linear regression capabilities of a graphing calculator to do defensible forecasting in real world settings. Students reach this goal by mastering the algebra of first degree equations and the coordinate geometry of straight lines, gaining familiarity with graphing calculators.

> Chapter 1. Turning Facts into Ideas
> Chapter 2. Welcome to Algebra
> Chapter 3. The Algebra of Straight Lines
> Chapter 4. Graphical Estimation

Book 1b generalizes and expands the ideas of *Book 1a*. It begins with techniques for solving two linear equations in two unknowns and interpreting such solutions in real world contexts. Functional relationships in everyday life are identified, generalized, brought into mathematical focus, and linked with the algebra and coordinate geometry already developed. These ideas are then linked to an examination of the fundamental counting principle of discrete mathematics and to the basic ideas of probability. Along the way, *Book 1b* poses questions about correlating two sets of data.

> Chapter 5. Using Lines and Equations
> Chapter 6. How Functions Function
> Chapter 7. Counting Beyond 1, 2, 3
> Chapter 8. Introduction to Probability: What Are the Chances?

Year 2 — Shapes in Space

Book 2a starts with the most basic ways of measuring length and area. It uses symmetries of planar shapes to ask and answer questions about polygonal figures. Algebraic ideas from Year 1 are elaborated by providing them with geometric interpretations. Scaling opens the door to similarity and then to angular measure, which builds on the concept of slope from Year 1. Extensive work with angles and triangles, of interest in its own right, also lays the groundwork for right angle trigonometry, the last main topic of this book. Standard principles of congruence and triangulation of polygons are developed and employed in innovative ways to make clear their applicability to real world problems.

Chapter 1. The Building Blocks of Geometry: Making and
            Measuring Polygons
Chapter 2. Similarity and Scaling: Growing and Shrinking Carefully
Chapter 3. Introduction to Trigonometry: Tangles with Angles

Book 2b begins by exploring the role of circles in the world of spatial relationships. It then generalizes the two dimensional ideas and thought patterns of *Book 2a* to three dimensions, starting with foldup patterns and contour lines on topographical maps. This leads to some fundamental properties of three dimensional shapes. Coordinate geometry connects this spatial world of three dimensions to the powerful tools of algebra. That two way connection is then used to explore systems of equations in three variables, extending the treatment of two variable equations in Year 1. In addition, matrices are shown to be a convenient way to organize, store, and manipulate information.

Chapter 4. Circles and Disks
Chapter 5. Shapes in Space
Chapter 6. Linear Algebra and Matrices

Year 3 — Mathematical Modeling

Book 3a examines mathematical models of real world situations from several viewpoints, providing innovative settings and a unifying theme for the discussion of algebraic, periodic, exponential, and logarithmic functions. These chapters develop many ideas whose seeds were planted in Years 1 and 2. The emphasis throughout this material is the utility of mathematical tools for describing and clarifying what we observe. The modeling theme is then used to revisit and extend the ideas of discrete mathematics and probability that were introduced in Year 1.

> Chapter 1. Algebraic Functions
> Chapter 2. Exponential Functions and Logarithms
> Chapter 3. The Trigonometric Functions
> Chapter 4. Counting, Probability, and Statistics

Book 3b begins by extending the modeling theme to Linear Programming, optimization, and topics from graph theory. Then the idea of modeling itself is examined in some depth by considering the purpose of axioms and axiomatic systems, logic, and mathematical proof. Various forms of logical arguments, already used informally throughout Years 1 and 2, are explained and used to explore small axiomatic systems, including the group axioms. These logical tools then provide guidance for a mathematical exploration of infinity, an area in which commonsense intuition is often unreliable. The final chapter explores Euclid's plane geometry, connecting his system with many geometric concepts from Year 2. It culminates in a brief historical explanation of Euclidean and non-Euclidean geometries as alternative models for the spatial structure of our universe.

> Chapter 5. Optimization: Math Does It Better
> Chapter 6. Playing By the Rules: Logic and Axiomatic Systems
> Chapter 7. Infinity — The Final Frontier?
> Chapter 8. Axioms, Geometry, and Choice

Appendices

Appendix A. Using a TI-82 (TI-83) Graphing Calculator. This appendix appears in all the books because graphing calculators are important tools for virtually every chapter. It provides a gentle introduction to these machines, and also serves as a convenient student reference for the commonly used elementary procedures.

Appendix B. Using a Spreadsheet. This appendix also appears in all the books. Although a spreadsheet is not explicitly required anywhere in these books, it is very handy for doing many problems or Explorations. It should be considered as a legitimate, optional tool for anyone with access to one.

Appendix C. Programming the TI-82 (TI-83). This appendix also appears in all the books. Students can use it to learn useful general principles of programming, as well as techniques specific to the TI-82 (TI-83).

Appendix D. Linear Programming with Excel. This appendix, which appears in Years 2 and 3, is not primarily a tool for doing problems within the chapters. Rather, it describes a technological approach to ideas that come up from time to time in various chapters. Linear Programming itself is discussed in detail in Chapter 5 of *Book 3b*. This appendix can be used either as a precursor to that discussion, or as an extension of it.

The Features

Here is an overview of the distinctive features that combine to make MATH *Connections* unique.

Standards-based — The NCTM Standards were guiding principles in the development of MATH *Connections*. In particular, various creative devices were employed to implement the four process standards:

- **Problem Solving** — MATH *Connections* demonstrates how mathematics is primarily about asking your own questions and looking for patterns, rather than finding someone else's answers and calculating with numbers. Marginal notes called *Thinking Tips* guide students in specific problem solving strategies. Within sections, many of the *Discuss this* questions and *Explorations* focus students' attention on the problem solving process, while the Problem Sets that conclude virtually all sections further challenge and extend their problem solving skills.

- **Communication** — MATH *Connections'* reading level has been carefully tailored to be, on average, at least one year below grade level. In addition, definitions of all mathematical terms used are clearly identified in the text. The most important ones are displayed as *Words (or Phrases) to Know*. The marginal devices *About Words* and

About Symbols explain fundamental mathematical language and notation. The *Discuss this* and *Write this* questions expect students to talk and write about mathematics in every chapter. The Problem Sets offer many open-ended questions that are suitable for group discussions, or homework writing assignments.

- **Reasoning** — The habit of logical justification is begun in Chapter 1 of *Book 1a* and carried through the entire series. Evaluation of suppositions and arguments occurs throughout all chapters of all books. In Year 1, the reasoning required is largely informal, progressing in Year 2 to more formal justifications. The idea of a counterexample is introduced formally at the beginning of the year and used routinely after that. Expectations of logical justification become more rigorous in Year 3. Also, formal proofs within axiomatic systems (not just in geometry), including mathematical induction, are examined and required.

- **Connections** — This is the defining theme of the entire MATH *Connections* program. The mathematical ideas are connected to the real world by different real life applications — more than 50 professions, 20 job fields and most academic disciplines — within the problems, Explorations and situations of the texts. The most common applications are to science (physics, chemistry, astronomy, biology, archaeology), but there are also applications to theater, music, business and the social sciences, as well as to daily life.

Blended — MATH *Connections* provides a proven approach in which students learn the underlying structure and the skills of mathematics. By presenting mathematics as the subject is used and by blending ideas from traditionally separate fields, students learn not only the topics but also the connections between algebra, geometry, probability, statistics, discrete mathematics, Dynamic and Linear Programming and optimization. By bringing ideas from a wide range of areas to bear on a question, MATH *Connections* presents mathematics as a seamless fabric perhaps with different patterns and colors in different areas, but with no clear boundary lines. Here are some typical examples of the blending of mathematics in MATH *Connections*, one from each year.

- In *Book 1a*, the discussion of carbon dating in Section 4.4 uses first degree equations (algebra), straight line graphs (geometry), least-squares differences (algebra and statistics), and a graphing calculator (technology) to solve a problem in archaeology (natural science). All these tools come together naturally in this context.

- At the end of Section 5.4 in *Book 2b*, algebra, geometry (the volume of a cone) and Cavalieri's Principle (usually in precalculus) are combined to calculate the volume of a sphere.

- In Chapter 6 of *Book 3b*, properties of axiomatic systems (logic) are introduced by examining the rules of a simple card game and the probabilities of certain kinds of outcomes. In Section 6.5 axiomatic systems are related to physics via the Law of the Lever and also to the arithmetic properties of the number systems. The Problem Set for this section includes ideas about population growth, physics, and integer arithmetic.

Accessible — Every major idea in MATH *Connections* is introduced from a commonsense viewpoint using ordinary, nonmathematical examples. For instance, coordinate geometry starts with reading map coordinates, slope follows from building a wheelchair ramp, the concept of function begins with fingerprint files, etc. This helps students see that a mathematical view of the world is not so very different from their view of the world.

Real — This curriculum is firmly grounded in the real world of the students' present and future. The scenarios are realistic, the data sets are drawn from real world sources, the applications are real. Even the whimsical settings use realistic measurements and conditions. Students see for themselves how mathematics is actually used.

Flexible — MATH *Connections* provides for a wide variety of student backgrounds and learning styles. There is plenty of opportunity for student-teacher and student-student interaction in discussing mathematics, students' experiences, opinions, and judgments. Two types of materials are available for each chapter which enhance a teacher's ability to meet the individual needs of all students.

- Supplemental materials for students whose understanding would be strengthened by additional problems or for those who are absent or enter the MATH *Connections* program in midstream

- Extension units for students who want to investigate in greater depth the concepts introduced in a particular section, either individually or in groups

Technological — Your students' future success is likely to depend, at least in part, on their ability to cope with and use the fast-changing tools of electronic technology. We treat the basic technological tools available in the world of work, particularly graphing calculators, as an integral part of the course material. Students are encouraged to use these tools whenever they are appropriate to the work at hand. Students are introduced to the graphing calculator early and are expected to use it routinely as they go along.

The Appendices of every book are filled with simple, illustrative examples of the calculator's basic tasks and functions. In addition, this Teacher Commentary contains many suggestions of where and how you can incorporate other technological tools — spreadsheets, Geometer's Sketchpad, or other geometry software — into your teaching.

Organized — The organization of these books is student friendly:

- Every chapter starts from a point of view that makes sense in the students' world. The expected Learning Outcomes are explicitly stated at the beginning of each section.

- The definitions of all technical terms and the statements of all important facts (theorems) are typographically distinctive, making them easy to find when reviewing.

- Each chapter ends with a retrospective paragraph to help students reflect on their learning.

- Each book contains a glossary, an index, and appendices.

Cooperative Learning

Classes designed around small group and whole class interaction fit the MATH *Connections* curriculum very well. Many of the *Do this now* and *Discuss this* activities are best done in groups. They are particularly useful in traditional and block scheduling situations where there is time for both group work and for direct teacher instruction or discussions involving the entire class. When students work in cooperative settings, they benefit from the interaction and insights of classmates. Student discussions on specific textbook questions often yield surprising and unexpected results. Moreover, working with graphing calculators is greatly enhanced by the group process, as students reinforce their understanding.

MATH *Connection's* classroom experiences reinforce the understanding that learning is a social process and that cooperative learning activities are essential if students are to be able to construct their own knowledge.

Suggested Reading
How to Use Cooperative Learning in the Mathematics Class, Second Edition
By Alice F. Artzt and Claire M. Newman
From the NCTM Educational Materials and Products Catalog
ISBN 0-87353-437-9 #650E1
1-800-235-7566

Elements

About Words are margin paragraphs that explain technical terms and other words that are not part of everyday language, showing how the mathematical use of a word is related to its usage outside of mathematics. This feature facilitates the integration of mathematics with English, especially with English as a second language.

About Symbols are margin paragraphs that point out how specific symbolic conventions are reasonable shorthand ways of writing and communicating the ideas they represent.

Thinking Tips are margin paragraphs that highlight the use of specific problem solving techniques.

A Word to Know and **A Phrase to Know** signal particularly important definitions. Some words appear in boldface where they are defined or first described.

A Fact to Know signals an important mathematical result. In mathematicians' terms, they are significant theorems.

 Do this now identifies questions that students should deal with before moving ahead. Often, these questions are answered within the few pages that follow. If the students don't wrestle with the question when it appears, its value as a learning experience may be lost.

 Discuss this identifies questions intended to provoke interchange among students, open-ended Explorations, and/or a variety of opinions. They can be done as soon as they appear in the text, but you might want to delay some and omit others, depending on your sense of how they might work best with your students.

 Write this identifies student writing exercises and requires students to gather information or reflect upon a topic. These exercises are suitable for group work and reporting, and can be assigned individually as homework or completed in class.

Pacing

Each school district will need to evaluate the pacing of the curriculum according to the needs of their students and schedule it has adopted. The two models shown here are examples of how school systems have adapted MATH *Connections* to fit their needs. Each semester assumes 16 weeks of class time, allowing two weeks per semester for other activities such as projects, statewide testing, etc.

3-YEAR PACING MODEL

Year 1:

Book 1a	Chapter 1	4 weeks including assessments
	Chapter 2	5 weeks including assessments
	Chapter 3	4 weeks including assessments
	Chapter 4	3 weeks including assessments
Book 1b	Chapter 5	4 weeks including assessments
	Chapter 6	6 weeks including assessments
	Chapter 7	3 weeks including assessments
	Chapter 8	3 weeks including assessments

Year 2:

Book 2a	Chapter 1	5 weeks including assessments
	Chapter 2	6 weeks including assessments
	Chapter 3	5 weeks including assessments
Book 2b	Chapter 4	5 weeks including assessments
	Chapter 5	6 weeks including assessments
	Chapter 6	5 weeks including assessments

Year 3:

Book 3a	Chapter 1	6 weeks including assessments
	Chapter 2	5 weeks including assessments
	Chapter 3	4 weeks including assessments
	Chapter 4	3 weeks including assessments
Book 3b	Chapter 5	3 weeks including assessments
	Chapter 6	3 weeks including assessments
	Chapter 7	4 weeks including assessments
	Chapter 8	4 weeks including assessments

(Teachers sometimes assign Chapter 8 as independent study resulting in student projects and research papers.)

The sections within each chapter vary in length and each is built around a major idea. Several sections might take only one class period; others may take two, three or four class periods. MATH *Connections* gives you the flexibility to successfully match the capabilities of your students with the program. Higher level classes complete the *a* and *b* books in one year, making this a 3-year program. In other situations, however, MATH *Connections* can be taught over $3\frac{1}{2}$ to 4 years.

3 $^1/_2$ - YEAR PACING MODEL

Year 1:

Book 1a	Chapter 1	5 weeks including assessments
	Chapter 2	6 weeks including assessments
	Chapter 3	6 weeks including assessments
	Chapter 4	4 weeks including assessments
Book 1b	Chapter 5	5 weeks including assessments
	Chapter 6	6 weeks including assessments

Year 2:

Book 1b	Chapter 7	3 weeks including assessments
	Chapter 8	2 weeks including assessments
Book 2a	Chapter 1	9 weeks including assessments
	Chapter 2	8 weeks including assessments
	Chapter 3	3 weeks including assessments
Book 2b	Chapter 4	4 weeks including assessments
	Chapter 5	3 weeks including assessments (up to section 5.5)

Year 3:

Book 2b	Chapter 5	3 weeks including assessments (start at section 5.6)
	Chapter 6	4 weeks including assessments
Book 3a	Chapter 1	7 weeks including assessments
	Chapter 2	6 weeks including assessments
	Chapter 3	5 weeks including assessments
	Chapter 4	7 weeks including assessments

Year 4:

Book 3b	Chapter 5	4 weeks including assessments
	Chapter 6	4 weeks including assessments
	Chapter 7	5 weeks including assessments
	Chapter 8	5 weeks including assessments

Published by IT'S ABOUT TIME, Inc. © 2000 MATHconx, LLC

Materials Included With Teacher Edition

MATH *Connections* provides a full range of support materials. In addition to the Teacher Edition, each year includes:

Assessments Form A — Allows students to use a variety of formats to demonstrate their knowledge of mathematical concepts and skills. Assessments are tied to specific mathematics objectives and are linked to the Learning Outcomes for each section of the chapter. Form A was developed for each quiz and chapter test.

The Solution Key and Scoring Guide for each quiz and chapter test contains solutions to all questions as well as suggestions for scoring based on a 100 point scale.

Blackline Masters — Contain spreadsheets, graphs, tables and forms which can be copied and distributed to students in the class or used as transparencies.

Ancillary Materials

Assessments Form B — Provides teachers with an alternative but equivalent assessment for each quiz and chapter test along with corresponding Solution Keys. They can be used as alternate assessments or makeup tests, etc.

Supplements — Complement the MATH *Connections* curriculum for students whose understanding of mathematics would be strengthened by additional work or for those who have been absent or enter the MATH *Connections* program in midstream.

Extensions — Enable students to investigate in greater depth the concepts and activities introduced in a particular section.

Test Banks — Allow teachers to prepare tests such as Midterms and Final Exams that are specifically tailored to the particular needs of their students. They include sample formats and test questions, along with the corresponding Solution Keys.

Professional Development Videos — It is always preferable to attend Professional Development Workshops. But sometimes schedules and budgets make attendance difficult. We developed these videos to support the professional development of teachers as they implement MATH *Connections* in their classrooms.

NCTM 9-12 Standards Correlation

Integrated into all three years of MATH *Connections* are the four process standards:

1. **Mathematics as Problem Solving:** Explorations, Projects, Simulations and Problem Solving

2. **Mathematics as Communication:** *Do this now, Discuss this, Write this, About Words, About Symbols, Justify Opinions*

3. **Mathematics as Reasoning:** A thematic thread in all three years

4. **Mathematics as Connections:** Connections with more than 50 professions, 20 job fields and all academic disciplines

	Year 1	Year 2	Year 3
5. **Algebra**	Chap. 2, 3, 4, 5	Chap. 1, 2, 6	Chap. 1, 5
6. **Functions**	Chap. 5, 6	Chap. 1	Chap. 1, 2, 3
7. **Geometry from a Synthetic Perspective**	Chap. 5, 6	Chap. 1, 2, 4, 5	Chap. 1, 5, 6, 8
8. **Geometry from an Algebraic Perspective**	Chap. 3	Chap. 1, 4, 5	Chap. 1
9. **Trigonometry**		Chap. 3, 4, 5	Chap. 3
10. **Statistics**	Chap. 1, 4, 5, 8		Chap. 4
11. **Probability**	Chap. 7, 8		Chap. 4
12. **Discrete Mathematics**	Chap. 1, 2, 4, 5, 7, 8	Chap. 2, 5, 6	Chap. 4, 5, 6
13. **Conceptual Underpinnings of the Calculus**	Chap. 2, 5, 6	Chap. 2, 5, 6	Chap. 1, 7
14. **Mathematical Structure**	Chap. 2, 4	Chap. 2	Chap. 6, 7, 8

Planning Guide Book 2b

This book begins by exploring the role of circles in the world of spatial relationships. It then generalizes the two dimensional ideas and thought patterns of *Book 2a* to three dimensions, starting with foldup patterns and contour lines on topographical maps. This leads to some fundamental properties of three dimensional shapes. Coordinate geometry connects this spatial world of three dimensions (and beyond) to the powerful tools of algebra. That two way connection is then employed to explore systems of three variable equations, extending the treatment of two variable equations in Year 1. Matrices are introduced as a tool for solving systems of linear equations. In addition, matrices are shown to be a convenient vehicle for organizing, storing, and manipulating information.

Chapter Objectives	*Pacing Range	Assessments Form A (A)	Blackline Masters
Chapter 4 Circles and Disks In this chapter, a few extended real world problems provide the context for developing fundamental properties of circles, sometimes in unusual ways. This chapter covers many familiar topics, such as the formulas for area and circumference and properties of inscribed angles. It also explains the meaning and significance of the number pi, shows how to graph circles parametrically, and explores curves of constant width. Some justifications resemble traditional proofs; others rely on algebraic arguments or on similarity principles from Chapter 2.	4-5 weeks including Assessments	Quiz 4.1-4.2(A) Quiz 4.3(A) Quiz 4.4-4.6(A) Project: 4.7(A) Chapter Test	Student pp. 312, 318, 335, 345, 350, 356
Chapter 5 Shapes in Space Chapter 5 explores three dimensional geometry, using hands-on activities to introduce deep ideas in easy ways. It examines the construction and basic properties of figures in 3-space, using both constructive (fold, stack, revolve) and analytic (coordinate) methods. The coordinate analysis relates geometric ideas to algebraic techniques and provides a platform for the discussion of four or more dimensions.	6 weeks including Assessments	Quiz 5.1(A) Quiz 5.2-5.4 Quiz 5.5(A) Quiz 5.6-5.8(A) Chapter Test	Student pp. 382 (2), 390-392, 392
Chapter 6 Linear Algebra and Matrices Starting with planes in 3-space, Chapter 6 treats solutions of systems of linear equations in a way designed to promote understanding of the process. The geometrical representation is linked with algebraic manipulation by means of matrix operations, including Gaussian elimination and matrix multiplication.	4-5 weeks including Assessments	Quiz 6.1-6.2(A) Quiz 6.3 A) Quiz 6.4-6.5(A) Chapter Test	Student p. 505

*Pacing Range   Teachers will need to adjust this guide to suit the needs of their own students. Not all classes will complete each chapter at the same pace. Flexibility — which accommodates different teaching styles, school schedules and school standards — is built into the curriculum.

Ancillary Materials

Assessments Form B (B)	Supplements for Chapter Sections	Extensions	Test Banks
Quiz 4.1-4.2(B) Quiz 4.3(B) Quiz 4.4-4.6(B) Chapter Test	4.2 Points on Circles — 1 Supplement 4.3 Drawing Circles With a Graphing Calculator — 1 Supplement 4.4 Area and Circumference — 4 Supplements 4.5 Pieces of Circles — 2 Supplements 4.6 Inscribed Angles — 1 Supplement	4.2 Round Pegs in Square Holes 4.4 Ellipses and the Planets	To be released
Quiz 5.1(B) Quiz 5.2-5.4(B) Quiz 5.5(B) Quiz 5.6-5.8(B) Chapter Test	5.1 Another Dimension — 4 Supplements 5.2 Three-Space in Layers — 2 Supplements 5.3 Finding Volumes by Stacking — 2 Supplements 5.4 The Volumes of Spheres, Cones and Pyramids — 2 Supplements 5.5 Turning Around an Axis — 2 Supplements 5.6 Space by the Numbers — 5 Supplements 5.7 Stacks of Lines — 1 Supplement 5.8 The Algebra of Three Dimensional Shapes — 1 Supplement	5.9 A Glimpse of Four Dimensional Space Following 5.9 Equations for Parabolas Geometric Optimization Parabolas: Why Headlights Light Ahead	To be released
Quiz 6.1-6.2(B) Quiz 6.3(B)) Quiz 6.4-6.5(B) Chapter Test	6.1 Systems of Equations and Graphs — 1 Supplement 6.2 Consistent and Inconsistent Systems — 4 Supplements 6.3 Solving Systems of Equations — 2 Supplements 6.4 Solving Linear Systems Using Matrices — 4 Supplements 6.5 Matrix Operations — 4 Supplements	Following 6.5 The Inverse of a Matrix Systems of Linear Inequalities	To be released

Chapter 4

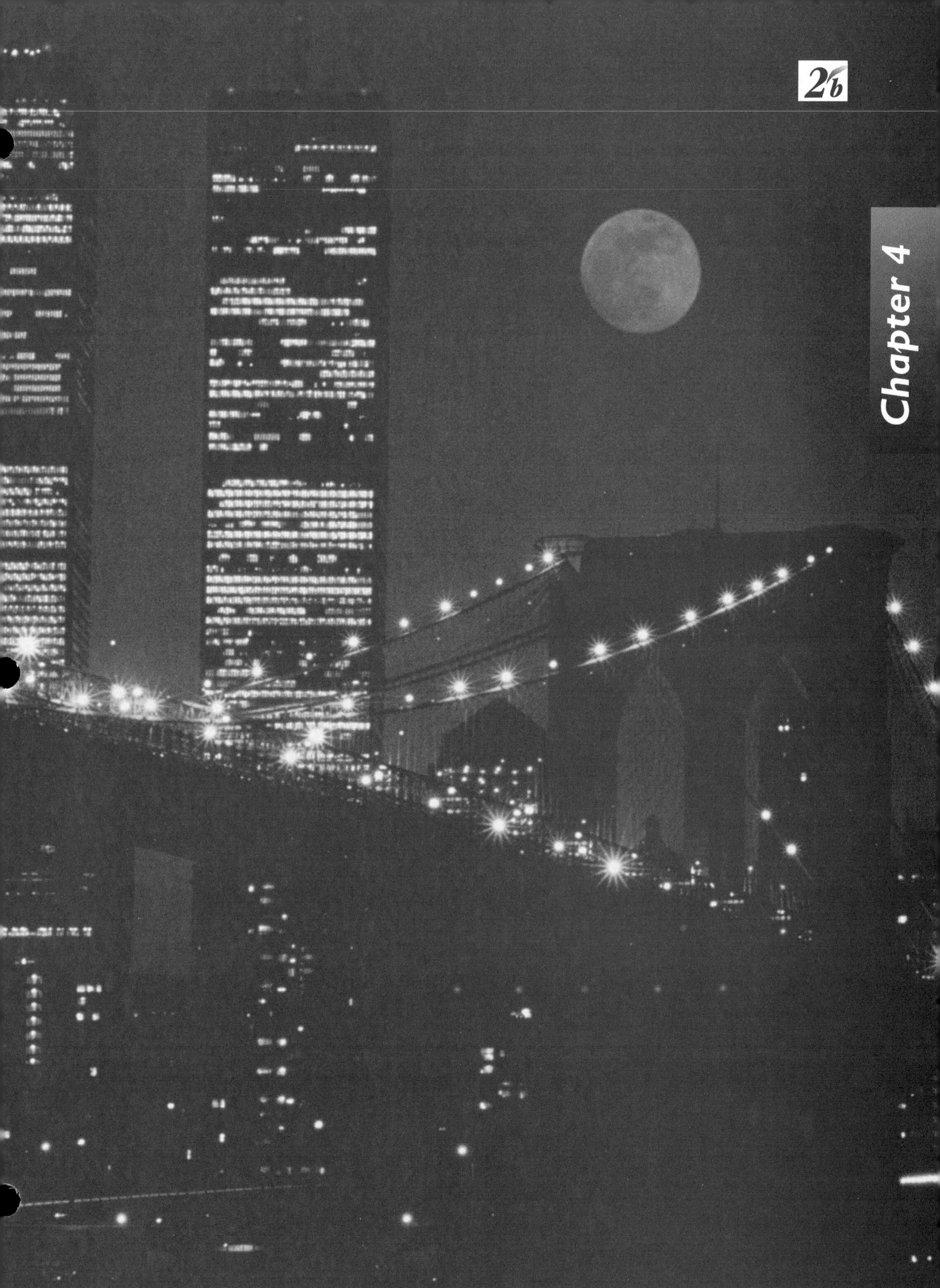

Chapter 4 Planning Guide

Chapter 4 Circles and Disks

In this chapter, a few extended real world problems provide the context for developing fundamental properties of circles, sometimes in unusual ways. This chapter covers many familiar topics, such as the formulas for area and circumference and properties of inscribed angles. It also explains the meaning and significance of the number pi, shows how to graph circles parametrically, and explores curves of constant width. Some justifications resemble traditional proofs; others rely on algebraic arguments or on similarity principles from Chapter 2.

Assessments Form A (A)	Assessments Form B (B)	Blackline Masters
Quiz 4.1-4.2(A) Quiz 4.3(A) Quiz 4.4-4.6(A) Project: 4.7(A) Chapter Test(A)	Quiz 4.1-4.2(B) Quiz 4.3(B) Quiz 4.4-4.6(B) Chapter Test(B)	Student pp. 312, 318, 335, 345, 350, 356

Extensions	Supplements for Chapter Sections	Test Banks
No Extensions for this Chapter	4.2 Points on Circles — 1 Supplement 4.3 Drawing Circles With a Graphing Calculator — 1 Supplement 4.4 Area and Circumference — 4 Supplements 4.5 Pieces of Circles — 2 Supplements 4.6 Inscribed Angles — 1 Supplement	To be released

Pacing Range 4-5 weeks including Assessments
Teachers will need to adjust this guide to suit the needs of their own students. Not all classes will complete each chapter at the same pace. Flexibility — which accommodates different teaching styles, school schedules and school standards — is built into the curriculum.

Teacher Commentary is indexed to the student text by the numbers in the margins (under the icons or in circles). The first digit indicates the chapter — the numbers after the decimal indicate the sequential numbering of the comments within that unit. Example:

4.9

4.9

Student Pages in Teacher Edition   **Teacher Commentary Page**

Observations

Errol Libby
Oxford Hills Comprehensive
High School, ME

"My favorite problem in the whole series is in this chapter — the Go-Kart problem. It goes on for several pages and eventually pulls together and blends many mathematical concepts — arc length, central angles, area of a sector, perimeters, etc.

"The students end up using and demonstrating many of the concepts that they have learned in earlier books and which have now become a part of their problem solving repertoire or tool box.

"In general, I really like the way concepts develop in this program. They go from simple ideas that are expanded upon and then expanded upon again. This is actually how geometry itself evolved and is a nice way to lead into circles."

Helen Crowley
Southington High School, CT

"Chapter 4 provides in-depth experiences with circles and disks, particularly their practical properties — the program tends to do a great deal of work with hands-on modeling and real world problems.

"Also in this chapter, students are introduced to various graphing calculator techniques for drawing circles. Parametric equations are introduced with the sine and cosine values of an angle. This reiterates earlier work with the trigonometric functions and leads into the extensive work done in the third year. Students gain a better understanding of the relationship between the trig functions and the unit circle, and also have another opportunity to work with values that transform graphs, relations, and functions."

Hugh M. Dunbar
Designing with Circles

At 13, Hugh Dunbar decided he would become an architect. He loved to draw and to build things. He also enjoyed geometry. "Then in 9th grade," Hugh states, "I met a teacher who taught me to love math." The teacher was a Peace Corps volunteer working in Hugh's hometown of Kingston, Jamaica. From him, Hugh learned about logic and how different approaches could be used to arrive at the same solution.

Hugh came to the U.S. and received his Bachelors of Architecture from Howard University. He now works for McKissack & McKissack, the oldest minority-owned architect firm in the U.S.

"Architecture is about space," explains Hugh, "especially how space can be used to make people's lives better. Circles are really important in defining space, whether you're designing a kitchen or a church. The Romans were the first to use the arch — an upside down, weight bearing circle. This new architectural element changed the way structures were designed from that time forward."

The proper use of circles is key to Hugh's work. "If we consider lighting for a room," he states, "we look at the circular area it covers. Light from the ceiling comes down as a cone. Then these cones intersect at certain points. If we are designing an office, the cones of light should intersect at the level of the desks."

Hugh's firm is well known for their outstanding designs in many major American educational and health facilities. But Hugh is also proud of his designs used in churches. "Circles play big roles in churches," he explains. "The pulpit is the center. The seats represent various perimeters, equally spaced from the pulpit. In a church, everyone needs to have the same view."

Published by IT'S ABOUT TIME, Inc. © 2000 MATHconx, LLC

282

Chapter 4   Circles and Disks

This chapter develops many of the fundamental properties of circles, sometimes in unusual ways. The fact that students have already seen some trigonometry allows us to use trigonometric functions in places. This is particularly important in Section 4.3, which uses parametric representations on graphing calculators. There is also some reliance on algebraic arguments in places, as well as on similarity principles from Chapter 2.

Chapter 4

Circles and Disks

CHAPTER

4

4.1 Points of View

Learning Outcomes

After studying this section you will be able to:

Describe the circle shape by its center-radius definition and by symmetry;

Identify some practical, symbolic, and aesthetic properties of circles.

Circles appear everywhere in art, nature, design, science, architecture, and technology. As you look around, you have no trouble finding circular objects. Did you ever wonder why so many things are circular?

Why do you think so many things are circular? Why do you think a circular shape often is more practical than, say, an oval or a rectangle or some other shape? Name some common objects that wouldn't work (or wouldn't work as well) if they were not circular.

4.1

The circle shape is practical for many uses. Moreover, its beauty and simplicity have made it an important symbol in many cultures and religions. Its mathematical properties have challenged and intrigued mathematicians for centuries. In this chapter we examine some of the properties that make circles so useful and interesting. We begin by comparing three very different views of the circle, from three people in very different walks of life.

283

4.1 Points of View

This first section is very rich. It contains a broad spectrum of material about circles, as well as some exploratory exercises that draw on ideas and techniques learned in previous chapters. Plan to spend more than one class on this section. The two most important descriptions of the circle (for our purposes) are the center radius definition and the statement about rotational symmetry. These ideas will recur throughout the chapter. The other ideas introduced here will arise from time to time, but without as much emphasis.

4.1

This question is intended to focus students on the fact that there is something to wonder about here. It need not take much class time.
Objects that come to mind easily are wheels of all sorts, as well as circular saw blades, gears, pulleys, etc. There are many other possibilities.

Chapter 4

Additional Support Materials:

Assessments	Qty
Form (A)	1
Form (B)	1

Blackline Masters	Qty

Extensions	Qty

Supplements	Qty

4.1 Points of View

Architect William Blackwell speaks of the circle in design and architecture like this.

The circle is the simplest of the two dimensional shapes and the easiest to draw on paper or inscribe on the ground. One can imagine the pleasure the ancients had in laying out the circles of Stonehenge. They needed only flat ground, a length of rope attached to a stake, someone to walk it around keeping it taut, and someone to mark the path . . .

The circle encloses a given area with the least perimeter or circumference and is the most compact of the plane geometric shapes. A complete circle appears to have no beginning and no end, and it has no corners . . . King Arthur's position at the round table is marked only by a special chair. The round table itself has neither head nor foot . . . The main point about a circular table is that everyone can see and be seen. It is the perfect shape for an assembly of equals . . .

When uniform pressure is exerted on a circle, either from the inside or from the outside, the circle is the strongest of shapes. For this reason, it is the shape of oil tanks, gun barrels, and shafts. For the same reason, masonry arches, possibly the most eloquent shapes of architecture, are semicircular . . .

In architecture, besides monuments, round tables, and cities, circles have been used in the plans of stadiums and arenas . . . stairways, dome structures, churches, museums, apartment houses, and parking garages . . .[1]

4.2

Can you summarize each of William Blackwell's four paragraphs with a sentence of six words or less? Try it.

4.3

1. **What is Stonehenge? Where is it? What does it have to do with circles?**

2. **Who was King Arthur? What was King Arthur's round table? Why do you think William Blackwell referred to King Arthur's round table in making his "main point about a circular table"?**

Published by IT'S ABOUT TIME, Inc. © 2000 MATHconx, LLC

284

Chapter 4

4.2

This is a reading comprehension question. The length limitation is deliberate. It forces students to distill the ideas, rather than just pick out some of Blackwell's sentences without understanding them. Possible (but not unique) answers are

(1) Circles are easy to make.

(2) Circles are compact and promote equality.

(3) Circular shapes withstand uniform pressure best.

(4) Architects use circles for many structures.

4.3

This is an encyclopedia research question that links the mathematics with a little history and geography.

Stonehenge is a large structure of standing stones on Salisbury Plain in Wiltshire, in southern England. It is surrounded by a circular ditch about 300 ft. in diameter. The stones are arranged in a circular pattern, too. It is believed to have been built about 1700 B.C. as a ceremonial monument.

King Arthur was a legendary king of Britain. Some stories about a King Arthur of Britain date back before 1000 A.D. The later, more popular form of this legend describes Arthur as the illegitimate son of King Uther Pendragon who showed his divine right to the throne by drawing a sword (Excalibur) from a stone. His court, with its Round Table, was in the legendary town of Camelot. The Round Table was for the formal meetings of King Arthur's knights. It was round so that there would be no quarrels about precedence; all the knights were to be considered equal in prestige. This is the point of Blackwell's reference.

Black Elk, a 19th century Native American leader, speaks of the circle's symbolic power.

You have noticed that everything an Indian does is in a circle, and that is because the Power of the World always works in circles, and everything tries to be round . . . This knowledge came to us from the outer world with our religion. Everything the Power of the World does is done in a circle. The sky is round, and I have heard that the earth is round like a ball and so are all the stars. The wind, in its greatest power, whirls. Birds make their nests in circles, for theirs is the same religion as ours. The sun comes forth and goes down again in a circle. The moon does the same, and both are round. Even the seasons form a great circle in their changing, and always come back again to where they were. The life of a man is a circle from childhood to childhood, and so it is in everything where power moves. Our tepees were round like the nests of birds, and these were always set in a circle, the nation's hoop, a nest of many nests, where the Great Spirit meant for us to hatch our children.[1]

What property of circles is Black Elk emphasizing in this quotation? Find a sentence or a phrase in Black Elk's words to support your opinion. Do other figures have this property, too? If so, why do you think Black Elk just refers to circles?

4.4

Professor Philip J. Davis, an artist who works in logic, comments as a mathematician. He describes some mathematical properties of the circle that are the basis for many of its practical, aesthetic, and symbolic uses.

The Greeks, who held some rather impressive notions of beauty and perfection, came to the conclusion that the circle was the most beautiful curve . . . What is the most perfect curve? . . . I'm not going to talk about visual beauty, but conceptual beauty . . . My answer is the Greeks' answer: the circle. I will defend my answer on the basis of the mathematical interest

---

[1]From *Black Elk Speaks, Being the Life Story of a Holy Man of the Oglala Sioux* as told through John G. Neihardt, (Flaming Rainbow), pages 194–196. Copyright © 1972 by John G. Neihardt. Reprinted with permission of University of Nebraska Press.

Published by IT'S ABOUT TIME, Inc. © 2000 MATHconx, LLC

4.4

Another reading comprehension question. The dominant property appears to be the repetitive, cyclic nature of a circular path, with no clear beginning or end. "Even the seasons form a great circle in their changing, *and always come back again to where they were.*" All closed curves have this property, but the circle (because of its symmetry and uniform curvature) is the simplest, clearest example of such a figure.

NOTES

of the circle. I will do it by pointing out a number of remarkable things that are true of it.

1. Every point on the circle is at the same distance from the center.

2. If you spin a circle about its center, the rotated circle always occupies precisely the same space as did the original circle.

3. Every diameter of the circle is an axis of symmetry.

4. A circle is a figure with constant width.

5. Every tangent to a circle is perpendicular to the radius drawn from the center to the point of tangency . . .

. . . There is no figure which provides simplicity when simplicity is sought, and profundity when profundity is sought, after the manner of the circle. There is no figure that can compare with it.[1]

4.5

Which of the properties listed by Professor Davis are familiar to you? Some may be unfamiliar. Jot down any terms or ideas that you don't understand and come back to them as you work through this chapter.

Professor Davis's five properties provide a good way to organize our investigation of circles. The first three will be the most important ones for our purposes. His first property is that every point on a circle is the same distance from its center. This is the simplest way of defining a circle; we'll call it the **center-radius definition**.

A **circle** is the set of all points in a plane that are a fixed distance (called the **radius**) from a particular point (called the **center**).

The radius specifies the size of the circle, and the center fixes its location. Since the size of a circle depends only on its radius, both the area enclosed by a circle and the distance around it (its **circumference**) can be calculated directly from its

---

[1]From *3.1416 and All That*, 2nd ed., by Philip J. Davis and William G. Chinn. Condensed from pages 88–93. Copyright © 1985 by Philip J. Davis. Reprinted with permission.

Published by IT'S ABOUT TIME, Inc. © 2000 MATHconx, LLC

4.5

It might be useful to have a brief class or small group discussion that allows students to compare the lists they make and iron out any trivial misunderstandings. You can also use the results of this discussion to see what needs to be stressed as you teach the chapter.

NOTES

radius. In other words, both the area and the circumference of a circle are *functions* of its radius. Simple formulas that describe these two important functions appear later in this chapter.

1. **When you draw a circle with a compass, which part of the compass positions the center? What determines the radius?**

2. **The center-radius definition says that all points must be in the same plane. Why is it necessary to say this?**

4.6

If you read the center-radius definition carefully, you can see that a circle consists *only* of the points of the curve, as shown in Display 4.1. The center is *not* part of the circle. The other points inside are not part of the circle, either. The circle together with all the points inside it is called a **disk.** In everyday speech we often say circle when we mean disk, allowing the context to make our meaning clear. Even in mathematics, we sometimes talk informally about finding the area of a circle when we really mean finding the area of the disk, the region enclosed by the circle.

About Words

The *diskettes* used in computers are little disks. Sometimes the word disk is spelled *disc*, as in compact disc (CD).

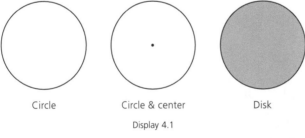

| Circle | Circle & center | Disk |

Display 4.1

Professor Davis's second and third properties are closely related. We begin by looking at the third one, since it is more familiar.

Every diameter of the circle is an axis of symmetry.

A **diameter** is a line segment that passes through the center and has both endpoints on the circle. (See Display 4.2.) You can easily find a diameter of a circle by cutting a disk out of a piece of paper. Fold the disk in half so that the two sides of the fold match, crease the fold, and open it up again. The crease shows you a diameter of the circular edge of the disk. It is an axis of symmetry for that circle. Each time you repeat this folding process you will get another diameter, another axis of symmetry. As you can see, a circle has an unlimited number of axes of symmetry.

Published by IT'S ABOUT TIME, Inc. © 2000 MATHconx, LLC

4.6

1. The pivot point of the compass determines the center of the circle; the opening (or the position of the pencil point) determines the radius.

2. If this restriction were not made, the definition would describe a sphere in space. Just such a description will appear in the chapter on three dimensional space.

NOTES

4.7

Cut a disk out of a piece of paper. Fold it in half so that the two sides of the fold match, crease it, and open it up again. Repeat the process, folding the disk in a different place. At what point do the two diameters intersect? Check your answer by folding the disk a third time, again in a different place.

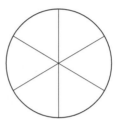

Diameters of a circle

Display 4.2

About Words

The plural of *radius* is *radii*.

A line segment with one endpoint on a circle and the other endpoint at its center is called a **radius** of the circle. The length of any such segment is *the* radius of that circle. This makes sense because all radii of a circle must have the same length. Thus, the word radius is used in two ways, one as a segment and the other as a length. That's not as confusing as it may seem. The intended meaning usually is obvious from the way in which the word is used in a sentence.

Any line segment with both its endpoints on a circle is called a **chord** of the circle. (See Display 4.3.) The diameters of a circle are the chords that pass through its center. The other chords are not axes of symmetry. Do you see why? Try folding your disk along one of them. Do the two sides of the fold match?

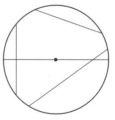

Chords of a circle

Display 4.3

288

4.7

This simple, hands-on activity reinforces two ideas: (1) a circle has infinitely many axes of symmetry (diameters), and (2) all diameters intersect at the center of the circle. The symmetry of each fold (equal halves) implies that each fold cuts any other diameter in half. Also, the definition of *diameter* requires all diameters to go through the center, so that must be the intersection point.

NOTES

When we talk about *the* diameter of a circle, we mean the length of any diameter of the circle, rather than the segment itself. This is just like the two uses of radius. However, the word chord is *never* used in this way. Why not?

4.8

The symmetry used to find diameters by folding is the kind you have studied before. It is symmetry about a line sometimes called *reflection symmetry*, and the line is called the *axis of symmetry*. Professor Davis's second property refers to a different kind of symmetry, called *rotation symmetry*. To understand this idea, let us look at a regular octagon.

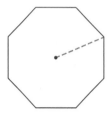

A regular octagon

Display 4.4

- Trace the regular octagon shown in Display 4.4 on two pieces of patty paper or other paper you can see through. Tape one to a piece of cardboard so that it will not slip.

4.9

- Now place the other piece on top of the first one so that the center points and the dashed lines coincide. Using a pencil point to keep the centers together, rotate the top copy around the center point, leaving the bottom copy in place.

- Use a protractor to find the smallest angle through which you can rotate the top copy so that it matches the bottom copy again. The dashed lines can help you do this.

- Continue to rotate the top copy and make a list of all the angles between 0° and 360° for which the top copy matches the bottom one.

Published by IT'S ABOUT TIME, Inc. © 2000 MATHconx, LLC

Chapter 4

4.8

The usage ideas here are important. Make sure they are clear now, to avoid confusion later. Since all diameters of a given circle have the same length, it is unambiguous to say "the diameter" as a shortened form of "the length of any diameter" of the circle. Similarly, "the radius" unambiguously refers to the (common) length of any radius of the given circle. However, different chords of a circle may have different lengths, so "the length of a chord" is ambiguous. Thus, "chord" always means a segment, never a length.

4.9

It is *important* that students actually do this exercise. Just reading about it will not have enough impact. There are eight distinct angles: 45°, 90°, 135°, 180°, 225°, 270°, 315°, and 360° (which is the same as 0°).

A figure has **rotation symmetry** if a copy of it can be made to coincide with the original figure by a positive rotation (around a point) of less than 360°. In such a case, we call any angle (including 360°) for which the tracing matches the original figure an **angle of rotation symmetry.**

Just as a reflection symmetry depends on a particular line (the axis of symmetry), so too does a rotation symmetry depend on the particular point around which the rotation occurs. That point is called the **point of symmetry.** Sometimes it is called the **center of symmetry.** Look at Display 4.4. If a tracing of the regular octagon is rotated around the center point, then the tracing will coincide with the original figure at some angles of rotation less than 360°. If you rotate it around any other point, the tracing will not match the original figure until it is rotated a full 360°. Try it.

4.10

1. How many angles of rotation symmetry does a regular octagon have? What are they? Does it matter how large the octagon is?

2. What if the octagon is not regular? In particular, what are the angles of rotation symmetry for the octagon in Display 4.5? What point of symmetry are you using? How could you locate that point using only a straightedge?

3. Draw an octagon with exactly four angles of rotation symmetry. Where is its point of symmetry?

4. Draw an octagon that does not have rotation symmetry around *any* point. Justify your answer.

5. Why do you think we are not considering angles of more than 360°?

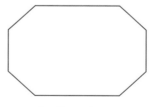

Another octagon

Display 4.5

Published by IT'S ABOUT TIME, Inc. © 2000 MATHconx, LLC

4.10

These questions reinforce and focus the idea of rotation symmetry and the point of symmetry. They also remind students that regular figures are more symmetric than irregular ones.

1. A regular octagon has eight angles of rotation symmetry. They are all multiples of 45°, as listed in the answer to 4.9. These answers do not depend on the size of the octagon.

2. An octagon that is not regular has fewer angles of rotation symmetry. The octagon in Display 4.5 has only two such angles, 180° and 360°. The point of symmetry can be found by drawing two diagonals from vertices – top left to bottom right, and top right to bottom left. Their intersection is the point of symmetry.

3. See Display 4.1T(a) for a typical answer.

4. See Display 4.1T(b) for one such octagon. Many different forms are possible.

5. For any angle of rotation larger than 360°, there is an angle less than or equal to 360° that puts the figure in the same position.

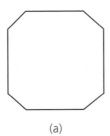   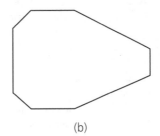

(a)   (b)

Display 4.1T

NOTES

_____

....

....

....

....

....

Chapter 4

1. How many angles of rotation symmetry does a regular hexagon have? What are they? *Hint:* Relate the number of vertices to the total number of degrees required to turn it all the way around. Where is its point of symmetry? Can you locate that point with a straightedge and a compass? If so, how?

4.11

2. How many angles of rotation symmetry does a regular pentagon have? What are they? Where is its point of symmetry? Can you locate that point with a straightedge and a compass? If so, how?

Notice that the more sides a regular polygon has, the more angles of rotation symmetry it has. What figure has the most angles of rotation symmetry? Professor Davis answers this when he writes, "If you spin a circle about its center, the rotated circle always occupies precisely the same space as did the original circle." In other words,

Every angle is an angle of rotation symmetry for a circle (around its center).

This is the main reason why circular shapes occur so often in machines, as gears, pulleys, or other rotating parts.

The other two properties of circles in Professor Davis's list deserve brief mention here.

Constant Width. Probably the most useful kind of circular object is the wheel. Before people used wheels, they moved heavy objects using rollers. When we imagine such rollers, we usually think of logs with circular cross sections, as in Display 4.6.

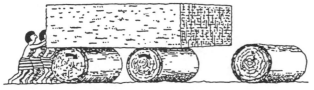

From *Mathematics Meets Technology*, by Brian Bolt. Copyright © 1991 by Cambridge University Press. Reprinted with the permission of Cambridge University Press.

Display 4.6

But rollers do not have to have circular cross sections! *All that is required is that the distance from the ground to the object being transported is always the same.* In other words, the cross sections of a roller must always have the same width, no matter where they are measured. We say that such a figure

4.11

These questions lead to the idea, developed in a problem at the end of this section, that the number and size of the angles of rotation symmetry of a regular *n*-gon are functions of *n*. This is a discussion question because how to find the point of symmetry (there is only one in each case) may not be obvious to students without talking it through.

1. A regular hexagon has six angles of rotation symmetry. They are all multiples of 60°, which is $\frac{360}{6}$: 60°, 120°, 180°, 240°, 300°, and 360°. The point of symmetry is the intersection of any two of the three main diagonals (the three segments that connect opposite vertices). This intersection point can be found with a straightedge only.

2. Both a compass and a straightedge are needed for these constructions. A regular pentagon has five angles of rotation symmetry. They are all multiples of 72°, which is $\frac{360}{5}$: 72°, 144°, 216°, 288°, and 360°. The point of symmetry (the center point) is a little harder to construct in this case. It can be found by constructing the angle bisectors of any two interior angles of the pentagon, all of which meet at the point of symmetry. Equivalently, drop a perpendicular from each vertex to its opposite side. All of these lines meet at a point, which is the point of rotation symmetry.

This would be a good place for a Geometer's Sketchpad (or other software) demonstration. For instance, you could set up a regular pentagon with the center point visible and ask students for suggestions about how much to rotate its tracing to get them to match. You'll probably have to have a tracing setup, too. Then ask them to pick any other point as the center of rotation and any angle, and observe how they can't get the tracing to match the original figure.

As a second, more difficult exercise, you might present a regular pentagon or hexagon *without* the point of rotation specified and ask students how to find that point. They should see, after a while, that the point they need is the center of the circumscribing circle, since vertices must coincide with vertices under the rotation. The question of how to find that center leads to the constructions above.

Chapter 4

has **constant width.** Circles are curves of constant width, *but they are not the only curves of constant width!* Does that surprise you?

 How can you design a roller that works smoothly but does not have a circular cross section? That is, how can you design a curve of constant width that is not a circle? Think about it.

4.12

Hint: One difference between wheels and rollers is that wheels have to pivot around their centers, but rollers do not.

Perpendicular Tangents and Radii. Do you remember seeing this property of circles earlier in **MATH** *Connections*? A problem outlined an argument showing that the angle made by a tangent line and the radius at the point of tangency is a right angle.

Problem Set: 4.1

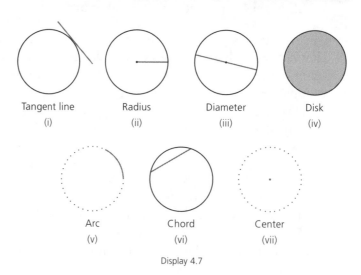

| Tangent line | Radius | Diameter | Disk |
| (i) | (ii) | (iii) | (iv) |

| Arc | Chord | Center |
| (v) | (vi) | (vii) |

Display 4.7

1. The diagrams in Display 4.7 illustrate some important terms related to circles. Match each of these written descriptions with one of those diagrams.

   (a) the circle together with its interior
   (b) a portion of the circle connecting two points on the circle

Published by IT'S ABOUT TIME, Inc. © 2000 MATHconx, LLC

Chapter 4

4.12

This question is intended to provoke thought, not to generate answers. If you plan to spend much time on it here (which is *not* recommended), you might find it helpful to look up the discussion of Reuleaux triangles in Section 4.7.

Problem Set: 4.1

1. (a) iv (b) v (c) iii (d) ii (e) i (f) vi (g) vii

NOTES

(c) a line segment that passes through the center and has both endpoints on the circle

(d) a line segment having one endpoint on the circle and the other at the center

(e) a line which touches the circle at only one point, and otherwise lies entirely outside the circle

(f) a line segment having both endpoints on the circle

(g) every point on the circle equidistant from this point

2. The following paragraph is from *Africa Counts: Number and Pattern in African Culture*, by Claudia Zaslavsky.[1]

> When a Chagga built his traditional beehive shaped house on the fertile slopes of Mount Kilimanjaro, he called upon the tallest man he knew. This neighbor would lie flat upon the site of the prospective home, with his arms outstretched. The span from the fingertips of one hand to those of the other is called a *laa*. To mark off the circumference, the builder tied a hoe to a rope of length equal to the desired radius, two to three *laa*. The rope was attached to a peg, and as he walked around this peg, he drew a circle with his hoe. The door height was equal to the span of the man's arms; its width was the circumference of his head, measured by a string.

(a) About how long do you think a *laa* is? Explain.

(b) Make a scale drawing of the circular floor of a house using a radius of three *laa*. Use the tallest person in your class or your family to determine the length of a *laa* for this house. Then pick a scale that will allow you to draw it conveniently. Estimate the circumference and area of the floor of the house. Explain how you made this estimate.

(c) Make a scale drawing of the door of this house, as described in the quote. Use the same person as you did for part (b).

---

[1]Copyright © 1990. Published by Lawrence Hill Books, an imprint of Chicago Review Press, Inc., 814 North Franklin Street, Chicago, IL 60610. Reprinted with permission.

Published by IT'S ABOUT TIME, Inc. © 2000 MATHconx, LLC

293

2. This exercise is self explanatory. The answers will vary. Look for "reason-ableness" in the responses.

NOTES

Chapter 4

3. The National Hurricane Center's July 10, 1996, bulletin about Hurricane Bertha said

"Hurricane force winds (75 mph or greater) extend 145 miles out from the center."

The structure of a hurricane is quite circular. You can use the concepts of this section to get an idea of how large an area would be affected by Bertha's damaging winds. Get a map of your area that shows the town your school is in and an area up to at least 150 miles away from you. A state or regional road map may work. Now assume that the eye (center) of Hurricane Bertha is over your school. In the eye of the hurricane you have calm, clear weather, but all around you the winds are howling and the rain is pouring down. When the eye of the hurricane passes, the wind direction reverses and the intensity of the winds increases.

(a) On your map, draw a circle with a scale radius of 145 miles to see how large an area the hurricane is affecting.

(b) List several places that are on or very near the outer edge of your circle. Have you ever been to these places? If so, how long did it take you to get there?

(c) If Hurricane Bertha is moving northeast at 25 miles per hour, how long will it be before Bertha is safely past your school? What if it's moving straight north or northwest?

Published by IT'S ABOUT TIME, Inc. © 2000 MATHconx, LLC

3. This activity problem shows a real world use of the center radius defini-
    tion of a circle. By overlaying the area of Hurricane Bertha on
    a familiar geographical region, the 145 mile radius of hurricane force
    winds becomes a visual, meaningful concept.

    (a) Look for proper scaling here. One challenge for students is to figure
        out how to draw a circle with a radius this large. For road maps or
        maps of similar scale, a compass is too small to be useful. Several
        pencils and a piece of string might work, or they might try some other
        approach. However they do it, the idea of radius is reinforced kines-
        thetically.
    (b) Answers will vary, of course. The second half of the question connects
        the scaled distance with the students' sense of real time and space.
    (c) A little less than 6 hours. Because the shape is circular, the clearing
        time is independent of direction. It's 145 miles from eye to edge in
        any direction.

NOTES

4. Display 4.8 illustrates a method for using Geometer's Sketchpad or other computer drawing software to draw regular polygons. It works because regular polygons have rotation symmetry.

    (a) Why were 45° rotations used to make a regular octagon?

    (b) What angle of rotation would be used to make an equilateral triangle? What about a regular pentagon?

    (c) Use Geometer's Sketchpad or any other geometry drawing tools to draw at least three other regular polygons by this method.

  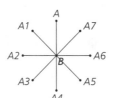

Original segment   The original segment *AB* is rotated 45° counterclockwise around point *B*.   Each segment created by rotating the previous segment 45° counterclockwise around *B*.

The endpoints of a segment are connected.   Extra points and segments are erased or hidden, leaving an octagon.

Drawing a regular polygon

Display 4.8

5. (a) Earlier in this section, you found the angles of rotation symmetry (greater than 0°, but not greater than 360°) for a regular pentagon and for a regular hexagon. What are they?

    (b) How many angles of rotation symmetry does an equilateral triangle have? What are they? Explain how you found them.

    (c) Which regular polygon has exactly four angles of rotation symmetry? Find the angles.

4. (a) Each interior angle of a regular octagon is 135°; to form such an angle, you need to turn off the straight path at an angle supplementary to it, which is 45°.

   (b) 120°; 72°

5. The first part of this problem reminds students of relevant results they have already found.

   (a) pentagon: 72°, 144°, 216°, 288°, and 360°
       hexagon: 60°, 120°, 180°, 240°, 300°, and 360°
   (b) Three: 120°, 240°, and 360°
   (c) A square. 90°, 180°, 270°, and 360°

Chapter 4

NOTES

(d) How many sides does a regular decagon have? How many angles of rotation symmetry does it have? What is the size of its smallest angle of rotation symmetry? What are the next three of these angles, in order of increasing size?

(e) Make a conjecture (a reasonable guess) about the angles of rotation symmetry of a regular n-gon (a regular polygon with n sides), where n can be any whole number greater than 2. How many are there? What are they?

6. This problem extends the results of problem 5.

(a) The smallest angle of rotation symmetry of a regular n-gon depends only on the number of sides, n. That is, it is a function of n. Explain this function in words. Then write a formula for it.

(b) The next-smallest angle rotation symmetry of a regular n-gon is also a function of n. Write a formula for this function.

(c) The next-to-largest angle rotation symmetry of a regular n-gon (largest less than 360°) is also a function of n. Explain this function in words. Then write a formula for it.

(d) Enter the three functions you found in parts (a), (b), and (c) into your graphing calculator. Then use them to find the smallest, next-smallest, and next-to-largest angles of rotation symmetry for each of these regular polygons.

(i) dodecagon

(ii) 16-gon

(iii) 20-gon

(iv) 100-gon

(v) 360-gon

(vi) 500-gon

Published by IT'S ABOUT TIME, Inc. © 2000 MATHconx, LLC

(d) Ten. Ten. 36°; 72°, 108°, 144°

(e) There are n such angles. They are the first n multiples of $\frac{360°}{n}$.

6. (a) Divide 360° by the number of sides of the polygon. $f(n) = \frac{360°}{n}$.
Of course, the function need not be called f; it could be a for angle, r for rotation, or whatever.

(b) $g(n) = 2 \cdot \left(\frac{360°}{n}\right)$

(c) There are two equally natural ways to do this: (1) Subtract the smallest angle of rotation symmetry from 360°: $h(n) = 360 - \frac{360°}{n}$, or (2) multiply the smallest angle by $n - 1$: $h(n) = (n - 1) \cdot \frac{360°}{n}$. Note that the algebraic expressions of these functions are equal, as one would expect.

(d) Using the graphing calculator here emphasizes the functional nature of these formulas in both senses.
(i) 30°; 60°; 330°
(ii) 22.5°; 45°; 337.5°
(iii) 18°; 36°; 342°
(iv) 3.6°; 7.2°; 356.4°
(v) 1°; 2°; 359°
(vi) 0.72°; 1.44°; 359.28°

NOTES

4.2 Points on Circles

To plan a circular garden or build a semicircular driveway or mark the 3-point line on a basketball court, you need to draw all or part of a circle. If you know the center of the circle and its radius, that's easy to do. Just cut a string to the length of the radius, then stand at the center with one end and have a friend hold the other end of the string taut and walk around you, marking points along the way, as in Display 4.9.

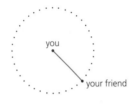

you

your friend

Display 4.9

Sometimes we know a few points that a circle must pass through, but we don't know where the center is. If we know some points, but don't know the center, how can we find more points on the circle? This section solves that problem and explores some patterns that relate circles to points and lines.

Each of the following explorations will be guided by these questions.

A. Is there a circle that satisfies the given conditions? Why or why not? How do you know?

B. If one or more exist(s), how do we find it (or them)? Will this method always work?

C. If more than one exists, what do they all have in common?

The directions for these explorations are given for doing them with pencil, paper, ruler, and compass. If you have Geometer's Sketchpad or other computer drawing software, you can use the program's tools to make the pictures.

Learning Outcomes

After studying this section, you will be able to:

Construct circles passing through one given point or two given points and describe what all such circles have in common;

Construct a circle passing through three given points and determine its center and radius;

Form the converse of a conditional statement;

Distinguish between the truth of a conditional statement and the truth of its converse.

Thinking Tip

To find good answers, you have to ask good questions. When looking at conditions of any sort, here are some good questions.

1. Does anything satisfy these conditions? (existence)

2. If so, how can I find it? (construction)

3. Does more than one thing satisfy these conditions? (uniqueness)

4. If so, what do they have in common? (pattern)

Published by IT'S ABOUT TIME, Inc. © 2000 MATHconx, LLC

4.2 Points on Circles

Besides its main geometric topic, the fact that three noncollinear points determine a unique circle, this section also has a habit of thought theme.
It serves as an extended example of how to think mathematically by posing a sequence of general questions that apply to any investigation of any set of mathematical conditions.

1. Does anything satisfying these conditions exist? How do we know?

2. If one or more exist, how do we find it or them? Will this method always work?

3. If more than one exists, what do they all have in common?

The Thinking Tip early in the section calls the students' attention to the generality of this approach.

This section begins in Exploration 1 by examining all circles that pass through a single given point. There are so many such circles that they have no interesting common properties when considered all together. However, by restricting the question to all *circles with a fixed* radius that pass through a common point, we do get some interesting results. The next step (Exploration 2) is to examine all circles that pass through two given points. In this case, something worthwhile can be said about all of these at once. Finally (in Exploration 3), we show that there is exactly one circle that passes through three given points, and we derive a general method for constructing it.

Additional Support Materials:

Assessments	Qty
Form (A)	1
Form (B)	1

Blackline Masters	Qty

Extensions	Qty
Round Pegs in Square Holes	1

Supplements	Qty
Points on Circles	1

EXPLORATION 1

4.13 Mark a point P on a piece of paper. How many circles with the same radius can you draw that will pass through P? Such circles are easy to find. Just mark another point a fixed distance (say, 3 centimeters) from P. Label it Q_1 and let it be the center of your circle. Then draw a circle with center Q_1 that contains P.

Now leave your compass setting the same and choose another point, Q_2. Draw the circle with center Q_2 (and the same radius) that contains P. Repeat this process until you have drawn six or seven circles, all with the same radius, all with different centers, and all passing through the point P. Your picture should look something like Display 4.10.

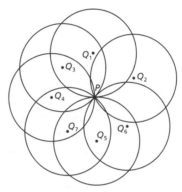

Circles through point P

Display 4.10

It is easy to see that there are lots of circles with the same radius through any single point, P. You can draw as many of them as you want, using a compass. That takes care of questions A. and B. The interesting question for this Exploration is question C. : What do all the circles through P have in common?

1. The centers (Q_1, Q_2, etc.) of the circles you drew might look as if they all lie on a single circle. Do they? How can you be sure? If so, what is the center of that circle and what is its radius?

2. Repeat this construction, this time choosing centers R_1, R_2, etc., farther away from P (say, 5 centimeters). That is, repeat the exercise using a larger compass setting. Do all the centers lie on a circle? If so, what is the radius of that circle? Explain your answer.

Published by IT'S ABOUT TIME, Inc. © 2000 MATHconx, LLC

298

EXPLORATION 1

1. The centers Q_1, Q_2, etc., all lie on a circle of the chosen radius (3 cm) centered at P. This must be the case (by the center radius definition of a circle) because all those centers were constructed at exactly the same distance from P.

2. The centers R_1, R_2, etc. all lie on a circle of the new chosen radius (5 cm) centered at P.

   The answers to questions 1 and 2 can be summarized and generalized as follows. The circles through a given point, P, can be subdivided into "families," with one family for each radius. For each set of circles with fixed radius, the centers of all the circles that contain P lie on another circle with center P and with the same fixed radius.

3. The hint for this question prompts the students to see that a circle of twice the chosen radius, centered at P, would be tangent to all the circles in the family. That is, regardless of which few circles the student drew originally, the larger circle would encompass them all, touching each one at exactly one point. There may also be other properties common to a circle family here, ones which we haven't anticipated. Be open to reasonable suggestions!

Published by IT'S ABOUT TIME, Inc. © 2000 MATHconx, LLC

3. Questions 1. and 2. present an example of the kind of common property we look for. Can you see any other properties common to all the circles with a particular radius that pass through P? *Hint:* Try drawing a circle centered at P with a radius twice that of the other circles in your version of Display 4.10.

EXPLORATION 2

Choose and mark two points, P and Q, at least an inch apart on a piece of paper.

1. Draw a circle that passes through both points, using any method and tools you choose. Mark the center of your circle and label it C. How did you choose this center point?

2. Draw $\triangle PCQ$. What kind of triangle is it? Explain.

3. Draw at least four more circles through P and Q. Mark the center of each circle. Now fold your paper so that points P and Q coincide. Crease your paper so that you can see the fold line clearly. What do you notice?

4. Draw a circle through P and Q with its center *not* on the fold line, if you can. If you cannot, explain why you think it can't be done.

5. Use your answers to parts 3 and 4 to make a conjecture about a common property of all circles through P and Q.

6. Draw the line segment PQ. Label as M the point where PQ crosses the fold line. Is M the midpoint of PQ? Justify your answer.

7. Is the fold line perpendicular to PQ? Justify your answer.

8. Choose any point F, except M, that is on the fold line. Draw FP and FQ. How are $\triangle FMP$ and $\triangle FMQ$ related? (See Display 4.11.) Justify your answer.

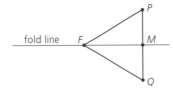

Display 4.11

Published by IT'S ABOUT TIME, Inc. © 2000 MATHconx, LLC

EXPLORATION 2

To make this Exploration effective, you must guide it carefully. The students must have a chance to do the drawing and some thinking on their own. The summary ideas in the text should not be allowed to preempt their own ideas. This may be difficult to orchestrate, but the ideal is that they discover for themselves that the centers of all the circles they draw lie along a straight line.

1. One easy way to find one is to take any disk (jar lid, coin, etc.) and slide it so that its rim just touches the two points. Of course, marking the center of such a thing isn't always easy.

2. It is an isosceles triangle because two sides are radii of the same circle, so they must be the same length.

3. One way to find as many such circles as you want is to cut different size disks out of paper and slide them around until they just touch the two given points. Any disk with diameter at least as large as the distance between the two points will work. If they are made with a compass, the center holes will allow you to mark the location of the center of each circle in relation to P and Q. All the center points should lie along the crease.

4. It can't be done. Explaining why cuts to the heart of this Exploration. If students don't give a very good explanation at this point, that's OK. The rest of this Exploration is designed to lead students to understand the key idea by making and refining conjectures about this fold line. The goal is to have them understand, *by the end of this Exploration*, the following fact.

   This fold line, which is the perpendicular bisector of the segment PQ, consists of all points that are equidistant from P and Q. Hence, it consists of the centers of all circles of any radius that pass through P and Q.

5. Conjecture: The center of any circle that passes through P and Q must be on the fold line.

6. Yes. By folding the paper so that P and Q coincide; the line segments PM and QM must coincide, too, so they must be equally long.

7. Yes. F is on the fold line, so, when the paper is folded over, ∠FMP and ∠FMQ coincide. That is, they are congruent. But their sum, ∠PMQ, is a straight angle because PQ was drawn as a straight segment, so they must add up to 180°. Thus, each angle must be 90°.

8. They are congruent. You might have to help out with some guidance here, but don't rush in too soon. The easiest justification is by SAS. By part 7, ∠FMP and ∠FMQ are equal (right) angles. By part 6, PM = MQ, and FM clearly equals itself.

9. If you draw a circle with the center at your chosen point *F* and passing through *P*, *must* it also pass through *Q*? Justify your answer.

10. What property do all the circles that pass through *P* and *Q* have in common? Make a conjecture. Then see if you can justify it.

(4.15)   Exploration 2 shows that you can draw as many circles containing *P* and *Q* as you want. All you have to do is choose as the center of the circle a point on the perpendicular bisector of *PQ*. This guarantees that the center will be as far from *P* as it is from *Q*. But do you know that there are no other circles containing both *P* and *Q*, circles with their centers *not* on the perpendicular bisector? Maybe you can see from Display 4.11 that a point above the fold line (the perpendicular bisector of *PQ*) appears to be closer to *P* than to *Q*, and a point below the line appears to be closer to *Q*. But can you prove it? That is, can you prove that a point which is the same distance from two other points *must* be on the perpendicular bisector of the segment that connects those other two points? Problem 9 will ask you to do that. Think about how you might use the ideas of Exploration 2 to help you.

About Words

The prefix *equi-* means equal or the same, so *equidistant* means equally distant or the same distance.

This discussion illustrates an important point of logic. Notice that there is a difference between the following two statements.

I. If a point is on the perpendicular bisector of *PQ*, then it is equidistant from *P* and *Q*.

II. If a point is equidistant from *P* and *Q*, then it is on the perpendicular bisector of *PQ*.

Knowing that statement I is true tells us that any point we pick on the perpendicular bisector will be equidistant from *P* and *Q*. Knowing that statement II is true tells us that, if a point is equidistant from *P* and *Q*, it *must* be on the perpendicular bisector. A point *not* on the perpendicular bisector won't have this property.

Published by IT'S ABOUT TIME, Inc. © 2000 MATHconx, LLC

9. Yes, because *PF* and *PQ* have the same length, by part 8. They are corresponding parts of congruent triangles.

10. This is an echo of part 5. Based on parts 6–9, a reasonable conjecture is: The center of every circle that passes through *P* and *Q* is on the perpendicular bisector of *PQ*. However, parts 6–9 do *not* prove this. They actually prove the converse, that every point on the perpendicular bisector is the center of some circle that passes through *P* and *Q*. Help students to make the conjecture *and* to see that it hasn't been proved (yet). This leads to the discussion that comes next.

(4.15) This might be a good place for a quick, simple Geometer's Sketchpad (or other software) demonstration, if that tool is available to you.

1. Draw a segment, *PQ*, anywhere on the display screen.

2. Construct its perpendicular bisector. This can be done in two steps. Construct the midpoint of the segment, then select both the segment and its midpoint and construct the perpendicular.

3. Pick any point, *A*, that is not on these two lines, construct segments *AP* and *AQ*, and display the lengths of these two segments.

4. Using the arrow tool, move *A* around the screen and watch the lengths of *AP* and *AQ* change. Have the students notice that the lengths are most nearly equal when *A* is on the perpendicular bisector of *PQ*. The sometimes inexact match of lengths results from the fact that the display pixels are a discrete grid, not a continuous sheet.

NOTES

Statements I and II are related to each other in a special way: Each is the *converse* of the other. We have used the word converse before; now it is time to take a careful look at exactly what it means. Notice the form of statement I.

IF *a point is on the perpendicular bisector of PQ,*

THEN *it is equidistant from P and Q.*

A statement that can be put in the form

If _____, then _____

is called a **conditional statement** (or sometimes simply a **conditional**). It is called a conditional statement because the "if" part is a condition that is supposed to guarantee the truth of the "then" part, which is called the **conclusion.** The condition itself usually is called the **hypothesis.**

About Words

The plural of *axis* is *axes*. The plural of *hypothesis* is *hypotheses*.

1. **What is the hypothesis and what is the conclusion of statement I?**

2. **What is the hypothesis and what is the conclusion of statement II?**

4.16

Each of these conditional statements, I and II, is made up of the same two parts. However, when we go from one statement to the other, the roles of the two parts are reversed. The hypothesis in the first conditional is the conclusion in the second, and vice versa. In such cases, each conditional statement is called the **converse** of the other one.

When both a conditional statement and its converse are true, we say that the two parts of these statements are **equivalent.** Each one guarantees the other. In such cases, we say that one part is true *if and only if* the other is also true.

In the case of finding the centers of circles containing *P* and *Q*, both conditionals play a role in settling the question. Statement I tells us where to find some centers that work; statement II tells us that there is no place else to look. We can combine these two statements into one by saying,

A point is on the perpendicular bisector of *PQ* if and only if it is equidistant from *P* and *Q*.

Published by IT'S ABOUT TIME, Inc. © 2000 MATHconx, LLC

Chapter 4

4.16

This reinforces the new vocabulary. There is one seemingly fussy, but important, detail to be careful about. The words "if" and "then" themselves are *not* part of the hypothesis or the conclusion, respectively. The hypothesis and the conclusion must be able to stand on their own as separate sentences, so that their truth values can be determined individually.

1. The hypothesis of statement I is
   A point is on the perpendicular bisector of PQ.

   The conclusion of statement I is
   It [the point] is equidistant from P and Q.

2. The hypothesis of statement II is
   A point is equidistant from P and Q.

   The conclusion of statement II is
   It [the point] is on the perpendicular bisector of PQ.

NOTES

1. What is the converse of each of these (true) statements?

4.17

(a) If a triangle has two angles of equal measure, then it is isosceles.

(b) If two triangles are congruent, then their corresponding angles have equal measure.

(c) If a quadrilateral has four axes of symmetry, then it is a square.

(d) Every square has four sides of equal length.

2. Which of the converses of these four statements are true? How do you know?

3. For which statements in question 1 are the hypothesis and the conclusion equivalent? Justify your answers.

From Exploration 1, you know that an unlimited number of circles can be drawn through any given point. In fact, you could choose any other point and there would be a circle centered at that point which passes through the given point. From Exploration 2, you know that an unlimited number of circles can be drawn through any given *pair* of points. The difference in this case is that the centers of all those circles must lie on a single line (on which line?). Now, what about three given points?

If the three points are on the same straight line, no circle passes through all three of them, of course. Points that lie on the same straight line are called **collinear**. The next exploration looks at the case of three points which are not collinear.

EXPLORATION 3

4.18 On a piece of paper, mark three points that are not collinear. Label them P, Q, and R.

1. Can you construct a circle through all three points, P, Q, and R? If so, where is its center? What is its radius?

Maybe that question is too much to handle all at once. Let's try some simpler ones.

2. Can you construct a circle through P and Q? How? Where must its center be? Remember Exploration 2.

302

4.17

These questions reinforce the formal logic of converses and equivalence. They also serve to review some basic facts from previous chapters. However, they can be covered lightly or skipped entirely if you are pressed for time or if you do not want to emphasize the formal logic at this stage.

1. Converses

   (a) If a triangle is isosceles, then it has two angles of equal measure.
   (b) If the corresponding angles of two triangles have equal measure, then the triangles are congruent.
   (c) If a quadrilateral is a square, then it has four axes of symmetry. (or, every square has four axes of symmetry.)
   (d) Every quadrilateral with four sides of equal length is a square.

2. The converses of (a) and (c) are true. The converse of (b) is false because AAA only guarantees similarity. The converse of (d) is false because a rhombus need not be a square.

3. The hypothesis and conclusion are equivalent in (a) and (c) because both of these true statements have true converses.

4.18

EXPLORATION 3

Again, some careful guidance is necessary for this Exploration. The result of this Exploration is the main idea of the section.

Three points that are not collinear determine exactly one circle.

Since both the result and the construction are so important, the text explains the process right after this Exploration. You will have to decide, based on your own students' interests and abilities, how much time to allow for independent or small group Exploration and how much to let them be led by the text and/or by you.

The approach here reflects two fundamental ideas that extend well beyond the geometry of circles. One is the problem solving strategy of asking simpler questions that can be answered fairly, then building up the answers to the more complex question at hand. The other idea mirrors a basic principle of linear algebra: the common solution for two linear equations lies at the intersection of the two lines.

1. Question 1 is answered by doing questions 2, 3, and 4.

2. Pick any point on the perpendicular bisector of PQ as the center. Use the distance between that point and either P or Q as the radius.

3. Can you construct a circle through Q and R? How? Where must its center be?

4. Now put the answers to these two questions together. Can you find a center point that works for *both* of the previous two questions at the same time? What about a radius?

5. Can you construct another circle that passes through all three points, P, Q, and R? If so, where is its center? What is its radius? If not, why can't it be done?

Let's review what Exploration 3 has shown.
- To find a circle that passes through three given points, we must first check to see if they are collinear. If they are, no such circle exists. If they are not collinear, we can *always* find a circle that passes through all three of them.

- The way to find the center and radius of a circle that passes through three non-collinear points, P, Q, and R, is illustrated in Display 4.12. It works like this.

  – Construct the segments between any two pairs of the three points, say PQ and QR.

  – Draw the perpendicular bisectors of these two segments. Their intersection point is the center, which we'll call C. Do you remember why this process works?

  – The radius of the circle is the distance from C to any one of the three given points. Why does it not matter which of the three points we use here?

- The center of any circle through all three points must be on both perpendicular bisectors. Since two straight lines cannot intersect at more than one point, there cannot be more than one such circle.

Published by IT'S ABOUT TIME, Inc. © 2000 MATHconx, LLC

303

3. Pick any point on the perpendicular bisector of QR as the center. Use the distance between that point and either Q or R as the radius.

4. The intersection point of the two perpendicular bisectors is on both lines, so it will work as a center point in both cases. The distance between that point and Q will also work in both cases, so there is a circle that passes through all three points!

5. This question requires a little deeper understanding of what was just done. No other such circle exists. By Exploration 2, the center of *every* circle passing through P and Q must lie on the perpendicular bisector of PQ, and the center of *every* circle passing through Q and R must lie on the perpendicular bisector of QR. Thus, the center of *any* circle that passes through all three of these points must lie on both of these lines. But two nonparallel lines have only one point in common (their intersection point), so there is only one possible center for such a circle. Once the center is determined, so is the radius because it must be the common distance between that center and each of the three points. Thus, there is only one such circle.

NOTES

These questions refer to Display 4.12.

4.19

1. How do we know that the perpendicular bisectors of *PQ* and *QR* *must* intersect? Why can't they be parallel?

2. Where will the perpendicular bisector of *PR* intersect the perpendicular bisectors of *PQ* and *QR*? How do you know?

Three points

Segments

Perpendicular bisectors

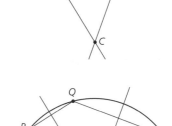

The Circle

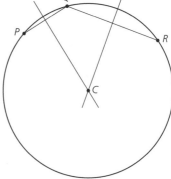

Display 4.12

304

4.19

These questions are worth a little discussion time to help tie up any loose ends of student confusion.

1. The noncollinearity of P, Q, and R means that segments PQ and QR do not form a straight angle. Therefore, the two perpendicular bisectors of these segments cannot be parallel. Make a sketch and look at the measures of the angles.

2. All three perpendicular bisectors must intersect at the same point. Any point equidistant from P and Q and also equidistant from Q and R must be equidistant from P and R. Hence, the point that is common to the perpendicular bisectors of PQ and QR must also be on the perpendicular bisector of PR.

NOTES

Chapter 4

Problem Set: 4.2

For some of these problems, you will need a compass and a straightedge, or a ruler and a protractor, for drawing perpendicular bisectors. You will need a compass for drawing circles.

1. Trace the three points of Display 4.13 on your paper. Then construct the circle that goes through all three of them.

For Problem 1

Display 4.13

2. Trace the four points of Display 4.14 on your paper. Draw as many circles as you can that go through three of these four points. How many possibilities are there?

For Problem 2

Display 4.14

3. Three woodsmen, Abner, Barney and Chester, live in separate cabins deep in the North Woods. Barney lives half a mile directly north of Abner, and Chester lives one mile directly west of Barney. A few days ago, a hunter who also sold TV satellite dishes sold the three woodsmen a satellite dish that could be hooked to all three cabins. They agreed to divide the cost equally, provided that the dish would be no nearer any one of them than it was to the other two. Where should the satellite dish be placed?

Published by IT'S ABOUT TIME, Inc. © 2000 MATHconx, LLC

305

Problem Set: 4.2

1. This mimics the construction in the text. Construct any two of the perpendicular bisectors of the three segments. Their intersection point is the center of the circle. The radius is the distance from that intersection point to any of the three given points.

2. The "how many" part of this problem is a combinatorial question that can be handled by common sense. Any three of these four points will determine exactly one circle. To get three points from four, you just have to leave out one. There are exactly four ways to do that, so there are four circles. Each is drawn as in problem 1.

    Note that there is a subtle problem with generalizing this problem to *any* four points that are not collinear. If all four points happen to lie on the same circle, then, no matter which three points are chosen, the same circle will be determined. Thus, the combinatorial approach gives you only a maximum number. In any given case, you have to draw at least one of the circles or find some other way to check that the four points are not all on the same circle.

3. A diagram of this situation begins with the three woodsmen's cabins, *A, B* and *C*, as the vertices of a right triangle with its right angle at *B*. The dish must be equidistant from all three cabins, so it must be the center of the circle through them. If students draw a sketch of this and solve it by constructing the perpendicular bisectors, they will see that the bisectors of *AB* and *BC* meet at the midpoint of *AC*. That's where the dish should be placed—exactly halfway between Abner's and Chester's cabins.

NOTES

4.2 Points on Circles

4. Ms. Lopez gave these instructions to her students.

Draw three points, A, B, and C. The distance from A to B should be 2 cm, and the distance from B to C should be 2 cm. Now draw the circle that contains all three points.

Minnie and Maxine both followed the teacher's directions completely and correctly. Then Minnie looked at Maxine's paper. "Wait, my circle is much smaller than yours! Why? Did I make a mistake?" Answer Minnie's question, using sketches.

5. Display 4.15 shows a diagram of one edge of a city park. The three trees are inside the park. The Public Works Department decides to put in a flower garden shaped like part of a circular disk bordered on one side by the wall. It wants the circular-arc border to include the three trees. Help them finish this plan. Trace Display 4.15 on a piece of paper. Then draw the outer rim of the garden on your copy and mark the center of its circle.

stone wall

○ tree

○ tree

○ tree

Display 4.15

6. (a) Draw a triangle. Make it whatever shape you want. Carefully construct the perpendicular bisector of each side. Do all three of these perpendicular bisectors meet at a single point? Are you sure?

   (b) Try to draw a triangle for which the perpendicular bisectors of the three sides clearly meet at three different points. Did you already do this in part (a)? Can it be done? Why or why not?

   (c) What do parts (a) and (b) have to do with circles?

306

Published by IT'S ABOUT TIME, Inc. © 2000 MATHconx, LLC

Chapter 4

4. The purpose of this question is to get students to see that the angle formed by the segments *AB* and *BC* is a determining factor in the size of the circle through the three points.

5. The intersection point of the perpendicular bisectors of the segments between the trees is the center of the circle. See Display 4.2T.

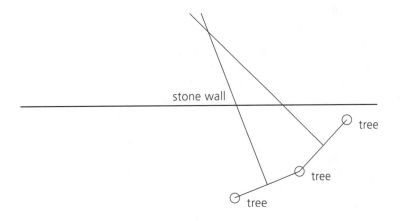

stone wall

tree

tree

tree

Display 4.2T

6. This echoes the second part of discussion 4.19.

   (a) Yes, the three perpendicular bisectors of the sides of a triangle will always meet at a single point. The answers given for this part might be, "Yes, because it looks that way" or "No; it looks like they don't exactly go through the same point." Unless a student has constructed a perpendicular bisector incorrectly or *very* carelessly, the picture should show the three lines coming very close to going through the same point.

   (b) This cannot be done. Each perpendicular bisector of a side is the set of all points equidistant from two vertices. The intersection of two of these bisectors is the point that is equidistant from all three vertices, so it must be a point on the third perpendicular bisector, too.

   (c) This common intersection point is the center of the circle that goes through the three vertices of the triangle. In more formal language, this point is the center of the circle that circumscribes the triangle; it is called the *circumcenter* of the triangle.

7. (a) Walking in the woods, Minnie and Maxine found the
piece of an old wagon wheel diagrammed in
Display 4.16(a). They made some measurements, as
shown in the diagram, which is drawn to scale. That is,
the drawing is in proportion to the real wheel. What
was the radius of the original wheel? You can assume
that it was circular, of course.

(b) Later on their walk, Minnie and Maxine found the
wagon wheel piece that is diagrammed to the same scale
in Display 4.16(b). Do you think it was from the same
wheel? Explain.

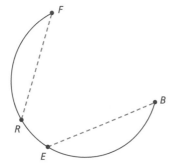

The straight distance from B to E is about 1 foot, 6 inches.
The straight distance from F to R is about 1 foot, 6 inches.

First Piece   Second Piece
    (a)   (b)

Display 4.16

Published by IT'S ABOUT TIME, Inc. © 2000 MATHconx, LLC

7. The scale of the drawing is intended to be 1 inch = 1 foot. Sometimes slight distortions occur in the printing process. You probably should have the students start this problem by carefully tracing the two parts of this diagram on a separate piece of paper.

   (a) The radius of the wheel is about 1 foot. The center is the intersection of the perpendicular bisectors of *FR* and *BE* (or any other two chords). In the diagram, the measured distance between this center point and any point on the wheel should be just about 1 inch if the students draw with some care.

   (b) It can't be from the same wheel. Although the distance between the endpoints of the arc seems to indicate that it would fit in just about right, the radius is too large! Using the endpoints and any point near the middle of the arc, students should be able to construct perpendicular bisectors with sufficient accuracy to see that this radius measures about $1\frac{1}{4}$ inches. That is, it's from a wheel with a 15 inch radius.

8. (i) *Hypothesis*: A point lies on the perpendicular bisector of a line segment.
   Conclusion: The point is equidistant from the two endpoints of the segment.
   Converse: If a point is equidistant from the two endpoints of a line segment, then it lies on the perpendicular bisector of the segment. True.
   A point lies on the perpendicular bisector of a line segment if and only if it is equidistant from the two endpoints of the segment.

   (ii) *Hypothesis*: A (particular) triangle is equilateral.
   Conclusion: The center of the circle through the vertices of a (particular) triangle lies inside the triangle.
   Converse: If the center of the circle through the vertices of a triangle lies inside the triangle, then the triangle is equilateral. False. Any acute triangle that is not equilateral works as a counterexample. Since the converse is false, the hypothesis and conclusion are not equivalent.

8. (a) State in clear and complete sentences the hypothesis and the conclusion of each of the following statements.

   (i) If a point lies on the perpendicular bisector of a line segment, then it is equidistant from the two endpoints of the segment.

   (ii) If a triangle is equilateral, then the center of the circle through its vertices lies inside the triangle.

   (iii) If every angle of a triangle is acute, then the center of the circle through its vertices lies inside the triangle.

   (iv) If three points are not collinear, then there is exactly one circle through all three of them.

   (v) If a quadrilateral is a square, then any circle through three of its vertices must also go through the fourth one.

   (b) What is the converse of each statement in part (a)? Which of these converses are true and which are false? Give a counterexample for each one that is false.

   (c) For which statements in part (a) are the hypothesis and the conclusion equivalent? For each one that is, rewrite the equivalent parts as a single "if and only if" statement.

9. Prove that a point which is the same distance from two other points *must* be on the perpendicular bisector of the segment that connects those other two points.

   Here's a start: Draw a diagram something like Display 4.11. Choose and label two points, say P and Q, draw the segment PQ, and label its midpoint M, and then choose a third point, A, which you assume to be equidistant from P and Q. Draw segments AP, AQ, and AM. Now what? What do you need to prove? How is that related to your diagram?

Published by IT'S ABOUT TIME, Inc. © 2000 MATHconx, LLC

(iii) *Hypothesis*: Every angle of a (particular) triangle is acute. *Conclusion*: The center of the circle through the vertices of a (particular) triangle lies inside the triangle.
Converse: If the center of the circle through the vertices of a triangle lies inside the triangle, then every angle of the triangle is acute. True (but maybe not obvious).
Every angle of a triangle is acute if and only if the center of the circle through its vertices lies inside the triangle.

(iv) *Hypothesis*: Three (given) points are not collinear.
Conclusion: There is exactly one circle through three (given) points.
Converse: If there is exactly one circle through three points, then the points are not collinear. True.
Three points are not collinear if and only if there is exactly one circle through all three of them.

(v) *Hypothesis*: A (particular) quadrilateral is a square.
Conclusion: Any circle through three vertices of a (particular) quadrilateral must also go through the fourth one.
Converse: If any circle through three vertices of a quadrilateral must also go through the fourth one, then the quadrilateral is a square. False. Any nonsquare rectangle works as a counterexample, but there are lots of others.
Since the converse is false, the hypothesis and conclusion are not equivalent.

9. See Display 4.3T for this diagram. It must be shown that line *AM* is perpendicular to *PQ*. By SSS, △*AMP* and △*AMQ* are congruent, implying that ∠*AMP* = ∠*AMQ*. But ∠*PMQ* is a straight angle, so the sum of these two equal angles is 180°, implying that each is a right angle. Thus, *A* is on the perpendicular bisector of *PQ*.

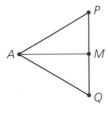

Display 4.3T

4.3 Drawing Circles With a Graphing Calculator

Making connections between algebra and geometry is a powerful, useful habit of thought in mathematics. For example, once you learned how to write algebraic equations for lines, you were able to use that knowledge for statistical forecasting. In this section, we use the graphing calculator to help you see how to describe circles algebraically.

Find out how to draw circles on your graphing calculator. If you have a TI-82 (TI-83), you can look this up in the Drawing Circles section of Appendix A. Then draw each of the following figures.

4.20

1. A circle with center (2, 1) and radius 6. If your graph does not look like a circle, how can you adjust the graph Window so that it does?
2. A circle with center (–7.2, 3.65) and radius $\sqrt{60}$. Adjust the graph Window so that the entire circle appears and looks like a circle.
3. Three circles with center (0, 0).
4. The five-circle Olympic Games logo.

Learning Outcomes

After studying this section, you will be able to:

Use a graphing calculator to graph a circle with a given center and radius;

Represent circles by parametric equations;

Graph circles parametrically;

Change the size and/or position of a circle by changing its equations.

Are you thinking something like, "OK, that's cute, but what is it good for?" A reasonable question. There are many situations in which you could use this drawing to solve a practical problem quickly. Here is an example. After we work through it with you, one or two others will appear as problems for you to think about and solve on your own.

Problem. Terry's Auto Repair has just moved into a 50 by 75 ft. garage that has compressed air outlets in three places along the walls. One is at the center of the front 50 ft. wall, and one is on each 75 ft. side wall, 55 feet back from the front wall. Terry wants to buy three air hoses for operating pneumatic tools long enough so that every place on the garage floor can be reached by at least one hose. Will 30 ft. hoses be long enough, or will Terry have to buy the more expensive 35 ft. size, or the even more expensive 40 ft. size?

Published by IT'S ABOUT TIME, Inc. © 2000 MATHconx, LLC

4.3 Drawing Circles With a Graphing Calculator

Up to now, we have looked at circles from a geometric point of view. This section combines the geometric and algebraic approaches to circles. The graphing calculator is used as a vehicle for examining how an algebraic description of a circle is related to its geometric description, and for recalling the unique image aspect of functions in the context of creating a graph of a circle.

A major goal of this section is to utilize the graphing calculator to introduce parametric equations in a relatively painless way. Parametric equations are useful tools in many areas of mathematics and its applications. This graphical approach makes the underlying ideas easily comprehensible.

4.20

These are mostly routine skill reinforcement exercises.

1. This is completely straightforward, except possibly for the Window adjustment. The standard Window settings for some calculators will make this circle appear elliptical. That can be fixed by adjusting the x and y settings for the Window display to account for the different number of pixels in these two screen dimensions. Most calculators have menu choices for doing this. For instance, on the TI-82 (TI-83), you can use either ZSquare or ZInteger from the ZOOM menu.

2. This exercise extends the ideas of the previous one in two ways.

   (a) It shows students that the center and radius need not be integers, and
   (b) It requires them to be a little creative in adjusting the dimensions of the Window.

3. Students can choose any radii for the three circles; they have to reenter the center point in each case.

4. This logo is five linked circles of equal size, three in an upper row linked by two in a lower row. In this logo, circles in the same horizontal row do not overlap. An example can be formed with circles of radius 5 centered at (10, 15), (22, 15), (34, 15), (16, 10), and (28, 10). On the TI-82 (TI-83), the ZInteger Window setting makes this look right automatically. Of course, students may find many other ways to draw this figure. Allow for some trial and error time as part of the skill reinforcement.

Assessments   Blackline Masters   Extensions   Supplements

For Additional Support Materials see page T-65

Solution. There are lots of different ways to solve this, of course. One way is to take a long string and mark out the different sizes of circles on the floor. But this is clumsy and time consuming, especially since some cars are in the way. Instead, Terry uses a graphing calculator to make some diagrams. Use your calculator to make Terry's diagrams.

- First, adjust the Window size to represent the 50 by 75 ft. floor by setting x between 0 and 75 and setting y from 0 to 50.

- Using these coordinates, the air hose outlets are at (0, 25), (55, 0), and (55, 50).

- Draw three circles of radius 30, one centered at each of the air hose outlets.

Finish the solution to Terry's problem by answering these questions.

4.21

1. With these coordinate settings, which edge of the calculator screen represents the front wall of Terry's garage? How do you know? Why do you think Terry set it up this way?

2. How does the graph show that the 30 ft. hoses are not long enough? How many separate areas of the garage floor cannot be reached by any of these hoses? Use your cursor to specify the coordinates of a point in each of these areas.

3. Are the 35 ft. hoses long enough? Clear the graph Window and draw three new circles to find out. If they are not, will the 40 ft. hoses be long enough?

4. Using the 35 ft. lengths, are there any places that can be reached by two hoses? Are there any places that can be reached by all three hoses? Use the calculator graph Display to help you justify your answers.

5. Can Terry save a little money by buying a 30 ft. hose for the front wall? How can you use your graphing calculator to find out?

Published by IT'S ABOUT TIME, Inc. © 2000 MATHconx, LLC

MATH *Connections*: A Secondary Mathematics Core Curriculum

4.21

Throughout this problem, the coordinates refer to measurements that match the Window settings. The lower left corner of the graph Window is the point (0, 0); the upper right corner is (75, 50).

1. The front of Terry's garage is at the left edge of the calculator screen. Since the maximum x setting is 75, the long sides are horizontal on the screen. The given coordinates of the airhose outlets show that the front wall is at the left, where the x-coordinate is 0. This setup allows for a more faithful screen representation of the shop floor because the screen is wider horizontally than vertically. This setup gives students practice in adapting their tools to a preexisting reality.

2. No. Four different regions of the floor are outside all three circles. (See Display 4.4T(a).) The one in the middle contains the point (33, 25) (and lots of others near it). (See Display 4.4T(b).) The three other uncovered regions are typified by these points: (23, 47), (23, 3), and (74, 25). Points near each of these are unreachable.

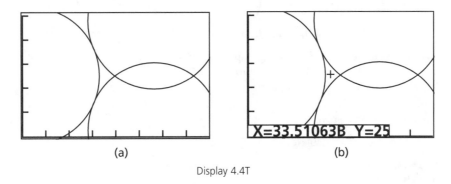

| (a) | (b) |

Display 4.4T

3. Yes, the 35 ft. hoses are long enough. The graph Window should look like Display 4.5T.

4. As Display 4.5T shows, there are three (overlapping) regions that can be reached by different pairs of hoses, and a small, almost triangular region that can be reached by all three.

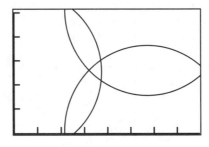

Display 4.5T

Chapter 4

The Navy is planning maneuvers for a 90 mile square area in the central Pacific Ocean. For safety, they want to station air-sea rescue helicopters so that every point in the area can be reached within 20 minutes. In 20 minutes, a helicopter rescue crew can travel up to 36 miles. Three different plans are proposed for the placement of these helicopters. Assuming that the lower left corner of the area is at $(0, 0)$ and the upper right corner is at $(90, 90)$ (miles), here are the helicopter placements for each of the three plans.

4.22

Plan A. Five helicopters—one in the middle of the square (at $(45, 45)$) and one at each corner.

$$(0, 0), (90, 0), (0, 90), (90, 90)$$

Plan B. Four helicopters—at

$$(20, 20), (20, 70), (70, 20), (70, 70)$$

Plan C. Four helicopters—at

$$(25, 25), (25, 65), (65, 25), (65, 65)$$

How can you use the circle drawing feature of your graphing calculator to evaluate these three plans? Do it. Which is the best? Which is the worst? Why?

Have you ever asked yourself

How does a calculator "draw" circles?

When we humans draw a circle, we usually push a pen or pencil around so that it makes a round mark on a flat surface. But a calculator doesn't actually *draw* anything; it *calculates*. That is, it applies the laws of arithmetic and logic to numbers. So, how does it turn numbers into the shapes we see in the graphics Window?

The graphics Window of a calculator is like part of a coordinate plane. It is a rectangular array of little dots, called *pixels*. The arrangement of the pixels may determine a picture. By setting the minimum and maximum Window values for x and y, you automatically give each of these dots a name, an ordered pair (x, y) of numbers that specifies the location of the dot in relation to the x and y axes. These numbers, x and y, are the *coordinates* of the dot. So, to draw a circle, the calculator must calculate the coordinates of all the dots to plot on its graph.

Published by IT'S ABOUT TIME, Inc. © 2000 MATHconx, LLC

311

5. A 30 ft. hose on the front wall would leave one small spot (around (31, 25)) unreachable. Moreover, this would be unwise from a practical standpoint because there would be no room to move tools around at the extreme edge of this hose span.

    We note, in passing, that Terry's gender is never specified in this story. You might spark a brief discussion about gender roles and stereotyping, if you want, by asking students whether they think Terry is male or female!

4.22

Initial discussion probably should focus on how to represent this problem on the graphing calculator. If the graph Window is set so that each coordinate goes from 0 to 90 (miles), then the display will represent the ocean area. Then circles can be drawn to represent the coverage of each helicopter in each case. Center the circle at the starting point for each helicopter and use their maximum range, 36 miles, as the radius. The three different scenarios are shown in Display 4.6T. Plan A doesn't work because some areas cannot be reached within the 15 minute time limit. Both Plan B and Plan C work, but Plan C is better because the circles overlap more. This means that there is more double coverage in the maneuver area and less wasted coverage outside it.

Note that because the area being studied is square and the display Window may not be, the circles may appear oval. It is important that students recognize this as a distortion of the picture, but *not* of the analysis. The calculator is providing accurate information about circular coverage. To remove this visual distortion on a TI-82 (TI-83), set the X range from 0 to 135 and the Y range from 0 to 90. You might also want to put a vertical line at X = 90 to show the limit of the area being studied.

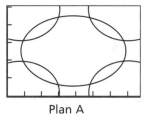
Plan A

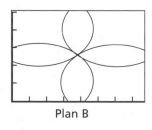
Plan B

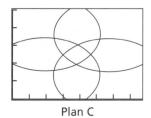
Plan C

Display 4.6T

Thinking Tip

Remember these?
Find or make examples.
Make a picture.
Look for a pattern.
Generalize your results.

To compute these coordinates from the radius and the coordinates of the center, the calculator must translate the geometric idea of a circle into some kind of algebraic recipe. There are several different ways to do this. In the rest of this section we shall explore one of those ways, based on some ideas you saw earlier in **MATH** *Connections*. By asking some easy questions and looking at picture patterns, we'll help the calculator show us how its inner workings behave. We begin with the **unit circle**, the circle of radius 1 centered at (0, 0).

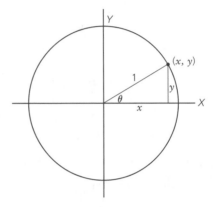

Display 4.17

4.23

Your teacher will give you a copy of Display 4.17, which illustrates the sine and the cosine of an angle, θ. Use it to answer these questions.

1. Express the x and y in Display 4.17 as functions of θ.

2. If $\theta = 30°$, what are x and y? Use your calculator and round your answers to two decimal places. Plot (x, y).

3. Draw a radius that makes an angle of 135° with the positive x-axis counterclockwise from the axis. Use your calculator to find to two decimal places the coordinates of the point where this radius meets the circle. Plot this point. Is it where you expect it to be?

312

4.23

A clear recollection of sine and cosine in the unit circle context is essential for the rest of this section. These questions, which are very much like problems from Section 3.5, serve to remind students of those ideas from the end of Chapter 3.

1. $x = \cos \theta, \quad y = \sin \theta$

2. $x = 0.87, \quad y = 0.5$

3. $x = \cos(135°) = -0.71, \quad y \sin(135°) = 0.71$. The point $(-0.71, 0.71)$ should be right where the radius meets the circle.

Chapter 4

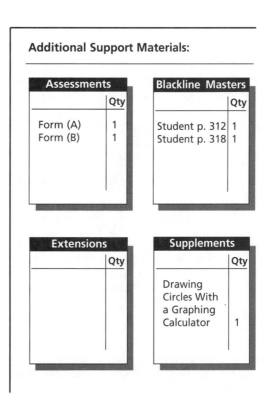

Additional Support Materials:

Assessments	Qty
Form (A)	1
Form (B)	1

Blackline Masters	Qty
Student p. 312	1
Student p. 318	1

Extensions	Qty

Supplements	Qty
Drawing Circles With a Graphing Calculator	1

4. Use your calculator to find (to two decimal places) the coordinates of the point where the radius that makes a counterclockwise angle of 238° with the positive x-axis meets the circle. Plot this point. Check by drawing the radius at that point and measuring the angle with a protractor.

5. Plot the point (0.34, 0.94). Does this point look as if it is on the circle? It should. Assuming that these values have been rounded to two decimal places, find (to the nearest degree) the counterclockwise angle between the positive x-axis and the radius at this point. If you can, find it first by using your calculator; then check your answer with a protractor.

6. Pick any point on the unit circle. Describe two different ways of finding the coordinates of that point, including one that uses the radius drawn from that point.

The questions above show how the coordinates of each point on the unit circle are *functions* of the angle made by a radius and the positive x-axis. If (x, y) is any point on the unit circle and if θ is the counterclockwise angle between the radius at (x, y) and the positive x-axis as in Display 4.17, then

$$x = \cos \theta \quad \text{and} \quad y = \sin \theta$$

The angle θ is called a **parameter** in this case; it is a variable that is used to describe other variables. Equations that contain parameters are called **parametric equations**. These two equations for the coordinates of a point on the unit circle are parametric equations.

A graphing calculator can draw circles by using such parametric equations. To see how it works, follow these steps to draw the unit circle.

4.24

1. Set your calculator to parametric mode.

2. Make sure that it is set to measure angles in degrees and to connect the points that it actually plots.

3. Now look at the function list. It should be waiting for functions *in pairs*, describing X and Y in terms of some parameter, which calculators often call T. In this case, the parameter is angle size, measured in degrees.

Published by IT'S ABOUT TIME, Inc. © 2000 MATHconx, LLC

313

4. (−0.53, −0.85)

5. 70° This question asks students to recognize $\cos^{-1}$ and $\sin^{-1}$ as inverse functions—that is, as ways of undoing what the cos and sin functions do.

6. One way is to drop a perpendicular to each axis, getting x and y directly. The other way is to draw the radius from that point, measure the counterclockwise angle it makes with the positive x-axis, and find the cosine and sine of that angle. Help students to see that this latter process always works.

4.24

Here are the specific settings and commands for drawing the unit circle parametrically using a TI-82 (TI-83). Their numbering corresponds to the more generic steps described in the student text.

1. In the fourth line of the MODE menu, choose Par (for parametric mode).

2. In that same menu, check that Degree is chosen on the line above and that Connected is chosen on the line below.

3. Now press Y= . This list is waiting for functions *in pairs*, describing X and Y in terms of some parameter, which the calculator always calls T.

NOTES

Chapter 4

4. To draw the unit circle, fill in the first two equations as follows.

$$X_1(T) = \cos(T) \text{ and } Y_1(T) = \sin(T)$$

Remember that T has to be entered as a variable.

5. To complete the picture, you have to tell the calculator three things about how to handle the parameter. Go to the calculator menu that lists the settings for T, X, and Y this is the Window menu on the TI-82 (TI-83) and set them as follows.

- T is the angle of rotation around the circle. Since you want a complete circle, let T go from 0° to 360°.

- Set the number of degrees between each plotted point and the next equal to 5 for now. This is Tstep on the TI-82 (TI-83.)

- Set X between −3 and 3 and set Y between −2 and 2, so that the unit circle fits well on the screen.

6. Now let the calculator do the rest. Tell it to graph these equations.

Call the number of degrees between plotted points the step of the parameter.

4.25

1. What happens if you set the step to 45? What if you set the step to 90? How does your calculator connect plotted points?

2. What happens if you set the step to 1? Why? Is that setting better or worse than setting it to 5? Explain.

3. What happens if you set both X and Y to a minimum of −2 and a maximum of 2?

EXPLORATION 1

4.26 Make sure that your calculator is in parametric mode. Set the parameter, T, and the variables X and Y as follows.

- T varies between 0° and 360° in 5° steps.

- X ranges from −6 to 6 with its scale marked at every unit.

- Y ranges from −4 to 4 with its scale marked at every unit.

Published by IT'S ABOUT TIME, Inc. © 2000 MATHconx, LLC

4. To draw the unit circle, fill in X₁ᴛ= by pressing COS and X,T,θ. The calculator will display our parameter θ, as T; the equation on the screen will be X₁ᴛ=cos T. Now put in the second equation, Y₁ᴛ=sin T. The TI-83 uses parentheses after the sin and cos functions: sin(T) and cos(T).

5. Press WINDOW. In parametric mode, the Window menu begins with three settings for the parameter, T.
   • Set Tmin=0 and Tmax=360.

   • Tstep sets the number of degrees between each plotted point and the next; set this equal to 5 for now.

   • Finally, set X between –3 and 3 and set Y between –2 and 2. The 3:2 ratio between these X and Y settings makes the circle look round, instead of oval.

6. Now press GRAPH.

4.25

These questions help familiarize students with some of the Window options for drawing circles parametrically.

1. The result is an octagon. You get a square (more or less). These things happen because the calculator connects plotted points linearly.

2. The circle is drawn much more slowly because the calculator is computing the coordinates of 360 points, rather than 72 points. If we were making a really large copy of the unit circle, this might be better. However, there are so few pixels on the calculator screen that the better theoretical accuracy is lost. The picture is no better, and the process is much slower. Thus, in this situation, step = 1 is actually worse than step = 5.

3. The circle looks oval, rather than round.

EXPLORATION 1

4.26

The detailed instructions of this setup serve to reinforce the parametric equation techniques just described. By following along, step by step, the students should see how to handle the details of such a setup in the future. SETUP for subsequent Explorations gradually become more terse.

Now enter these parametric equations into your equation list.

$$X_1(T) = \cos(T) \qquad Y_1(T) = \sin(T)$$

$$X_2(T) = 2\cos(T) \qquad Y_2(T) = 2\sin(T)$$

$$X_3(T) = 3\cos(T) \qquad Y_3(T) = 3\sin(T)$$

$$X_4(T) = 4\cos(T) \qquad Y_4(T) = 4\sin(T)$$

Do not graph these equations until you answer the questions in part 1.

1. When you graph these equations, you should get some circles. How many circles should you get? Where do you think they are centered? What size(s) are they? Now graph the equations and check your answers.

2. If you were to enter the equations

$$X_5(T) = 5\cos(T) \qquad \text{and} \qquad Y_5(T) = 5\sin(T)$$

   how do you think the picture would change? Try it.

3. Why do the circles appear to be flat near the axes?

4. If you were to enter the equations

$$X_6(T) = 10\cos(T) \quad \text{and} \quad Y_6(T) = 10\sin(T)$$

   what would you get? Try it. Is there a problem with your picture? There should be. How can you fix it?

5. What parametric equations would you use to draw a circle of radius 8 centered at (0, 0)? What about a circle of radius 25? What about a circle of radius 53.87?

6. Complete this statement.

   The parametric equations for a circle of radius r centered at (0, 0) are _____ and _____.

Exploration 1 shows us what happens when you multiply the parametric functions for the unit circle by a fixed number. It's natural to ask what happens when, instead of multiplying, you add a fixed number to these two functions. Exploration 2 answers this question.

Published by IT'S ABOUT TIME, Inc. © 2000 MATHconx, LLC

1. You get four circles, all centered at (0, 0), of radii 1, 2, 3 and 4.

2. A fifth circle is added to the picture. It is of radius 5 and is also centered at (0, 0).

3. This happens because the pixels of the calculator screen are not fine enough to reflect accurately the small change in curvature near the axes. The pixels are lined up in a horizontal/vertical array, so, although the curvature is the same everywhere on the circle, it is most difficult to represent near the horizontal and vertical axes. Some students might suggest fixing this by changing Tstep from 5 to something smaller. Let them try it. It's a good theoretical insight, but it doesn't work here because the trouble lies with the size of the pixels, not the size of the steps.

4. This question reinforces the connection between the radius and the constant multiple. The problem is that nothing new appears on the screen. That's because the X and Y ranges are too small. It can be fixed so that the sixth circle appears by setting the X and Y ranges large enough to accommodate a circle of radius 10. Students can fix the problem by choosing both the X and Y ranges between -10 and 10, but the circles will look oval. A better remedy is $-15 \leq X \leq 15$ and $-10 \leq Y \leq 10$.

5. $X = 8 \cos(T)$ and $Y = 8 \sin(T)$
    $X = 25 \cos(T)$ and $Y = 25 \sin(T)$
    $X = 53.87 \cos(T)$ and $Y = 53.87 \sin(T)$

    It might be good at this point to reintroduce the noncalculator notation for the variables.

    $x = 8 \cos \theta$ and $y = 8 \sin \theta$
    $x = 25 \cos \theta$ and $y = 25 \sin \theta$
    $x = 53.87 \cos \theta$ and $y = 53.87 \sin \theta$

6. The parametric equations for a circle of radius r centered at (0, 0) are
    $$x = r \cos \theta \text{ and } y = r \sin \theta$$
    In calculator notation: $X(T) = r \cos T$ and $Y(T) = r \sin T$.

EXPLORATION 2

4.27 Stay in parametric mode. Set X to vary between -9 and 9, and set Y to vary between -6 and 6. Then change the parametric equations to these.

$$X_1(T) = \cos(T) \qquad\qquad Y_1(T) = \sin(T)$$

$$X_2(T) = 2 + \cos(T) \qquad Y_2(T) = 2 + \sin(T)$$

$$X_3(T) = 3 + \cos(T) \qquad Y_3(T) = 3 + \sin(T)$$

$$X_4(T) = 4 + \cos(T) \qquad Y_4(T) = 4 + \sin(T)$$

Now graph these equations.

1. You should get some circles. How many? Where are they centered? What size(s) are they?

2. If you think of the obvious pattern formed by these circles, one seems to be missing. Fill it in by writing its parametric description as $X_5(T)$ and $Y_5(T)$. Graph again to check.

3. How do you think the picture would change if you put in

    $$X_6(T) = 10 + \cos(T) \quad \text{and} \quad Y_6(T) = 10 + \sin(T)$$

    Try it. If you don't get any change, try adjusting the Window settings.

4. Write parametric equations to make Display 4.18.

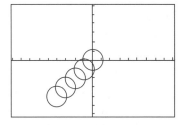

Display 4.18

5. Write parametric equations to make Display 4.19.

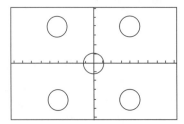

Display 4.19

316

4.27 It might be useful to emphasize the "What if... ?" habit of thought here. Asking about changing multiplication to addition (or some other operation) and seeing what happens exemplifies a thought pattern that should be encouraged. This Exploration shows that it can pay handsome dividends. The question right after this Exploration is generated by a similar thought pattern. Investigating it will pay off, too, but not until later in the section.

1. You get four circles, all of radius 1, centered at (0, 0), (2, 2), (3, 3) and (4, 4). (See Display 4.7T.)

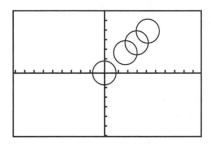

Display 4.7T

2. The missing circle is the circle of radius 1 centered at (1, 1). The pattern in changing from multiplication to addition should have added 1 in the first equations.

$$X_5(T) = 1 + \cos(T) \text{ and } Y_5(T) = 1 + \sin(T)$$

3. It doesn't change at all with these Window settings. The equations describe a circle of radius 1 centered at (10, 10), so the ranges of X and Y need to be increased at least enough to show points with one or the other coordinate equal to 11. Keeping the X to Y ratio at 3 to 2 (to preserve the appearance of circularity), one might simply extend the maximum values of X and Y to 15 and 12, respectively. Of course, there are other adjustments that are at least as good. For instance, setting $-1 \leq X \leq 14$ and $-1 \leq Y \leq 11$ still shows all the circles, but their slightly larger size makes the visual impression a little better.

4. Displays 4.18, 4.19, and 4.20 use the Window settings from the SETUP of this Exploration. This can be seen by counting the "tick marks" on the axes. Equations for Display 4.18.

$$X_1(T) = \cos(T) \qquad Y_1(T) = \sin(T)$$
$$X_2(T) = -1 + \cos(T) \qquad Y_2(T) = -1 + \sin(T)$$
$$X_3(T) = -2 + \cos(T) \qquad Y_3(T) = -2 + \sin(T)$$
$$X_4(T) = -3 + \cos(T) \qquad Y_4(T) = -3 + \sin(T)$$
$$X_5(T) = -4 + \cos(T) \qquad Y_5(T) = -4 + \sin(T)$$

6. Write parametric equations to make Display 4.20.

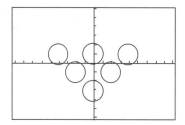

Display 4.20

7. Complete this statement.

The parametric equations $x = a + \cos \theta$ and $y = b + \sin \theta$ describe a circle of radius ___ centered at ____.

4.28

1. What would happen if you added a fixed number to the parameter, instead of to the function? In particular, what do you think the graph of

$$x = \cos(5 + \theta) \quad \text{and} \quad y = \sin(5 + \theta)$$

looks like? Try to figure it out before using your calculator.

2. What would happen if you multiplied the parameter, instead of the function, by a fixed number? In particular, what do you think the graph of

$$x = \cos(5\theta) \quad \text{and} \quad y = \sin(5\theta)$$

looks like? Try to figure it out before using your calculator.

If you are puzzled by what your calculator displays in these two situations, don't worry. Just try to make a conjecture (an intelligent guess) about what is going on in each case. How can you check your guesses? What other equations might you try, and what would you expect to happen?

Published by IT'S ABOUT TIME, Inc. © 2000 MATHconx, LLC

317

5. In the previous examples, there was an implicit (but not necessary) natural order in which the pairs of equations appeared. In this example and the next, the order in which the pairs of equations appear is purely a matter of personal choice. Equations for Display 4.19.

$$X_1(T) = \cos(T) \qquad Y_1(T) = \sin(T)$$
$$X_2(T) = 4 + \cos(T) \qquad Y_2(T) = 4 + \sin(T)$$
$$X_3(T) = -4 + \cos(T) \qquad Y_3(T) = 4 + \sin(T)$$
$$X_4(T) = -4 + \cos(T) \qquad Y_4(T) = -4 + \sin(T)$$
$$X_5(T) = 4 + \cos(T) \qquad Y_5(T) = -4 + \sin(T)$$

6. Equations for Display 4.20.

$$X_1(T) = \cos(T) \qquad Y_1(T) = 1 + \sin(T)$$
$$X_2(T) = 4 + \cos(T) \qquad Y_2(T) = 1 + \sin(T)$$
$$X_3(T) = -4 + \cos(T) \qquad Y_3(T) = 1 + \sin(T)$$
$$X_4(T) = 2 + \cos(T) \qquad Y_4(T) = -1 + \sin(T)$$
$$X_5(T) = -2 + \cos(T) \qquad Y_5(T) = -1 + \sin(T)$$
$$X_6(T) = \cos(T) \qquad Y_6(T) = -3 + \sin(T)$$

7. The parametric equations $x = a + \cos\theta$ and $y = b + \sin\theta$ describe a circle of radius 1 centered at (a, b).

4.28

This is a habit of thought inquiry. It should be natural to ask what happens if you apply the same processes (addition and multiplication) to different parts of the equations. When students try this out, the visual results probably will be unclear, if not downright confusing. In the first case (add 5), assuming that students leave the Window settings unchanged, the result will look like the unit circle again. In the second case (multiply by 5), the figure will also be the unit circle, but it will seem to appear much faster and then fatten up a little before the calculator finishes its graph. Allow the students to puzzle over these results and analyze them as best they can. Do not bring these issues to closure now; they will be the subject of a later Exploration.

In Exploration 1 you learned how to change the size of a circle in the calculator Window, and in Exploration 2 you learned how to change its location. Now we do both at the same time.

(4.29) EXPLORATION 3

Stay in parametric mode. Change the Window settings so that X varies between -12 and 12 and Y varies between -8 and 8. Your teacher will give you a copy of Display 4.21 to fill in during this Exploration. Enter these parametric equations as the first two pairs in your calculator.

– For circle C_1: $x = 3 \cos \theta$ and $y = 3 \sin \theta$

– For circle C_2: $x = 7 + 3 \cos \theta$ and $y = 4 + 3 \sin \theta$

Circles C_3 and C_4 appear in questions 6–10.

θ	Pt on C_1	Pt on C_2	Pt on C_3	Pt on C_4
0°		(10, 4)		
5°	(2.99, 0.26)			
45°				
57°				
120°				
143.5°				
180°				
245°				
273.84°				
325°				

Display 4.21

1. What is the center of circle C_1? What is its radius?

2. What is the center of circle C_2? What is its radius?

3. Use the Trace feature of your calculator to fill in the C_1 and C_2 columns of Display 4.21 with the coordinates of the points that correspond to the angle measurements in column 1. Round each number to two decimal places. One point for each circle is done for you; do as many more as you can. Why can't you do them all in this way?

4. How can you find the coordinates of points formed by angles that are not exact multiples of 5°? Test your answer by filling in the remaining entries in the table.

Published by IT'S ABOUT TIME, Inc. © 2000 MATHconx, LLC

The purpose of this Exploration is to explain the general parametric equations that describe a circle with center (a, b) and radius r. The first equation pair harks back to Exploration 1; C_1 is a circle centered at the origin, but with a radius larger than 1. C_2 is formed from C_1 by moving the center, as in Exploration 2. The equations are deliberately presented using θ, rather than T, as the variable, so that students get used to translating usual algebraic notation into calculator notation (and vice versa).

Note that there is a Blackline Master for this Exploration.

1. C_1 is centered at the origin, $(0, 0)$, and has radius 3.

2. C_2 is centered at $(7, 4)$, and has radius 3.

3. The table of Display 4.21, filled in, appears below. You might have to remind some students how the TRACE feature works.

   (a) The right and left arrows move the cursor around the circle counter-clockwise and clockwise, respectively.

   (b) The down and up arrows move the cursor from each figure to the next, in sequential and reverse sequential order, respectively.
   The 4th, 6th, and 9th lines of the table cannot be completed in this way.

θ	Pt on C_1	Pt on C_2	Pt on C_3	Pt on C_4
0°	(3, 0)	(10, 4)	(−1, 2)	(10.5, 4)
5°	(2.99, 0.26)	(9.99, 4.26)	(−1.02, 2.44)	(10.49, 4.31)
45°	(2.12, 2.12)	(9.12, 6.12)	(−2.46, 5.54)	(9.47, 6.47)
57°	(1.63, 2.52)	(8.63, 6.52)	(−3.28, 6.19)	(8.91, 6.94)
120°	(−1.5, 2.60)	(5.5, 6.60)	(−8.5, 6.33)	(5.25, 7.03)
143.5°	(−2.41, 1.78)	(4.59, 5.78)	(−10.02, 4.97)	(−2.41, 1.78)
180°	(−3, 0)	(4, 4)	(−11, 2)	(3.5, 4)
245°	(−1.27, −2.72)	(5.73, 1.28)	(−8.11, −2.53)	(5.52, 0.83)
273.84°	(0.20, −2.99)	(7.20, 1.01)	(−5.66, −2.99)	(7.23, 0.51)
325°	(2.46, −1.72)	(9.46, 2.28)	(−1.90, −0.87)	(9.87, 1.99)

The table of Display 4.21 completed

Display 4.8T

4. This question serves to reinforce the important idea that these parametric equations are functions of θ. There is more than one way to do this. The easiest way is to calculate the images. For the TI-82 (TI-83), use 1:value from the CALCULATE menu (2nd CALC)). Both parametric functions will be evaluated at once, for each circle. It is like using TRACE, except that you must enter the value of θ, instead of moving the cursor to it.

5. For the first four rows of the table, subtract the x and y coordinates in the C_1 column from the corresponding x and y coordinates in the C_2 column. What pattern do you see? Check to see if your pattern works by computing the coordinate differences for the rest of the rows. Are there *any* rows for which it doesn't work? Explain why your pattern works.

6. Write the two parametric equations that describe the circle, C_3, with radius 5 centered at $(-6, 2)$. Enter them as the third pair of equations in your calculator. Then check the graph to see if this circle appears where you think it should.

7. Fill in the C_3 column of Display 4.21 with the coordinates of the points that correspond to the angle measurements in column 1. Round each number to two decimal places.

8. Write the two parametric equations that describe a circle, C_4, which encloses circle C_2 inside it, but doesn't touch either axis. Then use your equations to fill in the C_4 column of Display 4.21.

9. Complete this statement.

   The parametric equations for a circle of radius r centered at (a, b) are _____ and _____.

10. Find and describe at least one pattern in your filled-out copy of Display 4.21 that we have not already discussed. Try to relate your pattern to your answer to part 9.

EXPLORATION 4

Begin with all settings and equations for circles C_1, C_2, and C_3 the way they were at the end of Exploration 3. Delete circle C_4.

1. Change the maximum T setting to 180°. Explain what effect you think this will have on the graph and why you think so. See if you are right by graphing.

2. Explain how to get each of the graphs in Display 4.22.

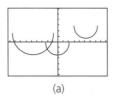 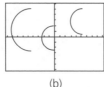 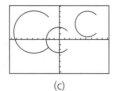

(a)	(b)	(c)

Display 4.22

Published by IT'S ABOUT TIME, Inc. © 2000 MATHconx, LLC

319

If students are having trouble seeing past the mechanics of button pushing, these questions might help to focus them on a transitional idea. Why does TRACE take 5° steps? Because the step for T was set that way. How could you make it take 2° or 1° steps? Reset the step for T to 2° or 1°. This might help them see that *they* control the function evaluation process.

5. Besides emphasizing a fundamental concept about representing circles, this question gives students practice in dealing with signed numbers. The pattern is that the difference is *always* (7, 4). If a student finds a row that "doesn't work," check the arithmetic! The pattern works because these are the constants that are added to each parametric equation to translate C_1, which is centered at (0, 0), to C_2, which is centered at (7, 4).

6. $x = (-6) + 5\cos\theta$ and $y = 2 + 5\sin\theta$

7. See Display 4.8T. Let students find these values by whatever method(s) they choose.

8. There are many ways to do this. One easy way is to use the same center as C_2, but increase the radius by a little (say 0.5). In this case, the parametric equations are

$$x = 7 + 3.5\cos\theta \quad \text{and} \quad y = 4 + 3.5\sin\theta$$

The entries in Display 4.8T reflect this choice, but there are infinitely many other circles that will satisfy these requirements.

9. The parametric equations for a circle of radius r centered at (a, b) are

$$x = a + r\cos\theta \quad \text{and} \quad y = b + r\sin\theta$$

10. This question is completely open-ended. There are *lots* of other patterns in this table; some of them are related to the parametric descriptions of the circles, some are not.

This final Exploration of the section leads to the questions raised in discussion 4.28. Beginning with the simple idea of restricting the domain of the parameter (θ), it leads students to discover that adding a fixed quantity to the parameter (rather than to its image under sin and cos) rotates the arc by that fixed amount. Moreover, multiplying the parameter by a fixed amount expands or contracts the size of the arc by that multiple. Since we have not discussed arcs or arc length formally yet, no explicit conclusions or specific results are established here. However, students should come away with the general idea of how the graph of a circle is changed when the parameter is altered by addition or multiplication of a constant.

1. The second (bottom) half of each circle will not be drawn because the angle of rotation must lie between 0° and 180°.

2. Together, these examples show that either the maximum or the minimum settings for the parameter, T, or both, may be altered, and they need not be altered by the same amount.

   (a) Set minimum $T = 180°$ and maximum $T = 360°$
   (b) Set minimum $T = 90°$ and maximum $T = 270°$
   (c) Set minimum $T = 45°$ and maximum $T = 315°$

3. Together, these examples show that the maximum value of T may be greater than 360° and that the range need not be symmetric about the 360° point.

   (a) Set minimum $T = 270°$ and maximum $T = 450°$
   (b) Set minimum $T = 315°$ and maximum $T = 405°$
   (c) Set minimum $T = 315°$ and maximum $T = 540°$

NOTES

Chapter 4

3. Explain how to get each of the graphs in Display 4.23.

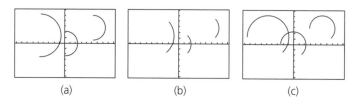

(a) (b) (c)

Display 4.23

4. Explain how to get the graph in Display 4.24.

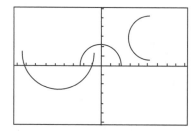

Display 4.24

5. Explain how to get the graph in Display 4.25.

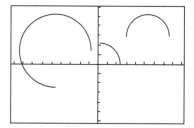

Display 4.25

How are parts 4 and 5 of Exploration 4 related to the discussion questions in Exploration 2?

4.31

Published by IT'S ABOUT TIME, Inc. © 2000 MATHconx, LLC

320

Chapter 4

4. This is the first difficult part. Because the three circles are changed in different ways, it is not possible to get this graph just by changing the parameter limitations. Students should be encouraged to talk their way through this problem, realizing (perhaps only gradually) that something different must be done to the T for each circle. For C_1, the strategy from part 1 suffices; if the range of T is set from $0°$ to $180°$, this circle will be altered in just the correct way. The other two will be the right size (i.e., semicircles), but they will be oriented incorrectly. Help students to see that C_2 needs to start at $90°$ and end at $270°$, and *that's the same as adding $90°$ to every angle between $0°$ and $180°$.* That is, the equations for C_2 should become

$$X = 7 + 3\cos(T + 90) \quad \text{and} \quad Y = 4 + 3\sin(T + 90)$$

This is the critical insight they need. Then it will be easier to see that the C_3 equations should become

$$X = -6 + 5\cos(T + 180) \quad \text{and} \quad Y = 2 + 5\sin(T + 180)$$

5. Once again, the solution to this part begins by setting the proper parameter limits to get C_1 correct: If the range of T is set from $0°$ to $90°$, this circle will be altered in just the right way. Help the students to see that the displayed part of C_2 starts at the same place ($0°$) but it is twice as long; that is, for every degree the parameter changes around C_1, it must change twice as far around C_2. Thus, the equations for C_2 should become

$$X = 7 + 3\cos(2T) \quad \text{and} \quad Y = 4 + 3\sin(2T)$$

Then it is fairly easy to see that the C_3 equations should become

$$X = -6 + 5\cos(3T) \quad \text{and} \quad Y = 2 + 5\sin(3T)$$

Note that the thinking pattern in the solution to this part parallels that of the previous part. This is a deliberate attempt to influence students' problem solving strategies by example.

4.31

The principles are exactly the same. The puzzlement in question 1 of the discussion 4.28 comes from the fact that adding only $5°$ to θ shifts the starting point of the circle drawing so little that it is difficult to see in the small scale Window. And since the parameter was allowed to vary through a full $360°$, the entire circle was drawn, making it appear that nothing had changed.

In question 2 of the discussion 4.28, the puzzlement arises from the fact that multiplying θ by $5°$ and allowing it to vary through a full $360°$ meant that the calculator actually plotted a point every $25°$, instead of every $5°$, but it went around the circle 5 times. This made the initial plotting appear to be much faster. Again, much of the confusion was due to relatively large pixels in a relatively small Window.

Problem Set: 4.3

1. Draw each of the following designs with your graphing calculator.

   (a) a circle with center (2, –5) and radius 7
   (b) a circle with center (4, 3.6) and radius 5.68
   (c) a circle centered at (–3, 1) that passes through (–3, 10)
   (d) a circle centered at (5, 8) that passes through (–7, 17)
   (e) four circles of radius 10 that pass through the point (7, –3)

2. These pairs of parametric equations describe circles. In each case, specify the center and the radius of the circle; then graph it on your calculator. Choose whatever Window settings you think best.

   (a) $x = 8 \cos \theta$ and $y = 8 \sin \theta$
   (b) $x = (-3) + 5 \cos \theta$ and $y = 5 \sin \theta$
   (c) $x = 5 + \cos \theta$ and $y = 6.25 + \sin \theta$
   (d) $x = 5.8 + 4 \cos \theta$ and $y = (-2.3) + 4 \sin \theta$
   (e) $x = (-2) + 0.7 \cos \theta$ and $y = (-3.5) + 0.7 \sin \theta$
   (f) $x = 1.25 + \frac{1}{3} \cos \theta$ and $y = (-1) + \frac{1}{3} \cos \theta$

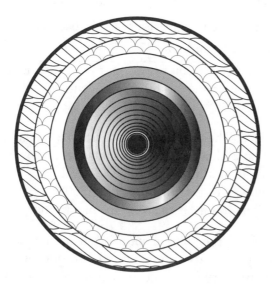

4.3 Drawing Circles With a Graphing Calculator

Chapter 4

Problem Set: 4.3

1. The two simplest methods of drawing circles with the calculator are (i) by using the cursor directly on the graph screen, and (ii) by entering the center and radius into the circle drawing function chosen from a Draw or Sketch menu. The implicit message of these examples is that neither method is better *all* the time.

   (a) This one is straightforward. It should be easy by either method.

   (b) This one cannot be done precisely by moving the cursor; the decimal coordinates of the pixels will permit only approximations. On the other hand, the function approach simply requires entering the three numbers.

   (c) This one requires a little arithmetic when done with the center radius function. Students must observe that the radius is given by the difference in the *y*-coordinates and subtract.

   (d) This one is easy with the cursor, but a nuisance with the function. In fact, unless students are skilled at computing the distance between two points from their coordinates (a skill that we don't expect at this stage), the function method will be difficult for them. The coordinates in this case have been chosen to work out to an integral radius, 15, but finding it requires applying the Pythagorean Theorem and computing or recognizing a square root.

   (e) This problem requires a little ingenuity. It has many different solutions. The simplest ones begin by observing that it is easy to locate the center of a circle through the given point by moving 10 units *in one of the coordinate directions*. Then both methods are fairly easy; the function approach requires a little more signed number arithmetic, but less button pushing, than using the cursor directly on the graph.

2. (a) Center is (0, 0); radius is 8

   (b) Center is (-3, 0); radius is 5

   (c) Center is (5, 6.25); radius is 1

   (d) Center is (5.8, -2.3); radius is 4

   (e) Center is (-2, -3.4); radius is 0.7

   (f) Center is (1.25, -1); radius is $\frac{1}{3}$

For problems 3 through 8, adjust your calculator Window settings and enter the parametric equations that will form the designs shown.

3.
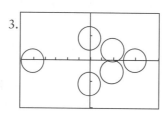

4.
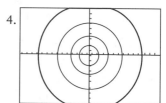

5.
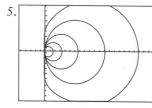

6.
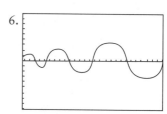

7.
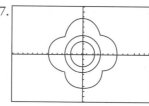

8.
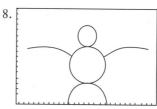

9. Here are three different ways to make a semicircle of radius 1 centered at (0, 0). The minimum for T is 0° and its step is 5° in all three cases.

(a) Set maximum $T = 180°$; use the equations $X = \cos(T)$ and $Y = \sin(T)$.

(b) Set maximum $T = 360°$; use the equations $X = \cos(0.5T)$ and $Y = \sin(0.5T)$.

(c) Set maximum $T = 90°$; use the equations $X = \cos(2T)$ and $Y = \sin(2T)$.

How do these different descriptions affect the way the graphing calculator plots the points? Which is the fastest? Which is the most accurate? Explain your answers.

Published by IT'S ABOUT TIME, Inc. © 2000 MATHconx, LLC

3. This problem focuses only on setting different center points.

   Window settings: $-6 \le X \le 6$, $-4 \le Y \le 4$

Equations	
$X = 4 + \cos(T)$	$Y = \sin(T)$
$X = 2 + \cos(T)$	$Y = 1 + \sin(T)$
$X = 2 + \cos(T)$	$Y = -1 + \sin(T)$
$X = \cos(T)$	$Y = 2 + \sin(T)$
$X = \cos(T)$	$Y = -2 + \sin(T)$
$X = -5 + \cos(T)$	$Y = \sin(T)$

4. This problem focuses only on setting different radii.

   Window settings: $-15 \le X \le 15$, $-10 \le Y \le 10$

Equations	
$X = \cos(T)$	$Y = \sin(T)$
$X = 2\cos(T)$	$Y = 2\sin(T)$
$X = 4\cos(T)$	$Y = 4\sin(T)$
$X = 7\cos(T)$	$Y = 7\sin(T)$
$X = 11\cos(T)$	$Y = 11\sin(T)$

5. This problem uses the same circle sizes as in the previous problem, but shifts the centers. It also requires students to adjust the X scale.

   Window settings: $-5 \le X \le 25$, $-10 \le Y \le 10$

Equations	
$X = \cos(T)$	$Y = \sin(T)$
$X = 1 + 2\cos(T)$	$Y = 2\sin(T)$
$X = 3 + 4\cos(T)$	$Y = 4\sin(T)$
$X = 6 + 7\cos(T)$	$Y = 7\sin(T)$
$X = 10 + 11\cos(T)$	$Y = 11\sin(T)$

6. This problem involves all the previous types of adjustments, as well as restriction to semicircles and 180° rotation of some of them. It is set along the positive x-axis to give students a handy reference scale for centers and radii as they figure out the necessary parametric adjustments.

   Window settings: $0 \le T \le 180$, $0 \le X \le 24$, $-8 \le Y \le 8$

Equations	
$X = 1 + \cos(T)$	$Y = \sin(T)$
$X = 3 + \cos(T + 180)$	$Y = \sin(T + 180)$
$X = 6 + 2\cos(T)$	$Y = 2\sin(T)$
$X = 10 + 2\cos(T + 180)$	$Y = 2\sin(T + 180)$
$X = 15 + 3\cos(T)$	$Y = 3\sin(T)$
$X = 21 + 3\cos(T + 180)$	$Y = 3\sin(T + 180)$

7. This design is made primarily from semicircles, centered in different places and rotated different amounts. If the students try to make the middle circles out of semicircles, they won't have enough room for the necessary equations. The trick is to double the parameter for those two circles, allowing each full circle to be drawn with a single set of parametric equations.

Window settings: $0 \leq T \leq 180$, $-12 \leq X \leq 12$, $-8 \leq Y \leq 8$

Equations

$X = 3 \cos T$	$Y = 3 + 3 \sin T$
$X = -3 + 3 \cos(T + 90)$	$Y = 3 \sin(T + 90)$
$X = 3 \cos(T + 180)$	$Y = -3 + 3 \sin(T + 180)$
$X = 3 + 3 \cos(T + 270)$	$Y = 3 \sin(T + 270)$
$X = 3 \cos(2T)$	$Y = 3 \sin(2T)$
$X = 2 \cos(2T)$	$Y = 2 \sin(2T)$

8. The "wings" of this one are a real challenge. The diagram can be made in several different ways. This is the one we used.

Window settings: $0 \leq T \leq 180$, $0 \leq X \leq 15$, $0 \leq Y \leq 10$

Equations

$X = 8 + 2 \cos(2T)$	$Y = 4 + 2 \sin(2T)$
$X = 8 + 2 \cos(T)$	$Y = 2 \sin(T)$
$X = 8 + \cos(2T)$	$Y = 7 + \sin(2T)$
$X = 3 + 6 \cos(0.25T + 60)$	$Y = 6 \sin(0.25T + 60)$
$X = 13 + 6 \cos(0.25T + 75)$	$Y = 6 \sin(0.25T + 75)$

9. The issue here is that multiplying T by a constant changes the step for T by the same multiple. (c) is the fastest because the calculator is actually plotting points on the semicircle in $10°$ steps, connecting the dots linearly in between. (b) is the most accurate, but also the slowest, because the calculator is plotting points on the semicircle in $2.5°$ steps.

NOTES

4.4 Area and Circumference

In your earlier schooling you probably have learned two very important formulas about circles:

$$C = 2\pi r$$
$$A = \pi r^2$$

The first formula tells you how to find the circumference of a circle from its radius. The second tells you how to find its area from its radius. Use these two formulas and the fact that the diameter of a circle is twice its radius to solve the following problems. Let your calculator supply the value of π.

At Pietro's Pizza, an 8 inch cheese pizza sells for $4.25, the 12 inch size sells for $8.50, and the 14 inch size sells for $10.25. All these pizzas are round. Which is the best deal? Why?

a
4.32

Jack the Tracker has 1200 feet of sturdy fencing for a pen to protect his flock of emus from the dingoes while he is away on walkabouts in the outback. Jack wants his emus to have as much space as possible for running around. He recalls that a square is the rectangular shape with the largest area, but wonders if a circular pen with the same perimeter might enclose even more area.

b
4.33

1. Should Jack build a square pen or a circular pen with his 1200 feet of fencing? Justify your answer.

2. What's an emu? What's a dingo? Which country do you think Jack lives in?

Now that you recall how these two circle formulas work, let's look at them a little more closely. Both depend on a strange number known as π. That symbol is the Greek letter which corresponds to our letter p. This Greek letter stands for some value—but what value? Your calculator says that π equals 3.141592654, but how does it know? You may have learned before that π equals $\frac{22}{7}$. Is that exactly the same as the calculator value? If you check, you'll see that it's not. They can't both be right. Is either one right? What's the *real* value of π?

Come to think of it, how do you *really* know that these two formulas always work? They say that you can find both the area and the circumference of *any* circle by using a number

Published by IT'S ABOUT TIME, Inc. © 2000 MATHconx, LLC

Learning Outcomes

After studying this section, you will be able to:

Describe the connection between the area and the circumference of the unit circle;

Express the area and the circumference of a circle as functions of its radius;

Find the radius, diameter, circumference, and area of a circle from any one of these measures;

Explain the meaning and the origin of the number π.

4.4 Area and Circumference

This section begins with the formulas for the circumference and the area of a circle. Your students should have learned them as computational devices in elementary and/or middle school, but probably are unaware of why they work. Moreover, students often misunderstand what is π. Computationally, they confuse it with one of its convenient approximations—3.14 or $\frac{22}{7}$ or the like—and conceptually they accept without surprise that a single constant is the key to the areas and circumferences of *all* circles.

To appreciate the power of these ideas, we take a naive, historical point of view, asking students to rediscover with us how these formulas came to be and how the number is actually defined. To carry off this approach, you will have to remind students from time to time that they don't yet know some of these facts that they have taken for granted. We shall try to help you by pointing out situations in which this is particularly important and suggesting ways to help students see the underlying questions.

There are three things that students should see as surprisingly nice, all of which underscore the importance of the number π.

- The ratio of the area of *any* circle to the square of its radius is the same number (i.e., a constant).

- The ratio of the circumference of *any* circle to its radius is the same number (i.e., a constant).

- One of these constants is exactly twice the other!

The fact that these two constants are related so nicely, even though neither of them is a particularly nice number in its own right, is a surprise that often is taken too much for granted. The development in this section seeks to present π with a touch of wonder at the remarkable role it plays in the universe.

Here is an outline of the main ideas of this section.

1. Since all circles are similar, the area and circumference of *any* circle can be found from the area and circumference of a unit circle, neither of which we know yet.

2. The area of the unit circle can be approximated in various ways. We call this the "mystery number" M.

3. The circumference of the unit circle can also be approximated in various ways. We call this the "unknown number" U.

4. $U = 2M$. This is a surprisingly convenient connection! It means that by pinning down the value of only one of these two "strange" numbers we'll also have the other one.

Assessments   Blackline Masters   Extensions   Supplements

For Additional Support Materials see page T-95

with a strange name and a value that you don't exactly know. Doesn't that seem peculiar? In this section you will see where these formulas come from and why they always work. You will also learn more about the powerful, elusive π.

To see why the area and circumference formulas work, let's reinvent them. In your imagination, travel back in time to about 1700 B.C., to a place in Egypt where these formulas are unknown. You are a curious, young Egyptian, trying to figure out how the radius of a circle is related to its circumference and the area it encloses. The key is this simple, very important idea.

A Fact to Know: All circles are similar.

If you think of *similar* as meaning "the same shape," then it should be easy to believe that all circles are similar. But to use this fact mathematically, we need a more careful definition of similarity. Here are two helpful facts from an earlier chapter.

- Two figures are similar if there is a constant k (the scaling factor) such that the distance between *any* two points of one figure is k times the distance between the corresponding two points of the other figure.

- If a planar region is scaled by a factor k, then the area of the scaled region is k^2 times the area of the original region.

4.34

1. What scaling factor relates a circle of radius 1 foot to a circle of radius 3 feet?

2. What scaling factor relates a circle of radius 5 inches to a circle of radius 7 inches?

3. If a circle encloses an area of 12 square inches and you form a similar circle using a scaling factor of 5, what is the area inside the larger circle? Why?

4. If a circle encloses an area A and you form a similar circle using a scaling factor of s, what is the area inside the larger circle? Why?

5. If a circle of radius 1 foot encloses an area of M square feet, what is the area inside a circle of radius 7 feet?

6. If a circle of radius 1 foot encloses an area of M square feet, what is the area inside a circle of radius r feet? Explain. Do you see why this is a VERY BIG idea?

324

Chapter 4

5. *M* is the number π—an important, elusive number with a long history of better and better approximations.

6. Putting these pieces together, we get the familiar formulas for computing the area and the circumference of any circle as *functions* of the radius.

4.32

This problem and the next serve to remind students how to use the formulas for the area and circumference of a circle. They give students who have forgotten (or never learned) these formulas a chance to see how they work and apply them in meaningful contexts. Since round pizza sizes refer to diameters, this problem uses the fact that the diameter is twice the radius as a natural part of the solution.

An 8 inch pizza has a 4 inch radius, so its area is $\pi \cdot 4^2$ sq. in. Rounded to one decimal place, that's 50.3 sq. in. Similarly, the 12 inch size has area $\pi \cdot 6^2 = 113.1$ sq. in. (approx.), and the 14 inch size has area $\pi \cdot 7^2 = 153.9$ sq. in. (approx.). Assuming that "the best deal" means the most pizza for the money, there are two natural ways to get to the answer. Calculate either cost per unit of area (sq. in.) or amount of pizza per dollar. Here are the results.

　　8 inch size: $0.084 per sq. in.; 11.8 sq. in. per dollar
　　12 inch size: $0.075 per sq. in.; 13.3 sq. in. per dollar
　　14 inch size: $0.067 per sq. in.; 15.0 sq. in. per dollar
　　Thus, the 14 inch size is the best deal.

4.33

This problem highlights the fact that the link between circumference and area is the length of the radius.

1. A square with perimeter 1200 feet has side length 300 feet, so its area is 90,000 square feet. You might check to see if all students know how to find the side length of a square from its perimeter. To find the area enclosed by a circle with a 1200 foot circumference you need the radius, which is found by using the circumference formula "backwards." That is, solve $1200 = 2\pi r$ to get $r = 191.0$ feet (to one decimal place). Now use the area formula: $A = \pi \cdot 191^2 = 114,608$ sq. ft. (approx.). Thus, a circular pen provides a lot more area for the same perimeter.

2. This part is a lighthearted link with another continent. An emu is a large, ostrich-like, nonflying bird. A dingo is a wild dog. Both are native to Australia, where Jack takes his walkabouts (long hikes) in the outback (undeveloped, semi arid regions). Recently, some people in other countries, including the U.S., have begun to raise emus as a source of lowfat, low cholesterol meat.

4.34

These questions apply to the two similarity principles about circles that were just stated. They focus on a simple, important idea: Since the size of a circle depends only on its radius, the scaling factor is just the ratio of the radii of the two circles. Students must understand and be able to apply this fact before moving on. The answer to question 6 is the objective here.

The answer to question 6 is a BIG step in simplifying your problem. Here's what it really says.

> If you know the area inside a circle of radius 1 of any unit length, then the area inside *any* circle is just M multiplied by the square of the radius of that circle.

In modern terms, this means that the area inside a circle is a function of its radius. If a circle has radius r, then the area inside it is

$$A(r) = M \cdot r^2$$

where M is the area inside a circle of radius 1 of whatever unit length is used to measure r. A circle of radius 1 of any unit length is called a **unit circle**. Thus, your next job is to find the value of M, the mystery number that is the area inside a unit circle.

You can get a very rough estimate of M from Display 4.26.

> These questions refer to Display 4.26.
>
>
> 4.35
>
> 1. **What is the area of the large square that contains the circle? How do you know?**
>
> 2. **What is the area of the shaded square inside the circle? How do you know?**

Display 4.26 shows that the area M inside the unit circle is less than 4, but greater than 2. In symbols,

$$2 < M < 4$$

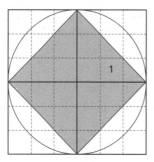

Display 4.26

To get a more accurate estimate of M, subdivide each 1 by 1 square in Display 4.26 into quarters, as in Display 4.27. Then the total of the small squares and half squares in the region that contains the circle and the total of the small squares and half squares inside the circle give us better upper and lower estimates

Published by IT'S ABOUT TIME, Inc. © 2000 MATHconx, LLC

Chapter 4

1. 3

2. $\dfrac{7}{5}$

3. $12 \cdot 5^2 = 300$ sq. in.

4. $A \cdot s^2$

5. $49M$

6. $M \cdot r^2$. Applying the ideas from questions 1 and 2, it should be clear that the scaling factor is $\dfrac{r}{1}$, which is just r. Now, as in questions 3, 4, and 5, the area inside the second circle must be r^2 times the area inside the first circle. The "big idea" question is intended merely to focus students on the fact that this result is important. The text immediately following these questions explains why.

4.35

1. The area of the large square is 4. The circle has radius 1, so it has diameter 2. This means that the square must be 2 by 2. Notice the four 1 by 1 squares in the picture.

2. The area of the shaded square is 2. Perhaps the easiest way to see this is to notice that half of each 1 by 1 square is shaded. Help students to look at the picture this way. Counting diagonal half squares is useful in understanding the next two figures, as well.

Additional Support Materials:

Assessments	Qty
Form (A)	1
Form (B)	1

Blackline Masters	Qty
Student p. 335	1

Extensions	Qty
Ellipses and the Planets	1

Supplements	Qty
Area and Circumference	4

for *M*. To get an even closer estimate, just subdivide the big square into smaller and smaller squares, and then count how many are inside and how many more it takes to make a region containing the circle.

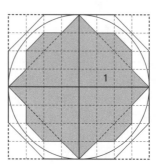

Display 4.27

a
4.36

These questions refer to Display 4.27.

1. What is the side length of each small square in the figure? What is the area of each small square?

2. How many of the small squares and diagonal half squares outside the circle can be removed without touching the circle? What area of the 2 by 2 square remains?

3. What is the total shaded area inside the circle? *Hint:* Count the number of small diagonal half squares that were added to the shaded square of Display 4.26.

4. Complete this statement.

According to Display 4.27, ___ < *M* < ___.

b
4.37

Take out a piece of graph paper and a compass. Use the side length of ten small boxes on the graph paper as your unit length. Place your compass pivot point at a crossing point of the lines on your graph paper somewhere near the center and draw a unit circle.

- By tracing along the lines of the graph paper grid, outline the largest region of small squares that is completely inside the unit circle.

- In the same way, outline the smallest region of little squares that encloses the unit circle. Your finished drawing should look like Display 4.28.

4.36

Besides illustrating how to approximate the area of the unit circle, these questions provide practice in understanding and manipulating simple fractions.

1. $\frac{1}{4}$; $\frac{1}{16}$ These answers can be found either arithmetically or by counting squares.

2. One full small square and two diagonal half squares can be removed from each corner, making a total of 8 small squares removed. The remaining area is $3\frac{1}{2}$.

3. Three half squares were added to each side of the shaded square, adding a total of $\frac{12}{16}$ to the area. Thus, the total shaded area is $2\frac{3}{4}$.

4. According to Display 4.27, $2\frac{3}{4} < M < 3\frac{1}{2}$.

Note that students may object to doing these approximations (or the next ones) by saying that they already know that M is π. If they do, remind them that is just another name for this quantity; it's a letter of the Greek alphabet, just as M is a letter of our alphabet. Neither name tells you anything about the *value* of this quantity! They "know" its value only because somebody told them something that they memorized. This investigation shows them where the value for this symbol comes from—and that the value they have learned for π is not exact.

4.37

This is a drawing exercise to give students a better sense of finding M by successive approximations. If you don't want to take the time to have your students do the drawing and counting, you should at least summarize the results of the process for them. See the previous Note regarding response to possible student objections.

Chapter 4

1. What is the side length of each small square of the grid? What is the area of each small square?

2. What is the area of the outlined region inside the circle?

3. What is the area of the outlined region outside the circle?

4. Use the preceding two answers to complete this statement.

$$\_\_ < M < \_\_$$

5. Refine your inner and outer regions by adding or removing small diagonal half squares. Use these regions to get a better estimate of the mystery number M.

$$\_\_ < M < \_\_$$

6. Do you think that M is *exactly* between these two numbers? Why or why not?

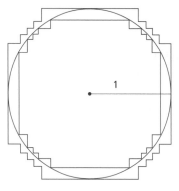

Display 4.28

Looking at the trends of the upper and lower estimates for M that you have found, a good guess for the actual value of M would be somewhere between 3 and $3\frac{1}{4}$. These examples show how you could sharpen this estimate further, but drawing and counting little squares is tiring work! This estimate is good enough for now. It's time to look for a connection between the radius of a circle and the distance all the way around it.

The distance all the way around a circle is called its **circumference.** The circumference is a length. One way to find out how the circumferences of circles are related to their diameters is to gather some data, as follows.

About Words

The Latin prefix *circum-* means around. When you *circumvent* something, you go around it. The *circumference* of a circle carries you around it.

Published by IT'S ABOUT TIME, Inc. © 2000 MATHconx, LLC

327

1. $\frac{1}{10}$; $\frac{1}{100}$

2. 3.52

3. 2.76

4. $2.76 < M < 3.52$

5. $2.88 < M < 3.42$

6. This last part is intended to alert students to the fact that finding the exact area is not as simple as this successive approximation process might suggest. The curvature of the circle makes it highly unlikely that M is exactly halfway between the last two approximations. Another clue to this is that the inner and outer approximations do not change by the same amount from part 4 to part 5.

NOTES

4.38

Collect at least six circular objects of different sizes, such as coins, jar lids, and pie plates. Measure the radius of each object as accurately as you can. Actually, the diameter is easier to measure than the radius because you don't have to find the center. Just measure the largest distance across the circle. Use a tape measure or a string and a ruler to measure the circumferences. Copy Display 4.29 and fill in the diameter, radius, and circumference of each circular object. The measurements of a U.S. quarter are filled in to start you off.

Now apply your data from Display 4.29 as follows.

1. Plot the data pairs of the last two columns on a piece of graph paper. Label the *x*-axis *radius* and the *y*-axis *circumference*.

2. Draw a straight line that seems to fit the data points of your graph. It should go through the point (0, 0). Why?

3. Estimate the slope of this line and write an equation for it. What is its *y*-intercept?

4. Check your estimate by entering the data into your graphing calculator and finding the line of best fit. Finding the line of best fit with your calculator is in Appendix A.

5. How can you use this line to estimate the circumference of other circles if you are given the radius? What if you are given the diameter? Use this line to add two more rows to your table, for these objects.
   - A bicycle wheel with radius 13 in.
   - A garbage can with diameter 21 in.

Object	Diameter	Radius	Circumference
quarter	2.4 cm	1.2 cm	7.5 cm
⋮	⋮	⋮	⋮

Display 4.29

Published by IT'S ABOUT TIME, Inc. © 2000 MATHconx, LLC

4.38

This is a multipurpose exercise suitable for small groups. Besides establishing an empirical reason for believing the functional relationship between area and circumference, it reviews the ideas of data gathering, measurement error, line of best fit, slope, and linear equations from **MATH** *Connections* Year 1.

2. It should go through $(0, 0)$ because a "circle" of radius 0 should have circumference 0.

3. The equation should be fairly close to $C = 6.3r$. The y-intercept is 0 because the point $(0, 0)$ should be on the line.

5. Just insert the radius for r in the equation or find it on the x-axis and use the line to find the corresponding y-coordinate. If given the diameter, divide it in half to find the radius, then proceed as before.

Chapter 4

NOTES

_____

Now let's think carefully about the results of this data gathering experiment. Does your line of best fit match your radius circumference data pretty well? If it does (and it should), then it tells you that the graph of the function that relates radius to circumference probably is a straight line through the origin. This means that it has the form $C(r) = U \cdot r$, where U is some "unknown" constant for which you have found an approximate value.

1. **What value for the unknown number U did you get from your line of best fit? Did everyone in your class get exactly the same value? Did everyone get approximately the same value?**

4.39

2. **Use the fact that all circles are similar to justify the claim that the function which relates radius to circumference must be a straight line.**

3. **In terms of U, what is the circumference of a unit circle? Justify your answer.**

4. **In terms of U, what is the circumference of a circle of radius 12 feet? Justify your answer.**

5. **Use Display 4.26 to justify the claim that $4\sqrt{2} < U < 8$. *Hint:* Use the Pythagorean Theorem.**

Here's what we have discovered so far.
* Because all circles are similar, we can calculate the area and the circumference of any circle if we know the area, M, and the circumference, U, of the unit circle.

* The area M is between 3 and $3\frac{1}{4}$.

* The circumference U is between $4\sqrt{2}$ and 8; in fact, judging from the slope of the line of best fit, it's most likely between 6 and $6\frac{1}{2}$.

Two natural questions remain.

1. Can we get better estimates of M and U?

2. Are M and U related in some convenient way?

It turns out that one way of approaching question 1 provides a surprisingly nice answer to question 2, as well! That approach

Published by IT'S ABOUT TIME, Inc. © 2000 MATHconx, LLC

329

Chapter 4

4.39

These questions should lead students to see why the circumference of any circle must be U times its radius.

1. As noted before, U should be about 6.3. Because of variations in the objects measured and the measurement process itself, not all students or student groups should have a measure of exactly the same value, but the values should all be pretty close to each other.

2. The important point to get across is that, if the measurements were all exactly accurate, then all the data points would be exactly on the line of best fit. We give here an algebraic argument. Less formal ways of expressing this same idea are acceptable, too, depending on the level of rigor you expect from your students.

   If O_1 and O_2 are similar circles, then there is some constant k such that every length measurement of O_2 is k times the corresponding length measurement of O_1. Both the radius and the circumference are length measurements, so (using subscripts in the obvious way) $r_2 = k \cdot r_1$ and $C_2 = k \cdot C_1$. Now, there must be *some* number—call it U—such that $C_2 = U \cdot r_2$. By substitution, $k \cdot C_1 = U \cdot k \cdot r_1$. Cancelling the common term k from both sides of the equation, we have $C_1 = U \cdot r_1$ for the same constant U.

3. U This is a simple, but very important, observation.

4. $U \cdot 12$ feet

5. Each side of the inner square is the diagonal of a unit square, so, by the Pythagorean Theorem, its length is $\sqrt{2}$. You might want to ask your students to approximate $4\sqrt{2}$ as a two place decimal, just to be sure that they have some intuitive idea of the lower number of this estimate.

begins with the rough estimate of U that you get from the inner and outer squares of Display 4.26. To get a better estimate, we can use regular polygons with more sides. Instead of looking at both inner and outer polygons, we'll just use inner polygons that have their vertices on the unit circle. They are called **inscribed polygons**. See Display 4.30 for some examples.

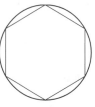  

Inscribed hexagon   Inscribed octagon   Inscribed decagon

Display 4.30

Display 4.30 shows that, the more sides the inscribed regular polygon has, the closer its perimeter is to the circumference of the circle. So the perimeter of one of these polygons with a large number of sides will be a good approximation of the circumference of the unit circle. But how can we find the perimeter of any of these polygons if we know only that the radius of the circle is 1? Suprisingly, by looking first at its area!

The area of an inscribed regular polygon can be found by adding up the areas of triangles. Start by connecting the center of the circle to each vertex of the polygon. Since the polygon is regular, all these triangles are congruent. Moreover, each segment that connects the center with a vertex is a radius of the unit circle. This means that all the triangles are isosceles, and the two equal sides both have length 1.

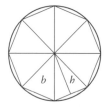

Display 4.31

NOTES

Let's look at an example. Display 4.31 shows a regular inscribed octagon with the triangles drawn in. The altitude of the lower right triangle is also drawn in. Now we put together several simple facts. Can you give a reason for each of these statements?

- The area of the triangle is given by the formula $\frac{1}{2}bh$, where b (the base) is one side of the octagon and h (the height) is the length of the altitude.

- The area of the octagon is 8 times the area of the triangle, that is,

$$\text{area of octagon} = 8 \cdot \left(\tfrac{1}{2}bh\right)$$

- The right side of this equation can be rewritten as

$$\tfrac{1}{2}h \cdot (8b)$$

- The perimeter of the octagon is $8b$.

- The equation for the area can be rewritten as

$$\text{area of octagon} = \tfrac{1}{2}h \cdot (\text{perimeter of octagon})$$

Now it's your turn. Suppose a regular decagon (10-sided polygon) is inscribed in a unit circle. Write a step-by-step explanation of

$$\text{area of decagon} = \tfrac{1}{2} \cdot h \cdot (\text{perimeter of decagon})$$

4.41

Hint: Mimic the steps for the octagon.

There's nothing magic about 8 or 10 as the number of sides of the inscribed polygon. If you have an inscribed regular polygon of 547 sides, the argument works exactly the same way.

- The area of each of the 547 triangles is given by the formula $\frac{1}{2}bh$, where b is one side of the 547-gon and h is the length of the altitude.

- The area of the 547-gon is 547 times the area of the triangle

$$\begin{aligned}\text{area of 547-gon} &= 547 \cdot \left(\tfrac{1}{2}bh\right)\\ &= \tfrac{1}{2}h \cdot (547b)\\ &= \tfrac{1}{2}h \cdot (\text{perimeter of 547-gon})\end{aligned}$$

The fact that this equation is true for inscribed regular polygons of *any* number n of sides is expressed as

$$(*) \qquad \text{area of } n\text{-gon} = \tfrac{1}{2}h \cdot (\text{perimeter of } n\text{-gon})$$

Published by IT'S ABOUT TIME, Inc. © 2000 MATHconx, LLC

Chapter 4

 4.40 Make sure that your students understand why each of these statements is true.

4.41 This writing exercise can be assigned as homework so that your students review the important argument we just gave. They need to see

(1) how this equation relates area to perimeter, and
(2) that such an equation holds for a regular inscribed polygon with any number of sides.
This latter fact will be stated explicitly right after this exercise; students need not have done this writing exercise (yet) to understand it.

- Triangulate the decagon into 10 congruent isosceles triangles by connecting each vertex with the center of the circle.

- The area of each triangle is given by the formula $\frac{1}{2}bh$, where b (the base) is one side of the decagon and h (the height) is the length of the altitude (the perpendicular distance from the side to the center).

- The area of the decagon is 10 times the area of the triangle; that is,
$$\text{area of decagon} = 10 \cdot (\tfrac{1}{2}bh)$$

- The right side of this equation can be rewritten as
$$\frac{1}{2} \cdot h \cdot (10b)$$

- The perimeter of the decagon is $10b$

- The equation for the area can be rewritten as
$$\text{area of decagon} = \frac{1}{2} \cdot h \cdot (\text{perimeter of decagon})$$

4.42 Some students may have seen a version of this approximation argument in middle school. If so, you can move through it rapidly. It is unlikely, however, that they will have seen or will remember what it establishes, namely, that the circumference of a unit circle in linear measure is twice its area in square measure. Notice that it does *not* establish what the actual amount is, either as a length or as an area. This argument only relates two unknown quantities to each other, so that, if one can be found somehow, the other will also be determined.

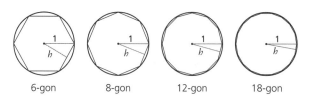

6-gon   8-gon   12-gon   18-gon

Display 4.32

Here's the last piece of this puzzle. Look at Display 4.32. Notice that, as n gets bigger, three things are happening,

- The area inside the polygon is getting closer to matching the area inside the circle.

- The height, h, is getting closer to matching the radius, 1.

- The perimeter of the polygon is getting closer to matching the circumference of the circle.

Even in the last circle of Display 4.32, with a polygon of only 18 sides, it is very difficult to see the differences between these measurements. Imagine what it would be like for a 360-gon or a 2000-gon! This means that, for large n, equation (∗) area of n-gon $= \frac{1}{2}h \cdot$ (perimeter of n-gon) is *very* close to

area of unit circle $= \frac{1}{2} \cdot 1 \cdot$ (circumference of unit circle)

But the area of the unit circle is M and the circumference of the unit circle is U. If we substitute M and U into the previous equation, we get

$$M = \frac{1}{2}U$$

Thus, our unknown values have been reduced from two to one.

If we find the value of either U or M, we immediately get the value of the other one.

4.43

1. In terms of M, what is the circumference of a unit circle?

2. In terms of M, what is the circumference of a circle of radius 10 meters? Justify your answer.

3. In terms of U, what is the area inside a circle of 6 meters radius? Justify your answer.

4. Is it possible for a circle to have area $2M$ and circumference $2U$? If so, find the radius of such a circle. If not, explain why not.

4.43

These exercises reinforce the relationship between M and U and remind students that we don't yet know the actual value of either of these numbers. Questions 1, 2, and 3 are routine; question 4 is a little more challenging. All four questions should be answered using the fact that $U = 2M$, without knowing the actual value of either M or U.

1. The circumference of a unit circle is U, which equals $2M$.

2. Circumference $= Ur$ and $U = 2M$, so the circumference of this circle is $20M$ meters.

3. Area $= Mr^2$ and $M = \frac{1}{2}U$, so the area inside this circle is $18U$ square meters.

4. No. $A = Mr^2$ and $C = Ur$. If there were such a circle, its radius, r, would have to satisfy the equations $2M = Mr^2$ and $2U = Ur$. The second equation implies that $r = 2$ by cancellation, but when r is replaced by 2 in the first equation, the result is $2 = 4$, a contradiction. Hence, no such radius exists.

We have called M a mystery number for good reason. Finding its value has been a mystery that people of many different civilizations have worked on and puzzled over for hundreds of years. Here are a few examples.

c. 1650 B.C. — The Ahmes Papyrus, from Egypt, declared M to be $4\left(\dfrac{8}{9}\right)^2$.

c. 240 B.C. — Archimedes showed that M is between $3\dfrac{10}{71}$ and $3\dfrac{10}{70}$.

c. 150 A.D. — Ptolemy, a Greek astronomer, used $\dfrac{377}{120}$ for M.

c. 480 — The Chinese scholar Tsu Ch'ung-chih used $\dfrac{355}{113}$ for M.

c. 530 — The Hindu mathematican Aryabhata used $\dfrac{62,832}{20,000}$ for M.

c. 1600 — A decimal value for M was computed to 35 places.

1737 — The great Swiss mathematician Leonhard Euler named M with the symbol by which it has been known ever since.

$$\boxed{\pi}$$

1873 — William Shanks of England computed, by hand, a decimal value for π to 707 places. It took him more than 15 years!

1949 — John von Neumann used the U.S. government's ENIAC computer to work out π to 2,035 decimal places (in 70 hours).

1987 — Professor Yasumasa Kanada of the University of Tokyo worked out π to 134,217,000 decimal places on an NEC SX-2 supercomputer.

1991 — Gregory and David Chudnovsky calculated π to 2,260,321,336 decimal places in 250 hours, using a home-built supercomputer in their New York City apartment. This many digits, printed in a single line of ordinary newspaper type, would stretch from New York City to Hollywood, California![†]

Yet *none* of these results is its *exact* value!

[†]For the story of the Chudnovskys and their amazing machine, see *"Profiles"* by R. Preston, in the March 2, 1992, issue of *The New Yorker*.

Published by IT'S ABOUT TIME, Inc. © 2000 MATHconx, LLC

333

NOTES

4.44

Which of the first five values in the historical list above come closest to the value for π in your calculator?

About 1765 (when America was working up to the Revolutionary War), a German mathematician named Johann Lambert proved that π is an **irrational number**; that is, it cannot be expressed exactly as a common fraction (the ratio of two whole numbers). Among other things, this meant that no decimal expression for π, no matter how far it was extended, would ever *exactly* equal π. We can find decimals as close as we want, if we're willing to be patient and do enough work, but they will always be just a little off the true value of the area of a unit circle. Such approximations are good enough for any real world application of this number, but no single one is good enough to use all the time. That's why the exact mystery number needs a name of its own — π.

π = 3.14159265358979323846264338327950288419716939937510582097494459230781640628620899862803482534211706798214808651328230664709384460955058223172535940812848111745028410270193852110555964462294895493038196442881097566593344612847564823378678316527120190914564856692346034861045432664821339360726024914127372458700660631558817488152092096282925409171536436789259036001133053054882046652138414695194151160943305727036575959195309218611738193261179310511854807446237996274956735188575272489122793818301194912983367336244065566430860213949463952247371907021798609437027705392171762931767523846748184676694051320005681271452635608277857713427577896091736371787214684409012249534301465495853710507922796892589235420199561121290219608640344181598136297747713099605187072113499999983729780499510597317328160963185950244594553469083026425223082533446850352619311881710100031378387528865875332083814206171776691473035982534904287554687311595628638823537875937519577818577805321712268066130019278766111959092164201989...

The first 1,000 decimal places of π

Display 4.33

The two most basic facts to remember about π are.
- The area of a unit circle is π square units.

- The circumference of a unit circle is 2π units.

With these facts you can find the area and the circumference of any circle, by similarity. Thus, the area and the circumference of a circle are functions of its radius. These functions can be written as formulas (algebraic recipes) that make it easy for you to find the

4.44

Calculator value: 3.141592654

Ahmes papyrus: $4\left(\dfrac{8}{9}\right)^2 = 3.160493827$

Archimedes: $3\dfrac{10}{71} = 3.14085 < \pi < 3.14286 = 3\dfrac{10}{70}$

Ptolemy: $\dfrac{377}{120} = 3.141666666...$

Tsu Ch'ung-chih: $\dfrac{355}{113} = 3.14159292$ (the closest of the five values)

Aryabhata: $\dfrac{62832}{20000} = 3.1416$

NOTES

Chapter 4

area and the circumference of any circle from its radius.

Area of a circle of radius r: $A(r) = \pi r^2$

Circumference of a circle of radius r: $C(r) = 2\pi r$

1. Explain what it means to say that the area and the circumference of a circle are functions of its radius.

4.45

2. Explain how to find the radius of a circle from its circumference or from its area. Illustrate your explanation with an example of each process.

3. (a) Can you write a function that gives a formula for finding the radius of a circle from its area? If you can, do it. If you cannot, explain why not.

   (b) Can you write a function that gives a formula for finding the radius of a circle from its circumference? If you can, do it. If you cannot, explain why not.

Problem Set: 4.4

1. Each line of the table in Display 4.34 contains one of the following items of information about a particular circle.

   radius, diameter, circumference, area

   Copy this table onto a piece of paper. Then fill in the other three items for each circle. Write your answers in terms of π; do not approximate.

Circle	Radius	Diameter	Circumference	Area
(a)	7 in.			
(b)		30 cm		
(c)			12π in.	
(d)				100π sq. mi.
(e)	π in.			
(f)		π in.		
(g)			1 m	
(h)				1 sq. m
(i)			36 in.	
(j)				64 sq. cm

Display 4.34

335

4.45

1. The radius of a circle determines the values of its area and its circumference. For each possible radius, there is exactly one area and one circumference.

2. We believe that this process is best taught if the students are allowed to see it first as a simple matter of common sense—reversing the recipe —rather than being presented initially as a formal algebraic manipulation. The next part formalizes the algebra.

3. From circumference: $r(C) = \dfrac{C}{2\pi}$ From area: $r(A) = \sqrt{\dfrac{A}{\pi}}$

   Students may object that the formula for finding radius from area is not a function because it is possible to have two square roots. That's a good observation, but it does not apply here because the radius, a length, must always be positive. Hence, the negative root can be ignored.

Problem Set: 4.4

1. This problem is simple, but not all parts of it are easy. In fact, some of the computations in (e)–(j) require quite a bit of care. Requiring that entries be written in terms of π emphasizes equation solving techniques, as well as the arithmetic of square roots and fractions. Display 4.9T shows the table of Display 4.33 with all the entries filled in. The entries given in the problem are boxed.

Circle	Radius	Diameter	Circumference	Area
(a)	7 in.	14 in.	14π in.	49π sq. in.
(b)	15 cm	30 cm	30π cm	225π sq. cm
(c)	6 in.	12 in.	12π in.	36π sq. in.
(d)	10 mi.	20 mi.	20π mi.	100π sq. mi.
(e)	π in.	2π in.	$2\pi^2$ in.	π^3 sq. in.
(f)	$\dfrac{\pi}{2}$ ft.	π ft.	π^2 ft.	$\dfrac{\pi^3}{4}$ sq. ft.
(g)	$\dfrac{1}{2\pi}$ m	$\dfrac{1}{\pi}$ m	1 m	$\dfrac{1}{4\pi}$ sq. m
(h)	$\dfrac{1}{\sqrt{\pi}}$ m	$\dfrac{2}{\sqrt{\pi}}$ m	$2\sqrt{\pi}$ m	1 sq. m
(i)	$\dfrac{18}{\pi}$ in.	$\dfrac{36}{\pi}$ in.	36 in.	$\dfrac{324}{\pi}$ sq. in.
(j)	$\dfrac{8}{\sqrt{\pi}}$ cm	$\dfrac{16}{\sqrt{\pi}}$ cm	$16\sqrt{\pi}$ cm	64 sq. cm

Display 4.33 completed

Display 4.9T

2. At Pietro's Pizza, an 8 inch cheese pizza sells for $4.25, the 12 inch size sells for $8.50, and the 14 inch size sells for $10.25.

   (a) Some students who had not studied **MATH** *Connections* thought that they ought to be able to get a 16 inch pizza for $8.50, since an 8 inch pizza costs $4.25. Explain to them why this would not be reasonable.
   (b) If price were proportional to area (based on the 8 inch size), what *should* a 16 inch pizza cost?
   (c) If price were proportional to area (based on the 8 inch size), what size pizza should cost $8.50?

3. Bicycle tires are usually classified by their diameters. For many touring bikes, $27\frac{1}{2}$ inches is a common size.

   (a) What is the circumference of a $27\frac{1}{2}$ inch tire? Express your answer in terms of π and also as a decimal, rounded to two places.
   (b) When a bicycle with tires of this size travels one mile, how many times does each wheel go around? Explain how you got your answer.
   (c) The size given here assumes that the tire is fully inflated. Suppose that the tire is a little soft, so that its diameter is actually $\frac{1}{4}$ inch less than its size rating. How many more times must this tire go around in a mile than the fully inflated tire of part (b)?

4. Lee and Avery work for Pi in the Sky Landscaping, a company that specializes in circular gardens.

   (a) How much fencing will they need to enclose completely a circular garden 20 feet in diameter? Round your answer to two decimal places.
   (b) How much fencing will they need to enclose completely a semicircular garden 20 feet in diameter? Round your answer to two decimal places. *Hint:* Draw a sketch.

5. A Doppler weather radar installation sits atop Talcott Mountain Science Center in Avon, Connecticut. This radar detects precipitation (snow, rain, etc.) by sending a microwave beam toward the horizon and waiting for reflections. If no precipitation is present, the microwave beam never comes back, but if rain, sleet, or snow is somewhere in the range, reflections come back a short time later, like an echo. The radar calculates the nearness of the

Published by IT'S ABOUT TIME, Inc. © 2000 MATHconx, LLC

2. It is reasonable to measure the amount of pizza in terms of area (square inches).

   (a) These students appear to be assuming that the price should reflect proportionally the size of the pizza. Under that assumption, twice the price of an 8 inch pizza should get you twice as much pizza. But the amount of pizza is measured in area. The area of an 8 inch (diameter) pizza is 50.27 sq. in. (see part (a)), but the area of a 16 inch (diameter) pizza would be $\pi \cdot 8^2 \approx 201$ sq. in., which is 4 times as much. The area increases with the *square* of the scaling factor.

   (b) Since the area is larger by a factor of 4, a 16 inch pizza should cost 4 times as much: $17.00.

   (c) A pizza twice as large in area as an 8 inch pizza would have an area of 100.52 sq. in. To find its radius, solve $100.52 = \pi r^2$.

   That is, $r = \sqrt{\dfrac{100.52}{\pi}} \approx 5.66$ inches. Thus, a pizza of diameter 11.3 inches is about the right size proportionally.

3. (a) $C = 27.5\pi = 86.39$ inches
   (b) There are 5280 feet in a mile, so there are 63,360 inches in a mile. $63,360 \div 86.39 = 733.4$ times around.
   (c) The circumference of the soft tire is $27.25\pi = 85.61$ inches, so this tire must go around $63,360 \div 85.61 = 740.1$ times in a mile. That's 6.7 more times than the fully inflated tire.

4. (a) $20\pi \approx 62.83$ feet
   (b) Half of the preceding answer plus the diameter gives
   $$10\pi + 20 \approx 51.42 \text{ feet.}$$

5. (a) This part asks for some intelligent guessing, rather than computation. High range permits greatest area coverage, so the radar could detect storms as far away as New Jersey or Maine. Low range restricts the total area covered, but permits much greater detail of precipitation location and intensity in the area covered. You would want to use high range to watch a distant storm approaching or leaving, but you would select low range during thunderstorm weather when one town can experience torrential rain, while no rain may be falling a short distance away.

precipitation by measuring the time it takes for the reflections to come back, and it calculates the intensity of the precipitation from the amount of microwave energy coming back. The beam rotates continually, scanning the entire horizon about once every 2.5 minutes.

(a) The operator can select the range (radius) of the radar. On low range, the radius of the beam is only about 10 miles; on high range the radius of the beam is up to 300 miles. What do you think are the advantages of operating the radar on high range? What are the advantages of operating the radar on low range? For which weather conditions do you think the operator would select the high range? For which weather conditions do you think the operator would select the low range?

(b) Calculate the total area covered (in square miles) when the radar is the range is 300 miles.

(c) Calculate the total area covered (in square miles) when the radar is the range is 10 miles.

(d) The radius at high range is 30 times the radius at low range. Is the area covered at high range 30 times the area covered at low range? Explain why or why not.

(e) Use a map to draw a circle of 300 miles radius centered in Avon, Connecticut. What states are included in the coverage? Use a map to draw a circle of 10 miles radius centered in Avon, Connecticut. What cities and towns are included in the radar coverage?

Published by IT'S ABOUT TIME, Inc. © 2000 MATHconx, LLC

(b) 282,743 square miles

(c) 314 square miles

(d) No, as can be seen from the answers to the preceding two parts. Since area varies with the square of the scaling factor (the radius), we should expect the area covered to be 30^2 times as large, and it is allowing for measurement roundoff error.

(e) This part may be omitted if a map is not available. Nothing in future sections depends on it. Avon, Connecticut, is 9 miles WNW of Hartford, along Route 44. It is marked on state maps of popular road atlases, but may not appear on maps of larger regions.

Chapter 4

NOTES
_____

6. Make a graph showing how the circumference of a circle varies with the radius. Put the radius on the *x*-axis, and the circumference on the *y*-axis. On another pair of axes, make a graph to show how the area varies with the radius. Put the radius on the *x*-axis and the area on the *y*-axis. Write an equation for each graph. We say that the circumference *varies directly as* the radius and that the area *varies directly as the square of* the radius. How are the graphs different? How are the equations different?

7. Mary makes and sells round butcher block table tops. The largest one she usually makes is 4 feet in diameter, but she takes custom orders for larger sizes. Mr. Washington wants to order a table top that would be 8 feet in diameter, but he is worried that it might be too heavy. If a 4 foot table top weighs 38 pounds, how much would an 8 foot table top of the same thickness weigh? Explain why your answer is *not* twice the weight of the 4 foot table.

8. The approximate value used for π affects almost all numerical work with circles. When you work with large circles or need very precise measurements, the value you choose can have a big effect on the results you get. However, after ten or twenty decimal places, changes in the digits rarely have any practical effect at all. The following questions illustrate these ideas.

   The diameter of the Earth at the equator is about 8000 miles. For this problem, assume that it is exactly 8000 miles.

   (a) Calculate the circumference of the Earth at the equator using the value of π proposed by each of the following people. Round your answers to the nearest tenth of a mile.

      (i) the Egyptian scribe Ahmes
      (ii) the Greek astronomer Ptolemy
      (iii) the Chinese scholar Tsu Ch'ung-chih
      (iv) the Hindu mathematician Aryabhata
      (v) the people who programmed your calculator

   (b) Most modern calculators express π to ten places: 3.141592654. Find the circumference of the Earth at the equator using this for π. Then change the last digit of this approximation of π to 0 and find that circumference again. What is the difference between your two values for the circumference, in miles? In inches?

Published by IT'S ABOUT TIME, Inc. © 2000 MATHconx, LLC

6 . The equations are the ones from the end of the section. In terms of x and y,

$$\text{(circumference) } y = 2\pi x \quad \text{and} \quad \text{(area) } y = \pi x^2$$

The first graph is linear, with slope 2π. The second graph is not linear. It is the positive half of a parabola.

7. The principle is that area does not vary directly as the linear scaling factor, but varies directly as its square. A table top of twice the diameter will have 4 times the area and hence, in this case, 4 times the weight. Thus, the 8 foot table top would weigh 152 pounds, which might be too heavy for some purposes, but not for others.

8. A major purpose of this problem is to drive home the dual message that (1) *all* values used for π are approximations, but (2) the decimal digits after the first few are irrelevant for most practical applications. Part (a) does this by comparing the effects of values for π actually used at various times in history.

   (a) (i) 25,284.0 miles
       (ii) 25,133.3 miles
       (iii) 25,132.7 miles
       (iv) 25,132.8 miles
       (v) 25,132.7 miles
   (b) Using $\pi = 3.141592654$, the circumference is 25,132.74123 miles. Using $\pi = 3.141592650$, the circumference is 25,132.7412 miles. The difference is 0.00003 miles. There are 63,360 inches (5280 feet) in a mile, so this difference is about 2 inches.

NOTES

4.5 Pieces of Circles

In this section you will see how the area and circumference formulas for a circle can be used to find lengths, perimeters, and areas of parts of circles and disks.

Since a circle is a closed curve, taking out a single point will not break it into two separate pieces, but taking out two points will. That is, any two points on a circle break it into two pieces, called **arcs**. (See Display 4.35.) If the pieces are not the same size, the larger one is called a **major arc** and the smaller one is called a **minor arc**. What are they called if they are the same size?

Learning Outcomes

After studying this section, you will be able to:

Use the measure of the central angle to find the length of any circular arc;

Use the measure of the central angle to find the area of any sector of a circle;

Find the area of an annulus and sectors of it.

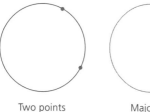 

| Two points | Major arc | Minor arc |

Display 4.35

The radii from the center to the two endpoints of an arc form an angle called the **central angle** of the arc. These radii and the arc enclose a region called a **sector** of the circle. The arc and the chord between its endpoints enclose a region called (unfortunately) a **segment** of the circle. The shaded regions in Display 4.36 are the sector and the segment of the minor arc; the unshaded regions inside the circles of parts (b) and (c) are the sector and the segment of the major arc. Sometimes, instead of saying major arc or minor arc, we describe an arc by the number of degrees of its central angle.

(a) Central angle   (b) Sector   (c) Segment

Display 4.36

Published by IT'S ABOUT TIME, Inc. © 2000 MATHconx, LLC

339

4.5 Pieces of Circles

The most important idea of this section is as simple as it is fundamental.

Arc length and sector area are proportional to central angle measure.

The section introduces the vocabulary needed to describe this idea efficiently, then looks at an extended example (the Go-Kart track) that provides a basis for student discovery and formalization of this "obvious" expectation, as well as illustrating a realistic setting in which the idea is useful. What is not so obvious, and is worth observing, is the fact that this proportionality depends on the fact that the circle is extraordinarily symmetric: It is symmetric with respect to any angle of rotation and any reflection line through its center. These ideas are explored *after* the students have a chance to formulate their intuitive sense of proportionality.

Although it would be ideal if all students were able to discover for themselves the proportionality principles here, this may be an overly optimistic expectation. You will have to gauge your students' ability to express these ideas in reasonably precise mathematical language and help them refine their thinking at critical places in the discussions. Specific comments to that effect appear in the commentary to various problems and Explorations in this section.

4.46 The arcs are equal, so they are (essentially) semicircles.

Additional Support Materials:

Assessments	Qty
Form (A)	1
Form (B)	1

Blackline Masters	Qty
Student p. 345	1
Student p. 350	1

Extensions	Qty

Supplements	Qty
Pieces of Circles	2

Chapter 4

4.47 The relationship between the area of a sector and the central angle of its arc is nicer than the relationship between the area of a segment and the central angle of its arc. What does nicer mean in this case?
Hint: Think about ratios of measures.

Here's a story to illustrate how knowing about arcs and sectors can be useful.

A Go-Kart Track (Part I)

Nip and Tuck, the Racer twins, have decided to build a Go-Kart track. They have found a simple design in their favorite Go-Kart magazine. Its diagram is shown in Display 4.37.

Notes with the diagram explain that the curves are circular arcs, to make the track easy to build. The dark line is the center of the track. The dashed lines are the edges, 10 feet away on each side of the center, all the way around. The notes go on to describe how to bank the turns, how to ease the curvature slightly when connecting the circular parts to the straightaways, etc., but these things will not concern us here.

The breaks in the straightaways, shown by the wavy lines, indicate that the length of the track may be adjusted, depending on how long these straight parts are made.

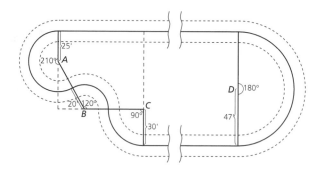

Go-Kart track

Display 4.37

Published by IT'S ABOUT TIME, Inc. © 2000 MATHconx, LLC

4.47

One major purpose of this discussion question is to get students familiar with the potentially confusing distinction between sector and segment. The other purpose is to bring out the intuitively natural idea that the area of a sector is proportional to its central angle, but the area of a segment is not. For instance, consider central angles of 45°, 90°, and 180°. This idea is of fundamental importance, but it is not essential that students discover it on their own at this point. It will be made explicit shortly.

Depending on how easily your class takes to this sort of discussion, there is a basic habit of thought to be pursued here.

> Given the center of the circle and two points on the circle, what are all the shapes and regions that are determined? How can you find their perimeters and areas just by knowing the locations of these three points?

These ideas might be pursued in an exploratory way for as long as class interest can be maintained. The discussion might touch on the following ideas.

- The distance between the center and either of the other points is the radius, which allows you to compute the circumference and the area of the entire circle.

- The two points on the circle determine two arcs. If you could find the length of either one, then you would know the length of the other, just by subtraction.

- A similar principle holds for sectors and segments. The area of either sector or either segment leads to the area of the other by subtraction from the area of the entire circle.

- The three points also determine an isosceles triangle by SAS, where the two sides are radii and the included angle is the central angle. How is the area of this triangle related to the areas of the minor arc sector and segment? Can we use it to find the length of the chord? Notice that the bisector of the central angle splits this triangle into two congruent right triangles.

The more you can help the students discover on their own, the easier the subsequent material will be.

Chapter 4

Published by IT'S ABOUT TIME, Inc. © 2000 MATHconx, LLC

As a first step in building their Go-Kart track, Nip and Tuck decide they need a larger scale drawing of the center line of Display 4.37. Make one for them. Draw it on a piece of $\frac{1}{4}$ inch ruled graph paper, to the scale of $\frac{1}{4}$ inch = 5 feet.

4.48

Hint: If you turn your $8\frac{1}{2} \times 11$ inch piece of graph paper sideways and start by placing point *A* about 3 inches down from the top left corner and 2 inches to the right, you should have just enough room.

Nip and Tuck like this track design because its length can be adjusted. They want to build a track exactly $\frac{1}{10}$ of a mile long. This will allow them to figure out easily each Go-Kart's average speed in miles per hour. Do you see how?

How long should they make the straightaways?

Nip and Tuck's question of how long to make the straightaways is difficult to answer all at once. Break it into smaller questions that are easier to handle. Then answer at least one of them.

4.49

Before deciding how long to make Nip and Tuck's straightaways, we have to know the lengths of the four circular arcs in the track. They are shown separately in Display 4.38. Let's take them one at a time.

Thinking Tip

Ask simpler questions. When faced with a question you can't answer, keep asking other questions about it until you find one you *can* answer. Then work from there.

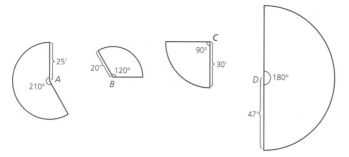

The circular arcs

Display 4.38

Chapter 4

4.48

Students will need a compass, a protractor, and a ruler for this drawing. It should be done on a standard size ($8\frac{1}{2} \times 11$ inch) piece of graph paper that is ruled in $\frac{1}{4}$ inch boxes.

Making this scale drawing is not a simple task, but spending some time on it should be a worthwhile activity. The critical step is the proper placement of points *B*, *C*, and *D*. The radius of the semicircle centered at *D* (47 feet) can be determined from the other radii; it depends on them. However, the computations required to find this length might distract students from the more immediate task of reconstructing the curve, so the measurement has been given.

This Go-Kart example will be used in several stages to illustrate most of the main ideas of this section. If you don't have your students do this scale drawing in class, you might assign it as homework. Make sure that they understand thoroughly how the center line of the track is formed.

4.49

This is a class or small group discussion that should be done before going any further. Guide your students to see that the question comes apart naturally into these pieces.

1. How long is $\frac{1}{10}$ of a mile? (528 feet)

2. How long are each of the four arcs, around *A, B, C,* and *D*?

3. How long is the part of the top straightaway between the two points directly above *A* and *C*? This one is less obvious. You may have to suggest that they extend the vertical radius from *C* up to the top straightaway and label that intersection point; it will be labeled *F* a little later.

4. What's the difference between the sum of all the lengths in parts 2 and 3 and the answer to part 1?

5. Where can this difference be added to the length of the track?

The first and last of these questions are easy to answer. For the first, if nobody in the class knows the length of a mile in feet, they can look it up in almost any dictionary. For the last question, if the required difference is known, half of it must be added to each of the two straightaways. Question 2 leads into the text's description of a general method for finding arc length.

4.50 Copy the table in Display 4.39. As you work through these questions, keep track of your answers by filling in your copy of Display 4.39. Use your calculator to help with the arithmetic, and round your answers to one decimal place.

1. Which of the four arc lengths of Display 4.38 do you think is easiest to find? Why?

2. Can you think of a way to find any of these lengths? If you can, do it, and explain your method.

3. If you completed the circle centered at D, what would its circumference be?

4. What is the length of the 180° arc centered at D? Explain.

5. If you completed the circle centered at C, what would its circumference be?

6. What is the length of the 90° arc centered at C? Explain.

7. If you completed the circle centered at B, what would its circumference be?

8. What is the length of the 120° arc centered at B? Explain.

9. Use your thinking from the previous questions to describe a process for finding the length of any arc from its radius and central angle. Explain your process in words and then in a formula. Use your process to find the length of the 210° arc centered at A.

Center	Radius (ft.)	Circumference	Central Angle	Arc Length
D	47		180°	
C				
B				
A				
?	r		$x°$	

Display 4.39

Chapter 4

4.50

This set of questions is intended to help students discover for themselves the proportionality of arc length to central angle measure. Students probably will pick C or D as the easiest case, but it doesn't matter which they choose. The rest of the questions start with the conceptually simplest of the four cases, D. Students should be comfortable with the intuitively obvious idea that the arc length of a semicircle is half the circumference of the circle. From there it's an easy step to a quarter circle (C), then to one third (B). By this time, it should be clear that the arc length depends on what fraction of the circle is represented. The arc at A is a stepping stone question to the general formula. The completed table of Display 4.39 appears in Display 4.10T.

Note that the fact that there is no column in this table for the ratio of central angle to 360° is deliberate. This is the key idea that we want students to discover for themselves, without having the table tell them what to look for.

Center	Radius	Circumference	Central Angle	Arc Length
D	47	295.3	180°	147.7
C	30	188.5	90°	47.1
B	20	125.7	120°	41.9
A	25	157.1	210°	91.6
?	r	$2\pi r$	$x°$	$(2\pi r) \cdot \frac{x}{360}$

Display 4.10T

NOTES

_____

Here is one way to describe the pattern in Display 4.39.

The length of a circular arc is proportional to the measure of its central angle.

For instance, to find the length, L, of a 54° arc of a circle with circumference 100π, we can solve the proportion

$$\frac{L}{100\pi} = \frac{54°}{360°}$$

Multiplying both sides of this equation by 100π (the circumference), we get

$$L = \frac{54}{360} \cdot 100\pi$$

In other words, to find the arc length, just express the measure of the central angle as a fraction of 360°, then take that fraction of the circumference.

Now that we have the lengths of all the circular arcs, let's see what else we need in order to figure out the lengths of the straightaways.

4.51

Look at Display 4.37 again. The top straightaway is longer than the bottom one. How much longer? To figure this out, it helps to draw in a few more points and lines, as follows.

- **Mark two more points, E and F, on your scale drawing of the track. E is the top end of the circular arc centered at A. Extend the vertical radius of the arc at C until it intersects the top straightaway; F is the intersection point. Do you see that the length of EF is the difference between the top and bottom straightaways?**

- **Extend AE downward and BC to the left until they intersect. Call the intersection point G.**

Published by IT'S ABOUT TIME, Inc. © 2000 MATHconx, LLC

4.51

Perhaps the most worthwhile part of this exercise is that it is another illustration of how to break a question into smaller parts whose answers are relatively easy to find. It is also useful for them to see that it sometimes is helpful to add lines to a figure.

NOTES

Chapter 4

Now your drawing should look like Display 4.40.

1. Why are the lengths of *EF* and *GC* equal?

2. How long is *BC*? Why?

3. How long is *AB*? Why?

4. How long is *GB*? Why?

5. How long is *GC*? Why?

6. How long is *EF*? Why?

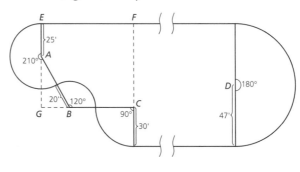

Display 4.40

Now we have all the pieces needed to answer Nip and Tuck's question. Here it is again.

4.52 Nip and Tuck want to make their Go-Kart track exactly $\frac{1}{10}$ of a mile long. How long should they make the straightaways?

A Go-Kart Track (Part II)

With the track design settled, Nip and Tuck are planning the cost of its construction. They want to cover the infield (the region inside the track) with sod (grass). The sod is sold by the square foot and is expensive, so they need an accurate measure of the area of the infield.

They begin by drawing the inner border of the track, 10 feet away from the center line all around. Then they divide the infield into 7 pieces and write in measurements for all the lengths and angles they know. Display 4.41 shows their drawing. The circled numbers mark the 7 regions. To find the area of the infield, they intend to calculate the area of each region separately, then add them up.

344

Chapter 4

1. Because they are opposite sides of a rectangle. Verifying that *EGCF* is a rectangle comes from tracing out the rotations of the various angles between the different radii. You might bypass this verification if everyone in the class agrees from the picture that *EGCF* is "obviously" a rectangle. The argument might be more distracting than enlightening at this point.

2. $20 + 30 = 50$ ft. (the sum of the radii from *B* and *C*).

3. $25 + 20 = 45$ ft. (the sum of the radii from *A* and *B*).

4. The easiest way to handle this is with trigonometry and a graphing calculator. $\triangle AGB$ is a right triangle with hypotenuse 45 ft. and $\angle GAB$ is 30°. Thus, the length of *GB* is

$$45\sin 30° = 45 \cdot \frac{1}{2} = 22.5 \text{ ft.}$$

5. $22.5 + 50 = 72.5$ ft. (the sum of *GB* and *BC*).

6. 72.5 ft. (by the answer to part 1).

4.52

This requires doing a little arithmetic with the answers from previous questions.

• The sum of the four arc lengths from Display 4.39 is 328.3 ft.

• Adding the length of *EF* from the previous question, we get 400.8 ft. as the total without the bottom straightaway and its corresponding piece in the top straightaway.

• Subtracting this sum from 528, the number of feet in $\frac{1}{10}$ of a mile, we get the total length to be added to the straightaways.

$$528 - 400.8 = 127.2$$

• Since half of this total must be in each straightaway, the bottom one must be 63.6 ft. long and the top one must be $72.5 + 63.6 = 136.1$ ft. long.

• If you want to emphasize the use of elementary algebra, the idea of these last two steps can be explained more efficiently as the solution of the equation

$$528 = 400.8 + 2x$$

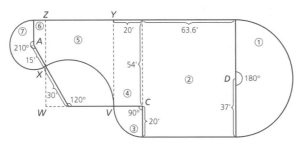

The infield of the Go-Kart track

Display 4.41

Region	Shape	Area (sq. ft.)
1	180° sector	
2	rectangle	
3	90° sector	
4	rectangle	
5	?	
6	quadrilateral	
7	210° sector	
	Total area	

Display 4.42

Copy the table in Display 4.42. As you work through these questions, keep track of your answers by filling in your copy of the table. Use your calculator to help with the arithmetic, and round your answers to one decimal place.

4.53

1. Compute the areas of the two rectangles; enter them into lines 2 and 4 of the table.

2. If you completed the circle centered at D, how much area would it enclose? What fraction of that area is in region 1? What is the area of region 1?

3. If you completed the circle centered at C, how much area would it enclose? What fraction of that area is in region 3? What is the area of region 3?

4. If you completed the circle centered at A, how much area would it enclose? What fraction of that area is in region 7? How do you know? What is the area of region 7?

5. The area of region 6 is approximately 161.2 square feet. How did Nip and Tuck find it?

Published by IT'S ABOUT TIME, Inc. © 2000 MATHconx, LLC

345

4.53

The completed table of Display 4.42 appears in Display 4.11T. The main purpose of this set of questions is to have the students observe and use the fact that sector area is proportional to central angle size. This fact is further emphasized in the next discussion question (4.54). Here are some other comments about these questions. Note that first of all, the radii in this figure differ by 10 ft. from those in Displays 4.36, 4.37, and 4.40 because that is the distance between the center of the track and its inside edge.

Region	Shape	Area (sq. ft.)
1	180° sector	2150.4
2	rectangle	4706.4
3	90° sector	314.2
4	rectangle	1080.0
5	?	1292.6
6	quadrilateral	161.2
7	210° sector	412.3
	Total area	10,117.1

Display 4.11T

1. These should be obvious from the diagram.

2. This is the first step in seeing the proportionality. The entire area is $\pi \cdot 37^2 \approx 4300.8$, so the area of the semicircular sector should be half of it, 2150.4 sq. ft.

3. This is the second step in seeing the proportionality. The entire area is $\pi \cdot 20^2 \approx 1256.6$ sq. ft. The area of this sector should be $\frac{1}{4}$ of it because a 90° angle is one-fourth of the way around a full 360° circle.

4. This is the third step in seeing the proportionality. The entire area is $\pi \cdot 15^2 \approx 706.9$ sq. ft. To find the area of the sector, we must first find out what fraction of 360° is represented by the 210° central angle: $\frac{210}{360} = \frac{7}{12}$. The area of the sector is this ratio times the area of the circle, 412.3 sq. ft. If the rounded full circle area is used, this will come out to 412.4.

5. One way is to split it into a rectangle (on top) plus a right triangle (below). The height of the rectangle is 15, the radius of the sector at A. Its width is $15 \sin 30°$, so the area of the rectangle is 112.5 sq. ft. The base of the right triangle is also $15 \sin 30°$, and its height is $15 \cos 30°$, so the area of the triangle is $\frac{1}{2}(15 \sin 30°)(15 \cos 30°) \approx 48.7$. The sum of these two areas is 161.2 sq. ft.

6. We have left the most difficult region for the last. Think of it as a rectangle, *ZYVW*, with a triangular corner, *XWB* cut off and then with a circular "bite" taken out of it.

   (a) How long is *ZY*? How do you know?
   (b) What is the area of rectangle *ZYVW*?
   (c) What is the area of triangle *XBW*? How did you find it?
   (d) What is the area of the circular bite, the 120° sector centered at *B*? How did you find it?
   (e) What is the area of region 5?

7. What is the total area inside the Go-Kart track?

4.54

Earlier in this section we described the pattern in Display 4.39 by saying that the length of a circular arc is proportional to its central angle. What is the analogous statement for area? Do you think it is always true? Why or why not?

A Go-Kart Track (Part III)

Now Nip and Tuck have to order the asphalt paving for their Go-Kart track. This, too, is sold by the square foot, so they have to figure out the number of square feet in their 20 foot wide track.

"I know how to do that," says Nip. "If we compute the area of the region bounded by the outside edge of the track and then subtract the area of the infield, we'll have the area of the track itself."

"Ugh!" says Tuck with a frown, "I can't bear to do another calculation with all those different pieces! I've got a better idea. Let's pretend that the entire track is straight. In that case, then all we'd have to do is multiply its length by its width, and we'd have our answer."

"But it's *not* straight," says Nip. "On every curve, the inner edge is shorter than the outer edge. Will all those differences balance out exactly?"

"Well, maybe not exactly," admits Tuck, "but it ought to be pretty close, don't you think?"

"I don't know," replies Nip, "but we'd better make sure before we're stuck with an extra truckload of asphalt or don't have enough to finish the track."

Published by IT'S ABOUT TIME, Inc. © 2000 MATHconx, LLC

6. (a) $72.5 - 20 - 15\sin 30° = 30 + 15 = 45$ feet.
    (b) $45 \cdot 54 = 2430$ sq. ft.
    (c) $\frac{1}{2} \cdot 30\sin 30° \cdot 30\cos 30° \approx 194.9$ sq. ft.
    (d) This is one more example of the proportionality of sector area to central angle size. The area of the entire circle would be $\pi \cdot 30^2$; the area of this sector is $\frac{1}{3}$ of that—approximately 942.5—because 120° is $\frac{1}{3}$ of 360°.

    (e) Thus, the approximate area of region 5 is
    $2430 - 194.9 - 942.5 = 1292.6$ sq. ft.

7. The sum of the region areas (the last column of Display 4.42).

4.54

This is simultaneously an exercise in analogy and in generalization. It asks students to connect what they have just done with what they did before, to recognize a similar pattern, and to formulate that pattern as a useful general rule. The statement is,

The area of a circular sector is proportional to its central angle.

This is precisely the principle that they should have used intuitively as they answered the sector area questions for the infield. It always holds for the same reason that the arc length principle always holds—because every angle is an angle of rotational symmetry for a circle. Thus, the area is evenly distributed around the circular region, so that the amount in a sector is proportional to the central angle of the sector.

NOTES

We can help Nip and Tuck out by applying some of the thought patterns we saw in this section in a creative way. First, let's simplify the problem in two ways.

- All the curves in the track are circular arcs, so, instead of trying to deal with the whole track at once, we'll look at a single circular arc.

- Since circles are symmetric with respect to all rotations about their center, if the balancing idea works for an entire circular path, it will work proportionally for any sector.

Thus, it makes sense to start with an example of a circular track, such as the one pictured in Display 4.43. In this case, the center line of the track has radius 30 feet and the width is 20 feet (just like the Go-Kart track). This track is the region between two concentric circles. It looks like a large washer. Such a figure is called an **annulus.** We'll compute the area of this circular track in two ways, Nip's way and Tuck's way.

About Words

Annulus is the Latin word for ring.

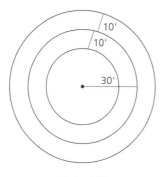

Display 4.43

Nip's way.

The area inside the outer circle is $\pi \cdot 40^2 = 1600\pi$. Subtract from that the area inside the inner circle, which is $\pi \cdot 20^2 = 400\pi$. The difference is 1200π.

Tuck's way.

If we straightened out this track, it would be like a rectangle 20 ft. wide and as long as the center line, which is the circumference of a circle with a 30 ft. radius: $2\pi \cdot 30 = 60\pi$. Thus, the area is $20 \cdot 60\pi = 1200\pi$.

The two answers came out *exactly* the same!

NOTES

a
4.55

Does this work all the time? That is, can you find the area of *any* annulus by taking the circumference of its middle circle and multiplying by its width? Try out some other cases. The first line of the table in Display 4.44 describes another annulus; the next three lines leave space for you to fill in examples of your own choice. The last line is for an algebraic description that applies to any annulus. In the last line, what does *s* represent? Copy and fill in this table.

Annulus		Area	
Middle radius	Width	Nip's way	Tuck's way
50 m	12 m		
r	2*s*		

Display 4.44

4.56

Does your work in filling out Display 4.44 convince you that the area of any annulus is the product of the circumference of its "middle" circle and its width? Why or why not? Do you think Display 4.44 *proves* this statement? If so, why? If not, what more is needed?

b
4.57

How many square feet of asphalt are needed to pave Nip and Tuck's Go-Kart track?

Here are three facts to remember about parts of circles.

1. The length of an arc is proportional to the measure of its central angle. Specifically, an $n°$ arc has length $\left(\frac{n}{360}\right)C$, where C is the circumference of the circle.

2. The area of a sector is proportional to the measure of its central angle. Specifically, an $n°$ sector has area $\left(\frac{n}{360}\right)A$, where A is the area of the circle.

3. The area of an annulus is the product of the circumference of its middle circle and its width.

348

Chapter 4

4.55

A partially filled in form of Display 4.44 appears as Display 4.12T. An intermediate form of each computation is included. We cannot anticipate your students' choices, but we supply one more case, in the event you should need one on short notice. No matter what the students choose, the last two columns should agree. Of course, the width cannot be more than twice the middle radius! The fact that the answer in the first line is 1200π, just like the text example, is an irrelevant coincidence that might, nevertheless, provoke some thought. The final line connects the geometric ideas with some algebra. (s represents half the width of the annulus.) It is important for the next discussion question (4.56) and also, in its own right, as an example of how algebra can be used to describe a pattern efficiently.

Annulus		Area	
Middle radius	Width	Nip's way	Tuck's way
50 m	12 m	$56^2\pi - 44^2\pi = 1200\pi$	$2\pi \cdot 50 \cdot 12 = 1200\pi$
27 ft.	40 ft.	$47^2\pi - 7^2\pi = 2160\pi$	$2\pi \cdot 27 \cdot 40 = 2160\pi$
r	$2s$	$(r + s)^2\pi - (r - s)^2\pi$	$2\pi \cdot r \cdot 2s = 4\pi rs$

The partially completed table of Display 4.44

Display 4.12T

4.56

This question focuses on the meaning of proof. It also provides an opportunity to illustrate a valuable algebraic technique, if you choose to pursue the question that far. Students might or might not say that they are convinced, but you should guide them to admit that a few examples do not constitute a proof. Some sort of general argument is needed, one that does not depend on particular choices of radius or width.

The last line of the table suggests such an argument. The algebra mirrors the process they used in each of the particular cases. You might have to help them see that, but it's not difficult. However, it has the advantage that r and k can be *any* numbers that describe an annulus. Proving that the entry in the third column always equals the entry in the fourth column is an exercise in algebraic simplification. One way to do it is by brute force; multiply out the terms of

$$(r + s)^2\pi - (r - s)^2\pi$$

collect and cancel terms, etc.

There is another way that illustrates the useful algebraic identity about the difference of two squares.

$$a^2 - b^2 = (a + b)(a - b)$$

Your students may not be familiar with this identity. If you are willing to teach it here, this provides an excellent example of its use, as well as an

Problem Set: 4.5

1. A windshield wiper pivots through an angle of 110°. The wiper is a 20 inch arm fastened to the middle of a 16 inch blade. (See Display 4.45.)

   (a) How many square inches of windshield does this wiper wipe? Round your answer to one decimal place.

   (b) The 16 inch blade is replaced by an 18 inch blade.

      (i) How many *more* square inches will this wiper wipe? Round your answer to one decimal place.

      (ii) What percentage increase is this over the 16 inch wiper? Is this increase the same percentage as the increase in length? Explain.

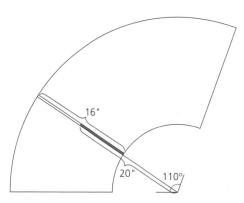

16"

20"   110°

Display 4.45

2. A landscape architect is planning the paved driveway diagrammed in Display 4.46. It has these design specifications.

   • All three arcs are circular.

   • The central angle of the arc centered at *A* is 100°; its radius to the center line of the drive is 48 feet.

   • The arcs centered at *B* and *C* are congruent and have the same radius: 16 feet to the center line of the drive.

   • The straight distance between the two small arcs is 98 feet.

   • The straight distance between the large and small arcs is 31.5 feet on each side.

Published by IT'S ABOUT TIME, Inc. © 2000 MATHconx, LLC

349

Chapter 4

example of letting one variable stand for a combination of others. Specifically, if we consider $a = r + s$ and $b = r - s$, then

$$
\begin{aligned}
(r+s)^2\pi - (r-s)^2\pi &= ((r+s)^2 - (r-s)^2)\pi \\
&= (a^2 - b^2)\pi \\
&= (a+b)(a-b)\pi \\
&= ((r+s) + (r-s))((r+s) - (r-s))\pi \\
&= (2r)(2s)\pi \\
&= 4\pi r s
\end{aligned}
$$

4.57

Because all the curves in the track are circular arcs, the annulus principle allows us to do this computation Tuck's way —straightening it out— and considering it as if it were a long rectangle. We know that the center line is 528 ft. ($\frac{1}{10}$ mi.) long and the track width is 20 ft., so the track area is
$528 \cdot 20 = 10{,}560$ sq. ft.

Problem Set: 4.5

1. (a) (Tuck's way) Take the length of the middle arc and multiply by the length of the blade (the width of the annular sector).
   $(\frac{110}{360} \cdot 40\pi) \cdot 16 \approx 614.4$ sq. in.
   (b) (i) The 18 in. blade will wipe $(\frac{110}{360} \cdot 40\pi) \cdot 18 \approx 691.2$ sq. in.
   Subtracting the answer to part (a), we get an increase of 76.8 sq. in.
   (ii) The percentage increase is $\frac{76.8}{614.4} = 12.5\%$. Yes, this is the same percentage increase as the increase in length. Even though we are dealing with areas, we have only increased *one* of the two lengths being multiplied to get the area, so the increase in area should be proportional to the increase in that one length.

2. (a) Since there are no reverse curves, the total of all the arc lengths is 360°, —all the way around once. The central angle at A is 100°, leaving 260° for the other two. Since the arcs at B and C are congruent, each must be 130° (half of 260°).
   (b) This requires finding the three arc lengths and adding them to the given straight lengths.
   Length of large arc: $\frac{100}{360} \cdot 2\pi \cdot 48 \approx 83.8$ ft.
   Length of each small arc: $\frac{130}{360} \cdot 2\pi \cdot 16 \approx 36.3$ ft.
   Total length of drive loop.
   $83.8 + 2 \cdot 36.3 + 2 \cdot 31.5 + 98 \approx 317.4$ ft.
   (c) The area of the driveway loop can be done Tuck's way because the curves are circular arcs: Multiply the length of the center line by the width of the driveway: $317.4 \cdot 14 = 4443.6$ sq. ft. To that, add the area of the rectangular entrance, which is $20 \cdot 18 = 360$ sq. ft. Thus, the total area to be paved is 4803.6 sq. ft.

- The driveway is 14 feet wide.

- The entrance to the driveway is a 20 by 18 foot rectangle.

Answer the following three questions. Be prepared to justify your answers.

(a) What size is the central angle of the arcs centered at *B* and *C*?
(b) How long is the center line of the driveway loop? Round your answer to one decimal place.
(c) How large is the total area to be paved?

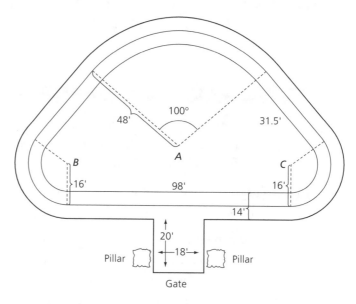

Display 4.46

3. Nip and Tuck designed their Go-Kart track to be exactly $\frac{1}{10}$ mile around, so that it would be easy to figure out the average speed of each Go-Kart from the time it took to go around the track. They use a stopwatch to get each Go-Kart's lap time in seconds. Write a function for Nip and Tuck's calculator that will automatically convert these times to miles per hour.

You might begin by answering these questions. How many seconds are in one hour? If a Go-Kart's lap time is 60 seconds, how long would it take for that Go-Kart to travel a mile at the same average speed? How many miles would it travel in one hour at that speed? What if the lap time is 90 seconds? What if it is *x* seconds?

3. This is a question about functions and simple algebra. The desired function is

$$y = \frac{3600}{10x}$$

where y is the speed in miles per hour and x is the lap time in seconds. The answers to the suggested preliminary questions are,

*   There are 3600 ($= 60 \cdot 60$) seconds in an hour.

*   600 seconds (10 minutes)

*   If it takes 10 minutes for 1 mile, in 60 minutes it would go 6 miles.

*   If its lap time is 90 seconds, it takes 900 seconds to go 1 mile, so it goes 4 miles ($= \frac{3600}{900}$) in an hour.

*   If its lap time is x seconds, it takes $10x$ seconds to go 1 mile, so it goes $\frac{3600}{10x}$ miles in an hour. This is the formula for the function.

4. At the beginning of this section, you saw that a *segment* of a circle is the part of the disk bounded by an arc and its chord.

   (a) What is a good strategy for finding the area of a segment?

   (b) Try out your strategy: Find the area of an 80° segment of a circle with radius 10 inches. Round your answer to two decimal places. The degree measurement here refers to the arc length of the segment. Start by drawing a rough sketch.

   (c) In part (b), how did you find the area of the isosceles triangle that has the chord as its base? Did you divide it into two right triangles and use trigonometric functions? If so, to which angle did you apply those functions?

   (d) Find the area of a 140° segment of a circle with radius 15 in. This time, use sin and cos of half the central angle to find the area of the isosceles triangle with the chord as its base. Round your answer to two decimal places.

5. This problem generalizes problem 4. Find the area of a $\theta°$ segment of a circle with radius r. Your answer should be a formula written in terms of the radius r and *half* of the central angle $\theta°$, which we shall call α (alpha).

   Begin by looking at Display 4.47, which shows a segment AB. Try to find a formula for the area of the segment bounded by chord AB and minor arc AB in terms of r and α. Work through the steps you used for part (d) of problem 4, substituting r for the radius and α for half of the central angle. Then see if you can put all those steps together into one fairly simple formula.

About Symbols

Alpha, α, is the first letter of the Greek alphabet, like our letter a. Greek letters often are used in mathematics as symbols for angles (and other things).

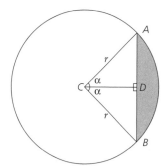

Display 4.47

4. This problem and the next lead from a computational exercise to an algebraic formula for the area of a sector in terms of the sin and cos of a related central angle. The four parts of this problem should be accessible to all students but are somewhat sterile if you don't intend to pursue the pattern. Problem 5 is a more challenging discovery exercise. Some students might need your guidance in discovering the general formula, after they have done parts (a)–(d) of this problem.

(a) Find the area of the corresponding sector, then subtract the area of the (isosceles) triangle that has the chord as its base.

(b) Area of sector: $\frac{80}{360} \cdot \pi \cdot 10^2 = \frac{200}{9}\pi \approx 69.81$ sq. in.

Area of triangle: $2 \cdot \left(\frac{1}{2} \cdot 10 \sin40° \cdot 10 \cos40°\right) \approx 49.24$ sq. in.

Approx. area of segment: $69.81 - 49.24 = 20.57$ sq. in.

Students who used sin and cos of the (50°) base angle of the isosceles triangle will get the same answer because $\sin50° = \cos40°$ and $\cos50° = \sin40°$.

(c) This question is self explanatory.

(d) Area of sector: $\frac{140}{360} \cdot \pi \cdot 15^2 = 87.5\pi \approx 274.89$ sq. in.

Area of triangle: $2 \cdot \left(\frac{1}{2} \cdot 15 \sin70° \cdot 15 \cos70°\right) \approx 72.31$ sq. in.

Approximate area of segment: $274.89 - 72.31 = 202.58$ sq. in.

5. Here are some questions to help you lead your students to the formula. It is best, of course, if they come up with these steps on their own, but you might have to coax them along from time to time.

(a) What is the area of the sector ABC?

Since the central angle is 2α, it is $\frac{2\alpha}{360} \cdot \pi r^2 = \frac{\alpha}{180} \cdot \pi r^2$.

Note that if we were using radian measure for angles, this would simplify very nicely: $\frac{2\alpha}{2\pi} \cdot \pi r^2 = \alpha r^2$.

(b) What is the area of one of the two right triangles, say $\triangle ACD$?

The area of $\triangle ACD$ is $\frac{1}{2} \cdot r \cos \alpha \cdot r \sin \alpha$

(c) What is the area of $\triangle ABC$?

It is twice the previous answer: $r \cos \alpha \cdot r \sin \alpha = r^2 \cos \alpha \sin \alpha$

(d) What is the area of the segment?

It is $\frac{\alpha}{180} \cdot \pi r^2 - r^2 \cos \alpha \sin \alpha$. This can be simplified slightly, by factoring out r^2, to $r^2 \left(\frac{\alpha}{180} \cdot \pi - \cos \alpha \sin \alpha\right)$.

Note that if we were using radian measure for angles, this would be even simpler: $r^2 (\alpha - \cos \alpha \sin \alpha)$.

4.6 Inscribed Angles

Learning Outcomes

After studying this section, you will be able to:

Define and identify inscribed angles;

Use the measure of the central angle to find the measure of an inscribed angle that intercepts the same arc;

Explain why any angle inscribed in a semicircle must be a right angle.

Much of the previous section depends on the fact that each arc of a circle corresponds to exactly one central angle, and the size of an arc can be described by the size of its central angle. Thus, a 40° arc is an arc with a central angle of 40°. Circular arcs are also related to another kind of angle. If you connect the two endpoints of an arc with some point on the circle but outside of the arc, you get an angle. Of course, you can get lots of angles from a single arc this way. (See Display 4.48.) Since they all are attached to the same arc, it makes sense to ask if they have anything else in common. Let's explore this question.

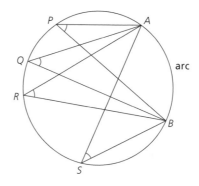

Display 4.48

About Words

To *inscribe* means to draw or write inside something. Think of an *inscription* on a ring or in a locket. To *intercept* something means to cut it off or to stop it at a particular place.

Two new terms will make our descriptions simpler to write. An angle formed by two chords with an endpoint in common is called an **inscribed angle.** We say that an inscribed angle **intercepts** the arc opposite its vertex. Display 4.48 shows four inscribed angles, at points *P, Q, R* and *S.* They all intercept the same arc, *AB.*

352

4.6 Inscribed Angles

This short section explains and verifies the measurement relationship between inscribed angles and their intercepted arcs. As an important special case, it shows why any angle inscribed in a semicircle must be a right angle.

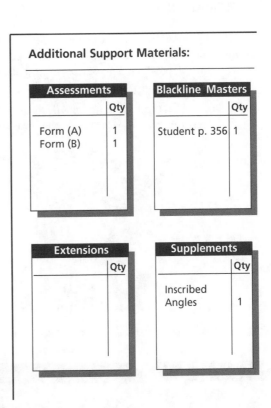

Additional Support Materials:

Assessments	Qty
Form (A)	1
Form (B)	1

Blackline Masters	Qty
Student p. 356	1

Extensions	Qty

Supplements	Qty
Inscribed Angles	1

Chapter 4

EXPLORATION 1
SETUP

You need four sheets of patty paper and a pen or sharp, soft pencil that will draw clear, dark lines.

- Draw four congruent circles, one on each sheet of patty paper. Mark the centers clearly.

- On one of these circles, clearly mark an arc of your choice. This will work best if your arc is between 45° and 120°.

- Put another piece of patty paper over the first one so that the circles match. Draw an inscribed angle that intercepts the arc.

- Repeat the previous step with the other two sheets of patty paper. Each time draw a *different* inscribed angle that intercepts the same arc.

QUESTION

How do the different inscribed angles compare in size?

To get some idea of this, put the sheets of patty paper in a stack so that the circles and the vertices of the three angles coincide. Then make a conjecture about the sizes of inscribed angles that intercept the same arc.

EXPLORATION 2
SETUP

For this Exploration you need a blank sheet of regular paper, a compass or other circle drawing tool, and a protractor.

- Draw a large circle on your paper and clearly mark an arc, *AB*, of your choice.

- Draw the central angle of arc *AB*, measure it to the nearest degree, and write down the measurement.

Published by IT'S ABOUT TIME, Inc. © 2000 MATHconx, LLC

353

EXPLORATION 1.

This Exploration gives students a direct, visual illustration of the fact that all inscribed angles of a given arc are congruent. The number of sheets of patty paper used can be varied, from at least 3 to as many as you would like. The activity can be done in small groups, with each student drawing one of the inscribed angles on his/her own sheet of patty paper. Then the group can match up all its sheets and see that, no matter who drew each inscribed angle, they all are the same size.

As an alternative to patty paper, you might set this up with overhead transparencies, have students draw inscribed angles on them, and then show the class how all the students' angles match. The next Exploration gets at the same idea, but by measurement, rather than by direct visual verification.

EXPLORATION 2.

Display 4.49 is included here in case you want your students to measure a second example without taking the time to make another drawing. It should not be regarded as the *best* way to draw this figure, just as one way to do it.

The responses to questions 1 and 2 should end up as two conjectures.

1. All the inscribed angles are the same size.

2. The measure of each inscribed angle is half that of the central angle.

If the students have measured with reasonable care, they should come up with these two ideas easily. These are important facts, but at this stage students might not be any more than pretty sure that they are true. After all, the statements are based on rounded measurement observations. Based on this evidence, it might be that all the inscribed angles must lie within a small *range* of values (a degree or two), but need not be exactly equal. The doubt here is useful because it invites a proof that exact equality is always true. The text provides a proof for a special case of this. The other cases appear as exercises.

- Draw five different inscribed angles that intercept arc *AB*. Display 4.49 is an example of a drawing made with these instructions.

- Measure each of the five inscribed angles to the nearest degree and write down your measurements.

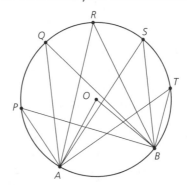

A central angle and five inscribed angles for arc *AB*

Display 4.49

QUESTIONS

1. How do you think the five inscribed angles are related to each other? Are you sure? Why or why not? Have you accounted for measurement error?

2. How do you think the five inscribed angles are related to the central angle? Are you sure? Why or why not?

3. If your arc is a semicircle, what is the measure of its central angle?

4. Based on your answers to questions 1 and 2, what do you think is the measure of an angle inscribed in a semicircle?

5. Will you get the same results if you begin with an oval instead of a circle? Try it and see.

The principles you have observed in these Explorations apply to a surprising variety of things. Here is just one example; this one is from show business.

Overhead stage lighting units come in three standard sizes: 20°, 30°, and 40°. The degree values refer to the largest cone of light that the unit produces. If we cut this three dimensional light cone down the middle, so to speak, we get a triangular cross section, as in Display 4.50. The three sizes refer to the size of the angle of this triangle at the light source.

354

Questions 3 and 4 lead to an important corollary.

> An angle inscribed in a semicircle is a right angle.

In question 5, the inscribed angles will come out with different measures easily enough, and that might be all you want to do. Finding a central angle requires first dealing with what an appropriate center point would be, which, in turn, raises questions of symmetry. This can lead to an interesting discussion with a curious, highly motivated class.

If you have Geometer's Sketchpad or other similar software, either for student use or as a demonstration tool, this is a good place to use it. Construct a circle, mark the endpoints of an arc, and draw the central angle. Then mark any point, P, on the circle and connect it to the endpoints of the arc. Display the measures of the central angle and the inscribed angle at P. Then move P around and see that the measure of the inscribed angle does not change until you move P from the major arc to the minor arc, or vice versa. You can also observe that the sum of the two different values for the inscribed angles is exactly 180°, which suggests the relationship between the central angle and the inscribed angles.

NOTES

Set designers place light units at different angles to the stage for special lighting effects. One of their problems is to place each unit so that the maximum light cone covers the entire performing area, but does not extend beyond it into the audience or other offstage areas. Here is a particular case.

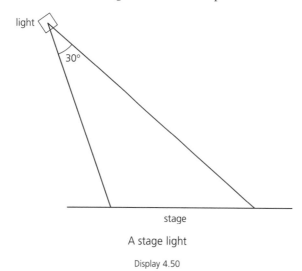

light

30°

stage

A stage light

Display 4.50

Mike MegaWatt is setting up the lighting for a rock concert. The stage measures 20 feet from front to back. He wants midstage lighting from the upper back, directly overhead, and the upper front. Three 30° units are available for this part of the job. Where can he put them?

Fortunately, Mike studied geometry (as all set designers do). He remembers that all the inscribed angles for a circular arc have the same measure, so he decides to think of his light units as 30° inscribed angles. But for which circle?

He thinks of the 20 foot stage depth as a chord of the circle because that's exactly what he wants the lighting angles to intercept. The central angle of this chord (and its arc) must be twice the inscribed angle, so the central angle must be 60°. He draws an isosceles triangle, using the 20 foot stage line as the base. Each base angle must be 60°, so he draws lines at a 60° angle from each end of the chord, sloping up toward each other. The place where they cross is the center of the circle he wants. Why? He draws the circle with this center point and the radius determined by one edge of the stage. (See Display 4.51.) He can place his lights anywhere on that circle!

Published by IT'S ABOUT TIME, Inc. © 2000 MATHconx, LLC

NOTES

Chapter 4

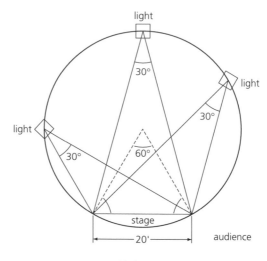

Display 4.51

4.60

1. How does Mike MegaWatt know that the triangle he wants is isosceles?

2. How does he know that its base angles are 60°?

3. How long is the radius of Mike's circle?

4. If Mike wants to put a light directly overhead, he will have to build scaffolding to hold it. How high above the stage must that light be?

5. If he doesn't want the lights to be so far from the stage, should he replace his 30° lighting units with 20° units or with 40° units? Why?

4.61

1. When Mike MegaWatt rechecks his lighting units, he finds that they are the 20° size. Redo Display 4.51 for this size. Specify the size of the angles of the triangle, the radius of the circle, and the height of the topmost light. Round your answers to one decimal place.

2. Mike doesn't like the design for 20° units; the scaffolding would have to be too high. He manages to trade them for three 40° units. Redo Display 4.51 for this size. Specify the size of the angles of the triangle, the radius of the circle, and the height of the topmost light. Round your answers to one decimal place.

356

Chapter 4

4.60

This discussion should review some earlier ideas and solidify students' understanding of this example.

1. A central angle is formed with radii, and all radii of a circle are equal.

2. The sum of the angles must be 180° and one of them (the central angle) is 60°. Thus, the sum of the other two must be 120°, and they are equal because the triangle is isosceles. In simple algebraic terms, solve $60 + 2x = 180$.

3. 20 feet. Since all three angles are 60°, this triangle is equilateral.

4. Its height is the height of the isosceles triangle plus the radius; that is,
$$20 \sin60° + 20 \approx 37.3 \text{ ft.}$$

5. With 40° units. The larger light cone will span more stage area from the same distance, so it can be moved in closer.

4.61

These questions are suitable for small group work. Redoing the figure in each case is straightforward and should be possible for all groups of students. More challenging for some students is finding the radius in these cases and the height of the top light, which depends on it.

1. The central angle is 40°, so the two base angles are 70° each. Finding the radius draws on some trigonometry and algebra. Bisect the isosceles triangle from top to bottom. Each half is a right triangle with its hypotenuse a radius of the circle and the horizontal leg 10 ft. long. If r is the radius, then
$$10 = r \cos70°$$
so
$$r = \frac{10}{\cos70°} \approx 29.2 \text{ ft.}$$
The height of the topmost light is approximately
$$(29.2 \sin70°) + 29.2 \approx 56.6 \text{ ft.}$$

2. The central angle is 80°, so the two base angles are 50° each. To find the radius, use the method from the previous question.
$$10 = r \cos50°$$
so
$$r = \frac{10}{\cos50°} \approx 15.6 \text{ ft.}$$
The height of the topmost light is approximately
$$(15.6 \sin50°) + 15.6 \approx 27.6 \text{ ft.}$$

Here are three important facts about inscribed angles.

1. All inscribed angles that intercept the same arc are the same size.

2. The measure of an inscribed angle is half that of the central angle of its intercepted arc.

3. Any angle inscribed in a semicircle is a right angle.

We haven't really justified these statements, except by looking at some examples. The key to all three of them is statement 2. If we know that, then statements 1 and 3 follow easily. We end this section by proving one case of statement 2 with you. The other cases can be proved from this one. They will appear in the problem set.

Here is what we shall prove now.

The measure of an inscribed angle *with one side passing through the center of the circle* is half that of the central angle of its intercepted arc.

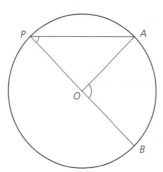

Display 4.52

Display 4.52 illustrates this case. Arc *AB* is intercepted by the inscribed angle at *P*. Angle *AOB* is the corresponding central angle. We want to show that

4.62

$$\angle APB = \frac{1}{2}\angle AOB$$

Answer these questions to complete the proof.

1. What kind of triangle is △ *POA*? Why?

2. How are ∠*APB* and ∠*PAO* related? Why?

3. What is the sum ∠*APB* + ∠*PAO* + ∠*POA*?

4. What is the sum ∠*AOB* + ∠*POA*?

Published by IT'S ABOUT TIME, Inc. © 2000 MATHconx, LLC

357

4.62

1. It is isosceles because *OP* and *OA* are radii.

2. They are equal because they are the base angles of an isosceles triangle.

3. 180°

4. 180°

5. Both sides of the equation equal 180°

6. The same quantity has been subtracted from both sides of the previous equation.

7. By 2. above, ∠*PAO* = ∠*APO*

8. Divide both sides in problem 7 by 2.

NOTES

5. Justify. $\angle APB + \angle PAO + \angle POA = \angle AOB + \angle POA$

6. Justify. $\angle APB + \angle PAO = \angle AOB$

7. Justify. $2 \cdot \angle APB = \angle AOB$

8. Justify. $\angle APB = \frac{1}{2}\angle AOB$

Problem Set: 4.6

1. Explain how the relationship between inscribed angles and their intercepted arcs can be used to prove that the sum of the angles of a triangle must be 180°.

2. A polygon is *inscribed* in a circle if all its angles are inscribed angles of that circle.

   (a) Prove or disprove. If a quadrilateral inscribed in a circle has one right angle, then it must have at least two.
   (b) Prove or disprove. If a quadrilateral inscribed in a circle has one right angle, then it must be a rectangle.

   Recall that to prove a general statement, you must give an argument to show that it is always true. To disprove it, you need only find a counterexample, one instance in which it is false.

3. Here is another case of the statement.

   The measure of an inscribed angle is half that of the central angle of its intercepted arc.

   In this case, the line through the vertex of the inscribed angle and the center of the circle crosses the intercepted arc, as in Display 4.53. Prove that the statement is true in this case.

   Plan: Draw *PO* and extend it until it hits the circle at some point, say *C*. Use *PC* to break the problem into two instances of the case we proved at the end of this section. Add the results.

Published by IT'S ABOUT TIME, Inc. © 2000 MATHconx, LLC

Problem Set: 4.6

1. Any triangle can be inscribed in a circle because three noncollinear points determine a circle. The vertices of the triangle divide the entire circle into three arcs that do not overlap. Thus, the sum of these arcs is 360°. Each of these arcs is the intercepted arc of one of the three angles of the triangle. Since the measure of each angle is half the sum of its intercepted arc, the sum of the three angles must be half the sum of the intercepted arcs. That is, the sum of the angles must be 180°.

2. (a) This is true. Suppose that $ABCD$ is such a quadrilateral (with its vertices labeled in alphabetical order the usual way) and that $\angle A$ is a right angle. Then the intercepted arc of $\angle A$, which is BD, must be a semicircle. But B and D are also the endpoints of the chords that form $\angle C$, so the intercepted arc of $\angle C$ must also be a semicircle. Thus, $\angle C$ must also be a right angle.

   (b) This is false. To make a counterexample, draw a diameter of a circle. If you use its two endpoints as two vertices of the quadrilateral, then it is easy to pick another point on each semicircle so that the resulting figure is obviously not a rectangle. For instance, a kite shape will work.

3. $\angle APB = \angle APC + \angle CPB$

   $\angle APC = \frac{1}{2} \angle AOC$ and $\angle CPB = \frac{1}{2} \angle COB$ by (4.71)

   Substituting into the first equation,

   $\angle APB = \frac{1}{2} \angle AOC + \frac{1}{2} \angle COB = \frac{1}{2} (\angle AOC + \angle COB) = \frac{1}{2} \angle AOB$

4. $\angle APB = \angle APC - \angle BPC$

   $\angle APC = \frac{1}{2} \angle AOC$ and $\angle BPC = \frac{1}{2} \angle BOC$ by (4.71)

   Substituting into the first equation,

   $\angle APB = \frac{1}{2} \angle AOC - \frac{1}{2} \angle BOC = \frac{1}{2} (\angle AOC - \angle BOC) = \frac{1}{2} \angle AOB$

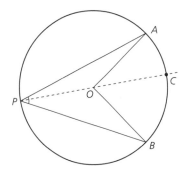

Display 4.53

4. Here is the remaining case of the statement.

> The measure of an inscribed angle is half that of
> the central angle of its intercepted arc.

In this case, shown in Display 4.54, the line through the
vertex of the inscribed angle and the center of the circle
does not cross the intercepted arc. Prove that the statement
is true in this case.

Plan: Draw *PO* and extend it until it intercepts the circle at
some point, say C. Use *PC* to break the problem into two
instances of the case we proved at the end of this section.
Subtract one result from the other.

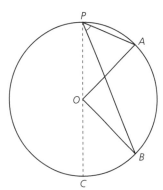

Display 4.54

Published by IT'S ABOUT TIME, Inc. © 2000 MATHconx, LLC

NOTES

4.7 Coming Around Full Circle

Learning Outcomes

After studying this section, you will be able to:

Use methods from this chapter to construct Reuleaux triangles and other non-circular curves of constant width;

Use methods from this chapter to find perimeters and areas of Reuleaux triangles and other non-circular curves of constant width.

We end this chapter by looking back at Section 4.1, to the properties of circles listed by Professor Philip Davis. Most of those properties have already been explored in some detail, but one of them—*constant width*—deserves a little more attention.

Loosely speaking, a figure of *constant width* is a closed curve that is the same width everywhere. That sounds simple, until you think about how to check a particular curve for this property. What does "everywhere" mean? With circles, it's easy enough. It means that the diameter is the same everywhere. But what if the curve is not a circle? What plays the role of the diameter? How do you measure the width of an isosceles triangle or an oval or the Go-Kart track in Section 4.5?

In Section 4.1 the idea of constant width was explained in terms of rollers. We said that there can be rollers which work smoothly but do not have circular cross sections. Picture how rollers work. Look back at the picture, in Section 4.1, if you need help visualizing this. For a roller to work smoothly, the distance between the flat surface on which it is rolling and the flat object rolling on it must always be the same.

 Display 4.55 shows the side view of some rollers. Which of these rollers will work smoothly? Which will cause an uneven ride? Which are you not sure about?

4.63

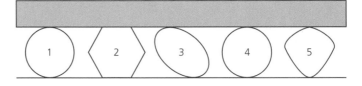

Display 4.55

Display 4.55 shows us a good way to think about the width of a figure: "Trap" the figure between two parallel lines and measure the distance between the lines. That measurement may vary, of course, depending on where you draw the parallel

360

4.7 Coming Around Full Circle

This section explores an idea introduced in Section 4.1—constant width. It uses machinery developed in the rest of the chapter to examine some peculiar planar shapes. In particular, it uses facts about arcs and sectors to find the perimeters and areas of Reuleaux triangles and related curves of constant width. This investigation provides a good conclusion for the chapter.

4.63

The purpose of this picture is to get the students to visualize curves between parallel lines. This provides the basis for the generalized idea of *width*. Curves 1 and 4, which are circles, clearly will work smoothly. It should be obvious that 2 and 3 will not, but there ought to be some uncertainty about 5.

Chapter 4

Additional Support Materials:

Assessments			Blackline Masters	
	Qty			Qty
Project 4.7 (A)	1			

Extensions			Supplements	
	Qty			Qty

lines. For example, the rectangle in Display 4.56 has many different widths, three of which are illustrated.

The width between lines l_1 and l_2 is 1 cm.

The width between lines m_1 and m_2 is 2 cm.

The width between lines n_1 and n_2 is 2.2 cm.

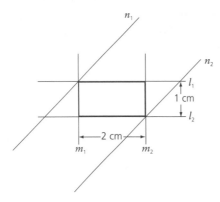

Display 4.56

1. What is the smallest width of the rectangle in Display 4.56? What is its largest width? Explain.

a
4.64

2. What is the smallest width of a square that is 3 cm on a side? What is its largest width? Explain.

3. What is the smallest width of a circle of radius 4 cm? What is its largest width? Explain.

Use a ruler to measure, to the nearest millimeter, the largest and smallest widths of the three curves in Display 4.57.

b
4.65

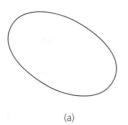

(a)   (b)   (c)

Display 4.57

Published by IT'S ABOUT TIME, Inc. © 2000 MATHconx, LLC

361

Chapter 4

4.64

This reinforces the fact that most figures have variable width. Don't look for profound explanations here; be content with anything that shows student understanding of the width concept.

1. Smallest is 1 cm, obviously. Largest is the length of the diagonal, $\sqrt{5}$ cm (approx. 2.236 cm).

2. Smallest is 3 cm, the length of a side. Largest is the length of the diagonal, $3\sqrt{2}$ cm (approx. 4.24 cm).

3. The circle is a curve of constant width. Its largest and smallest widths are the same—the diameter, 8 cm.

4.65

This exercise has two purposes.
(1) To illustrate the fact, with curve (c), that the width of a figure is not the same as the shortest distance across any middle of it, and

(2) To have students verify for themselves that a noncircular figure, (b), can have constant width.

These measurements presume no change in the size of these figures during the printing process. You might need to check them yourself.

(a) 22 mm, 37 mm   (b) constant 29 mm   (c) 21 mm, 30 mm

Now that you have a good idea of width, the term constant width should make sense. A figure (usually called a curve) has **constant width** if the distance between two parallel lines that trap (just touch, but don't cross or squeeze) the figure is always the same, no matter how the figure is placed between them.

Circles have constant width, of course. The distance between any two parallel tangent lines is always the same; it's the diameter of the circle. But *many* other curves have this property. One of them appears in Display 4.57. Did you notice that all the width measurements for (b) came out the same? That curve is one of the simplest non-circular curves of constant width. It is called a **Reuleaux triangle**, named after Franz Reuleaux, the German engineer who first investigated these shapes.

Reuleaux triangles are not really triangles, but they are closely related to equilateral triangles. This simplest type, shown again in Display 4.58, is made by starting with an equilateral triangle (also shown) and drawing three 120° arcs from vertex to vertex. Each arc uses one vertex of the equilateral triangle as its center and the side length of the triangle as its radius. In this way, all the points of each arc are the same distance from the vertex opposite that arc, guaranteeing that this figure will "roll smoothly" despite having corners.

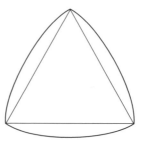

A Reuleaux triangle

Display 4.58

Chapter 4

NOTES

How can you be *sure* that a Reuleaux triangle is a curve of constant width? That is, how does its construction guarantee that parallel trapping lines are always the same distance apart? *Hint*: Look back at Professor Davis's property #5, which appears in Section 4.1.

4.66

Construct a regular pentagon 5 cm on a side. Use this pentagon to construct a curve of constant width with 5 vertices. Generalize the method for constructing a Reuleaux triangle. Will this method of construction work with any regular polygon? Explain.

4.67

You might think that Reuleaux triangles and other such things are just curiosities that some people play around with. Not so. Curves of constant width have been used in the design of many things, including the Wankel rotary engine, a drill for square holes, and coins to fit vending machines. One source for learning more about curves of constant width is *Mathematics Meets Technology*, by Brian Bolt (New York: Cambridge University Press, 1991).[†]

REFLECT

Here are some facts about circles that appeared in this chapter.

- Circles are the most symmetric of all two dimensional shapes, and all circles are similar.

- A circle can be determined by a point (its center) and a length (its radius), or by any three non-collinear points.

- Circles can be graphed in parametric form by using the sin and cos functions.

- The formulas for the area and the circumference of any circle are linked by a constant, π, which is fundamental to many, many parts of mathematics and science.

- π is an irrational number; it cannot be represented exactly by a decimal or a fraction. Nevertheless, π itself is an exact value—the area enclosed by a unit circle, the ratio of the circumference of any circle to its diameter, and the ratio of the area of any circle to the square of its radius.

- Because all circles are similar, their arcs can be measured in terms of their central angles and also by angles inscribed in the circle.

- Rollers don't have to be circular; there are other curves of constant width.

[†]In this book, they are called "curves of constant breadth."

363

4.66

This argument depends on the fact that every tangent to a circle (and hence to any circular arc) is perpendicular to the radius at the point of tangency. Given any pair of parallel "trapping lines" for a Reuleaux triangle, one of them will be a tangent to one of the three circular arcs. The radius of this arc at that point of tangency passes through the opposite vertex (which is the center of the arc). This must be the point where the other trapping line touches the Reuleaux triangle (by perpendicularity to the radius), and the length of the radius is the distance between the two parallels. But the radii of all three circular arcs have the same length: the side length of the equilateral triangle. Therefore, the distance between the parallel trapping lines must be this same length everywhere.

4.67

This is a good small group question. The construction is not difficult. The only part that requires a little ingenuity is the choice of arc radius. A little experimentation should make it clear that the proper radius is the distance between a vertex and one of the two nonadjacent vertices. A similar process works for any regular polygon *with an odd number of vertices,* but not for those with an even number.

NOTES

This list doesn't include everything you should know about circles, but it should remind you of the most important ideas you have seen. More importantly, it might suggest other questions that you can investigate on your own with the tools this chapter has given you.

Problem Set: 4.7

1. (a) Is a square a curve of constant width? Explain.
    (b) Is a regular pentagon a curve of constant width? Explain.
    (c) Extend your answers to parts (a) and (b) to the general case of any regular polygon.

2. (a) Construct a Reuleaux triangle from an equilateral triangle 3 in. on a side.
    (b) What is the perimeter of this Reuleaux triangle?
    (c) What is the area of this Reuleaux triangle?
    (d) What is the perimeter of a Reuleaux triangle constructed from an equilateral triangle of side length s?
    (e) What is the area of a Reuleaux triangle constructed from an equilateral triangle of side length s?

3. Follow these instructions to draw another non-circular curve of constant width:
    *   Construct an equilateral triangle 6 cm on a side.

    *   Extend each side 3 cm beyond each vertex.

    *   Using one vertex as center, draw these two arcs,

        – an arc of radius 3 cm connecting the nearer ends of the two sides that cross at the vertex, and

        – an arc of radius 9 cm connecting the farther ends of the two sides that cross at the vertex.

    *   Repeat the previous step at the other two vertices.

Published by IT'S ABOUT TIME, Inc. © 2000 MATHconx, LLC

Problem Set: 4.7

1. (a) No. (b) No. (c) No regular polygon is a curve of constant width. A variety of informal explanations will suffice here.

2. (a) This asks the student to follow the instructions accompanying Display 4.58.
    (b) It is the sum of three 60° arcs of radius 3 in., which is the same as one 180° arc (a semicircle) of that radius: 3π in.
    (c) There are several ways to do this, some more elegant than others. One way is to think of the Reuleaux triangle as three (overlapping) 60° sectors. They overlap precisely on the equilateral triangle, so the sum of the areas of the three sectors minus twice the area of the equilateral triangle is the desired result.

    $$\frac{1}{2}\pi \cdot 3^2 - 2 \cdot \left(\frac{1}{2} \cdot 3 \cdot 3 \, \sin 60°\right)$$
    $$\frac{9}{2}\pi - 9 \, \sin 60°$$

    (d) This part and the next generalize parts (b) and (c). $\pi \cdot s$.
    (e) $s^2 \cdot \left(\frac{\pi}{2} - \sin 60°\right)$

3. Encourage your students who do this problem to draw the figures with care. Part of the value of such exercises is the practice in making carefully constructed diagrams.

    The width of the figure is 12 cm, the length of each extended side of the triangle. The larger the triangle is, relative to the constant width, the more rounded the figure appears. The circle of diameter 12 cm is the "limiting" case as the triangle gets smaller and smaller. The Reuleaux triangle of width 12 cm is the limiting case at the other extreme. All the figures have the same perimeter: 12π cm. (!!!)

The resulting figure should look like Display 4.59.

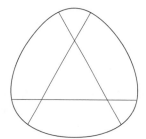

Another curve of constant width

Display 4.59

(a) What is the width of this figure you just drew?
(b) Draw a curve of the same constant width, starting with an equilateral triangle 3 cm on a side.
(c) Draw a curve of the same constant width, starting with an equilateral triangle 9 cm on a side.
(d) Describe the general pattern of your results in whatever way you can.
(e) How does a circle fit into this pattern? Which circle? What is its diameter or radius?
(f) How does a Reuleaux triangle as described in this section fit into this pattern? Which one? Draw it.
(g) Find the perimeter of each of these figures, including the circle and the Reuleaux triangle.

4. Write a careful, logical argument to prove that the figure constructed in problem 3 is a curve of constant width.

Chapter 4

4. This is very much like the argument for Discuss this 4.66. Observe that there are three special pairs of tangent lines—the ones at opposite ends of the extended triangle sides. Clearly, the distance between these pairs of lines is 12 cm. For all other parallel pairs of tangents, one of the lines will be tangent to a small arc and the other to a large arc. Each is perpendicular to a radius through the same vertex; the shorter radius is 3 cm, the longer one is 9 cm. Since these two radii are perpendicular to parallel lines and meet at a point, they must lie along a straight line. Thus, the distance between the parallel tangents is the sum of the lengths of the radii, 12 cm.

NOTES

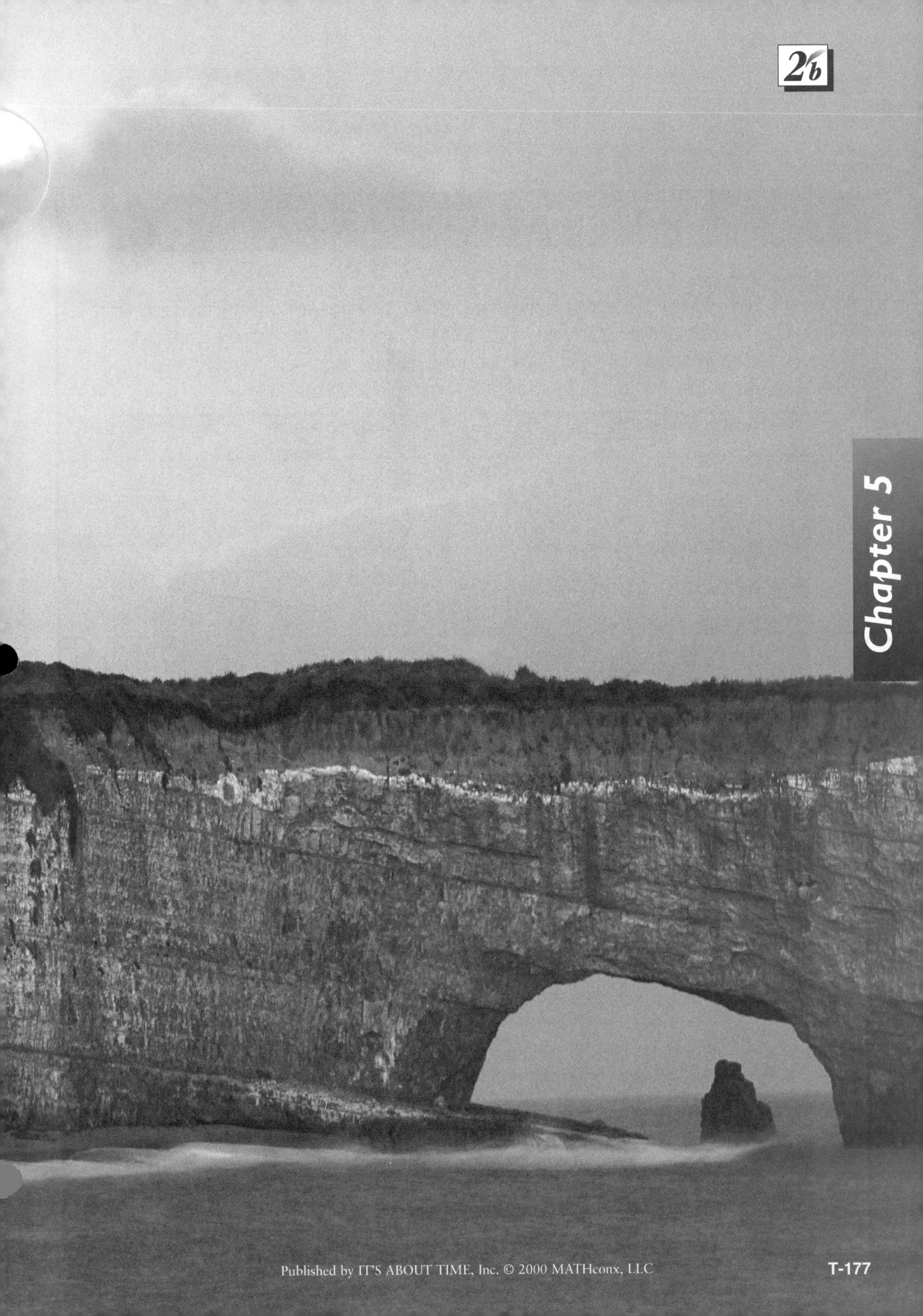

Chapter 5

Chapter 5 Planning Guide

Chapter 5 Shapes in Space

Chapter 5 explores three dimensional geometry, using hands-on activities to introduce deep ideas in easy ways. It examines the construction and basic properties of figures in 3-space, using both constructive (fold, stack, revolve) and analytic (coordinate) methods. The coordinate analysis relates geometric ideas to algebraic techniques and provides a platform for the discussion of four or more dimensions.

Assessments Form A (A)	Assessments Form B (B)	Blackline Masters
Quiz 5.1(A) Quiz 5.2-5.4(A) Quiz 5.5(A) Quiz 5.6-5.8(A) Chapter Test(A)	Quiz 5.1(B) Quiz 5.2-5.4(B) Quiz 5.5(B) Quiz 5.6-5.8(B) Chapter Test(B)	Student pp. 382 (2), 390-392, 392

Extensions	Supplements for Chapter Sections	Test Banks

5.9   A Glimpse of Four Dimensional Space
 	
Following 5.9 Equations for Parabolas
Parabolas: Why Headlights Light Ahead
Geometric Optimization
	5.1 Another Dimension — 4 Supplements
	5.2 Three-Space in Layers — 2 Supplements
	5.3 Finding Volumes by Stacking — 2 Supplements
	5.4 The Volumes of Spheres, Cones,` and Pyramids
	        —2 Supplements
	5.5 Turning Around an Axis — 2 Supplements
	5.6 Space by the Numbers — 5 Supplements
	5.7 Stacks of Lines — 1 Supplement
	5.8 The Algebra of Three Dimensional Shapes —
	        1 Supplement
		To be released

Pacing Range 6 weeks including Assessments
Teachers will need to adjust this guide to suit the needs of their own students. Not all classes will complete each chapter at the same pace. Flexibility — which accommodates different teaching styles, school schedules and school standards — is built into the curriculum.

Teacher Commentary is indexed to the student text by the numbers in the margins (under the icons or in circles). The first digit indicates the chapter — the numbers after the decimal indicate the sequential numbering of the comments within that unit. Example:

5.9

5.37

5.9

5.37

Student Pages in Teacher Edition   **Teacher Commentary Page**

Observations

Helen Crowley
Southington High School, CT

Errol Libby
Oxford Hills Comprehensive
High School, ME

Chapter 5

"Some of the material in this chapter may seem unusual to the typical high school mathematics teacher who has taught from a more traditional approach. Rotation symmetry, for example, is something that we did not spend a lot of time with in the past. In this series, however, it becomes an important building block for the students to have. It is critical, therefore, that teachers go through each problem carefully themselves before assigning it to students.

"In this chapter students are familiarized with cones and pyramids by constructing models. They use these models to measure and then calculate heights, base radii, and other dimensions. This hands-on approach is particularly helpful for students who have difficulty visualizing 3-D figures from 2-D drawings.

"What's intriguing is that students are introduced to several topics in this chapter that might previously have been saved for calculus, but they are able to work with them at this level. I've found that many students particularly enjoy this chapter."

"Throughout this chapter the students are asked to create three dimensional objects in a very hands-on way. This gives them a good idea of how solid figures evolve from flat plane geometry. One of the toughest things for most students is to be able to visualize real objects from abstract material, so having them build three dimensional objects is terrific and this three dimensional visualization component is actually a centerpiece to the chapter rather than just an aside."

Scott Veirs
Investigating the Ocean Depths

Scott Veirs is a graduate student at the University of Washington School of Oceanography. He studies Marine Geology and Geophysics. The best way to do this is to get out of the classroom and get into the water. Deep water. In a submarine.

"We're observing volcanoes on the sea floor in the Philippines," explains Scott by e-mail. "We're looking for hydrothermal vents. To help find them, I collect data on temperature, salinity and the transmission of light over the volcanoes."

Scott is exploring how the temperature around oceanic volcanoes changes over time, and how it affects the overlying ocean. His goal is to discover connections between these hydrothermal vents and their potential impact on earth systems like climate.

"There is certainly no shortage of basic math applied down deep," Scott writes. "We

calculate systems with plenty of variables and exponents. We do lots of averaging and differencing of raw data. We also design experiments and write computer programs to process our data."

The relationship between two and three dimensional space underlies all Scott's work. "We use bathymetric maps for our explorations," he explains. "These are just like topographical contour maps used on the earth's surface, only they're for underwater. It's very dark down there. So to make decisions like where to take the submarine, we need to constantly interpret and rework these maps. The hydrothermal plumes or vents of warm water we study are also in three dimensional space. But the temperature readings we take of them are search pattern projections in two dimensional images."

Scott has always loved the sea. Earlier, he spent a year in Malaysia as a balloon launcher on a Chinese oceanographic ship for atmosphere experiments. "For one whole summer, I sat elated atop the main mast of the ship 'Shenandoah' rolling 94 feet above the ocean's swell." Swell indeed!

368

Published by IT'S ABOUT TIME, Inc. © 2000 MATHconx, LLC

Chapter 5　Shapes in Space

This chapter examines the construction and basic properties of solid and hollow figures in 3-space, using both constructive (fold, stack, spin around) and analytic (coordinate) methods. The coordinate analysis relates geometric ideas to algebraic techniques and provides a platform for the discussion of four and more dimensions.

Many, many fascinating and useful topics in three dimensional geometry *could* be presented at this level. Choosing what to include in this all too brief chapter was partly a matter of personal taste, but we have been guided by these precepts.

- The three dimensional figures studied should be related to common shapes of everyday experience that can be held, modeled, or drawn.

- It is critically important for students to see how the principles of plane geometry can be applied to three dimensional figures. Many three dimensional problems can be solved by reducing them to two dimensional problems.

- The process of generalizing to higher dimensions is a natural extension of the step from 2-space to 3-space.

- Students' understanding of dimension should be as broad as possible, extending beyond visual intuition, allowing for the widest application of this fundamental concept.

Chapter 5

Shapes in Space

CHAPTER

5

5.1 Another Dimension

What does the word *space* mean? How many different meanings can you think of? Which meaning best fits the way it is used in the title of this chapter?

5.1

The space we live in is called **three dimensional space** (or simply **3-space**) because there are three independent directions for us to move around in: forward-backward, left-right, and up-down. The plane, in which we have done most of our geometry so far, is a kind of space, but it's not the kind we live in. It is the *two dimensional* world of forward-backward and left-right, the space of flat things.

5.2

"So why have we spent so much time studying flat things?" you may ask.

That's because flat shapes, which are simpler than thick or lumpy shapes, can tell us a lot about the more complex shapes of the space in which we live, if we think about them properly. Exploring how this generalization process works is the main theme of this

Published by IT'S ABOUT TIME, Inc. © 2000 MATHconx, LLC

369

5.1 Another Dimension

One purpose of the hands on exercises and explorations in this section is to get students comfortable with thinking about how 2-space diagrams can represent 3-space objects. Another, less obvious, purpose is to get them thinking about logical patterns. This latter purpose is suggested and reinforced by the kinds of questions asked and the order in which they are presented.

The habits of thought described after Discuss this 5.5 are strategies that apply to all sorts of problems and situations, not just in mathematics, but in just about everything one does. Mathematics is a convenient, uncluttered context in which to examine and practice them. You might not want to dwell on such things explicitly, but students should be "let in on the secrets" sometimes. Every now and then we ought to tell them what we think we are doing to/for them.

Chapter 5

Additional Support Materials:

Assessments	Qty
Form (A)	1
Form (B)	1

Blackline Masters	Qty
Student p. 382	2

Extensions	Qty

Supplements	Qty
Another Dimension	4

Published by IT'S ABOUT TIME, Inc. © 2000 MATHconx, LLC

5.1 Another Dimension

chapter. We begin with a familiar way of turning flat things into three dimensional objects: folding them up.

Let's look first at how to build a (hollow) cubic box from square pieces of a plane. To make a cutout pattern, think of one square in a plane as the base of the cube. On each of its sides, draw another square of the same size. These are the four "walls" of the box. Now draw a square on the outer edge of one of the side squares. This is the top of the box. The cutout pattern for a cube should look like Display 5.1.

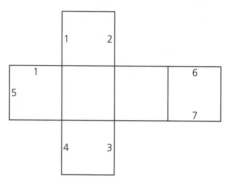

A cutout pattern for a cube

Display 5.1

5.3

1. Take a piece of heavy paper or light cardboard and using Display 5.1 to guide you, draw the cutout pattern for a cube 6 cm on a side. An old file folder will work fine.

2. Number the edges as shown. Then number the rest of the outer edges so that edges with the same number fit together when the cube is folded up. Edge 1 is done for you.

3. Cut out and fold up the cube as shown by Display 5.2. Use a little tape to hold the flaps in place. Do your edge numbers match properly?

5.4

1. Design a different cutout pattern for a cube and number its sides for folding. How is your pattern really different from the one in Display 5.1? What should *different* mean in this situation?

370

Published by IT'S ABOUT TIME, Inc. © 2000 MATHconx, LLC

5.1

There are many different meanings for "space"; *The Random House Dictionary* lists 19 different meanings, just for the noun! Let students talk about various meanings for a few minutes, then focus them on the title of the chapter. Guide them to see that the meaning intended here is some rough approximation of the idea that *The Random House Dictionary* describes as "the ... three dimensional realm in which all material objects are located and all events occur."

5.2

There is a possible, though unlikely, source of confusion in the sentence following Discuss this 5.1. Some students may object that there seem to be six, not three, directions listed here. That's true, in a sense, but the two in each pair depend on each other. Forward is "unbackward," left is "unright," and so on. For instance, you can travel forward and left at the same time, but you can't travel forward and backward at the same time. Thus, there are really only three *independent* directions. This informal description of dimensions should be good enough for now. Later in this chapter there will be a more careful definition of *dimension*.

5.3

You may have to supply heavy paper or light card stock for this; it can be done with an 8.5×11 in. sheet. This hands-on process is an essential step in helping students without good spatial intuition to visualize shapes in 3-space from their two dimensional descriptions. Please don't skip it. The proper numbering is shown in Display 5.1T.

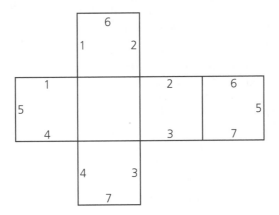

Display 5.1 with numbering completed

Display 5.1T

5.4

These questions are good for small group interaction.

1. Several, but not all, ways are shown by Display 5.2T which is not in the student text. The threshold question is the meaning of "different." This is an opportunity for a good, intuitive student discussion about basic geometric ideas. Don't worry about getting a precise definition of different, but help them to see that two things which should *not* count as differences in the pattern are (i) its size and (ii) its orientation.

2. Will *any* way of drawing six congruent squares together, edge to edge, form a cutout pattern for a cube? If you say Yes, give an argument to justify your answer. If you say No, draw a six-square pattern that *cannot* be folded into a cube.

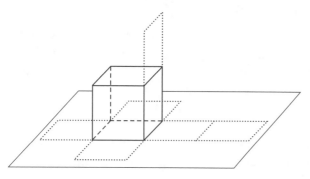

Folding up the cutout pattern for a cube

Display 5.2

These questions refer to Display 5.3, which shows a square with an equilateral triangle attached to each side.

5.5

1. What shape do you get when you fold the triangle up so that the same numbered edges go together? Can you describe it in words? Do you know a name for it?

2. Make one of these shapes out of a piece of paper. Start by drawing a square 6 cm on a side. Then draw (carefully) an equilateral triangle on each side of the square. Cut the figure out and fold the figure up. Do all the "top" points of your triangles come together? They should.

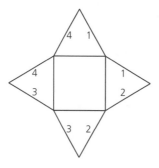

Display 5.3

Published by IT'S ABOUT TIME, Inc. © 2000 MATHconx, LLC

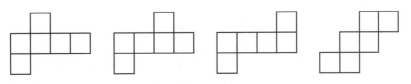

Display 5.2T

2. No. For example, six squares attached in a single row won't work.

A relatively difficult question is almost begging to be asked here—how many truly different cutout patterns for a cube are there? We have refrained from asking it in the text because it can be a time sink. However, if you have an inquisitive class that raises the issue, you might profitably spend a little time on it. Its investigation connects with some fundamental ideas of combinatorics. The truly difficult part is deciding which of the many possible designs are the same and which are different in a way that is organized enough to let you count things accurately.

5.5

These questions represent another step in the two to three dimension visualization process. They also begin a chain of analogous questions that lead the students to several important observations about how planar angles come together in 3-space. This could be done in small groups or individually, but it is far better for a student to make this object than just to watch it being made.

1. It is a pyramid with a square base.

2. A formal compass and straightedge construction is not necessary here, but the triangles must be drawn with some care. Otherwise, the "top" vertices will not come together at a single point when the figure is folded.

3. Let P stand for the point (in space) at which the top vertex of each of the four triangles meet. When you set this figure down on its square base, P is directly above some point, Q, of the square. Which point? Unfold your cutout and mark the location of Q. Then explain how you found that location.

4. When your figure is folded up, what is the distance between P and Q? Explain how to calculate this distance without measuring it. Then measure it and see if your calculation agrees with your measurement within a reasonable margin of error.

5. Do you notice anything interesting about your results?

You have just investigated some properties of a pyramid with a square base. The pyramid is a three dimensional figure; it has length, width and height. Nevertheless, you only needed two dimensional tools to find and describe its measurements in 3-space, including its height. This was a first example of how the ideas of plane geometry help us deal with the three dimensional space we live in.

What shall we do next? The thinking habits that we would like you to develop can lead in several different directions.

Generalize what you've done. Build on what you've already done by changing some condition just a little and seeing what effect it has. For instance, what if the triangles are not equilateral? What if they are not all congruent? Is this possible? We'll look at some of these ideas in the problem set.

Argue by analogy. What kinds of shapes are like Display 5.3 in some important ways, but are different enough to be interesting? Can we ask the same questions as we did before? Can we answer them in the same ways? If not, how close can we come to asking the same questions and using the same methods?

Look for a pattern. Can we change things in some kind of order or sequence and then look for predictable differences in results as we go from each case to the next?

Let's try to do all three of these things at once!

Published by IT'S ABOUT TIME, Inc. © 2000 MATHconx, LLC

3. This answer depends, in part, on an intuitive sense of symmetry in 3-space. The point is in the exact center of the square, 3 cm from each side. It is the point at which the perpendicular bisectors of two adjacent sides intersect.

4. This is the most difficult of these questions, and also the most important. It requires students to visualize a triangle that isn't there explicitly in the cutout figure. Some small group discussion might help here.

If you call one corner of the square A, then PQ is one side of the right triangle APQ that is "standing up" perpendicular to the plane of the square. It is standing on side AQ, and its hypotenuse, AP, is one ("glued") edge of two adjacent equilateral triangles. Once the students can visualize this triangle, the rest is relatively straightforward application of work with triangles from earlier in **MATH** *Connections*.

Since the hypotenuse is an edge of the equilateral triangles, it must be 6 cm long. Side AQ is half of the diagonal of the square.

$$AQ = \frac{1}{2} \cdot \sqrt{6^2 + 6^2} \approx 4.24 \text{ cm}$$

Note that $(AQ)^2 = \frac{1}{4} \cdot (6^2 + 6^2) = 18$. This is much simpler and more useful for what we need in the following computation. This is an instance when calculator approximation actually makes the computation unnecessarily complex. Now use the Pythagorean Theorem on triangle APQ.

$$(AP)^2 = (AQ)^2 + (PQ)^2$$
$$6^2 = 18 + (PQ)^2$$

This means that

$$(PQ)^2 = 36 - 18 = 18$$

so

$$(PQ) = \sqrt{18} \approx 4.24 \text{ cm}$$

5. This is an open-ended question that could have a variety of answers, of course. One interesting observation is that the two legs of triangle APQ are exactly equal! That is, the height of this pyramid is exactly the distance from one base corner to the middle of the base. This also means that the "seam lines" between the triangles make 45° angles with the plane of the base.

Chapter 5

EXPLORATION 1

The pyramid of Display 5.3 is particularly nice because all its triangular sides are equilateral and its base is square. That is, it is made up of regular polygons. What happens if you keep the idea of a regular polygon and just change the shape of the base? Suppose you use an equilateral triangle for the base, or a regular pentagon, or a regular hexagon, and so on? How do the results change? Is there a pattern to them?

1. Let's start with the simplest case.

   (a) Construct an equilateral triangle, $\triangle ABC$, 6 cm on a side. Then draw another equilateral triangle on each side of the first one. Cut out the figure and fold it up so that the three outside vertices of the triangles meet at a point, P. (See Display 5.4.)

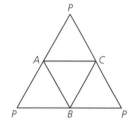

Display 5.4

   (b) When you set this figure down on its triangular base, P is directly above some point, Q, of $\triangle ABC$. Which point? Unfold your cutout and mark the location of Q. Then explain how you found that location.

   (c) When your figure is folded, what is the distance between P and Q? Explain how to calculate this distance without measuring it. Then measure it and see if your calculation agrees with your measurement within a reasonable margin of error.

Part 1 describes the first of a sequence of pyramids with different regular polygons as bases. Display 5.3 and the questions about it described the second pyramid in this sequence. Now it's your turn to extend the sequence, by analogy.

2. Rewrite questions (a), (b), and (c) of part 1 to apply to a pyramid with a base that is a regular pentagon 6 cm on a side. Then answer them.

Published by IT'S ABOUT TIME, Inc. © 2000 MATHconx, LLC

373

(5.6) This is a good, but difficult, small group Exploration. It can easily take up a class period or two. If you are pressed for time, it's important to cover this material in some fashion. The first three questions develop the tetrahedron, which is needed later, and the last few questions lead by example to the key principle of the section. You can bypass some of the details by giving students some of the intermediate information, if necessary.

1. (b) The same process for finding the point works here, but the reason is a little less obvious. It should be clear intuitively that the point must be equidistant from all the vertices of the base. This means that it is the center of the circle that passes through those three vertices. As we saw in Chapter 4, this center is the intersecting point of the perpendicular bisectors of the sides. Any two sides can be used to find this point.

(c) This question depends on visualizing the same kind of triangle—APQ, where A is a vertex of the base. However, finding the length of AQ is more difficult in this case. AQ is a radius of the circle in which the base equilateral triangle is inscribed. Here is a sequence of questions that you might use to help students who get stuck on this part.

(i) If B is another vertex of the base, what kind of triangle is AQB? How do you know? It is isosceles because AQ and BQ are radii of the same circle.

(ii) What is the measure of $\angle AQB$? How do you know? 120°. Call the remaining base vertex C and draw the third radius, QC. These three radii subdivide the base triangle into three conguent triangles (by SSS), so the three angles at Q must be equal. Since their sum is 360°, each one must be a third of that.

(iii) What is the measure of $\angle QAB$ and $\angle QBA$? Why? They are equal and total 60°, (180° − 120°), so each must be 30°.

(iv) How long is AQ? Explain. This part is really tough, unless your students have a good command of the right triangle trigonometry of Chapter 3. If you think the argument is too difficult for them, help them to see that the information gathered so far *determines* triangle AQB (by ASA), so AQ can be found somehow. Then give them its length. Here is the argument.

Let r denote the length of AQ. Now, AB, the base of isosceles triangle AQB, is a side of the original base triangle, so its has length 6 cm. Drop a perpendicular from Q to AB; it will bisect AB at some point D. Thus, ADQ is a right triangle with a 30° angle at A, its hypotenuse has length r, and side AD is 3 cm long. This means that $r\cos30° = 3$ cm, so $r \approx 3.46$ cm.

Chapter 5

(v) How long is PQ? Explain. This part mimics the argument for question 4 of (5.5). Triangle APQ is a right triangle and its hypotenuse, AP, is 6 cm long. Side AQ is 3.46 cm long by the previous part, so

$$6^2 = 3.46^2 + (PQ)^2$$

That is, $PQ = \sqrt{24} \approx 4.9$ cm

2. (a) Question: Draw a regular pentagon 6 cm on a side. Then draw an equilateral triangle on each side of the pentagon. Cut it out and fold it so that the five outside vertices of the triangles meet at a point.

   (b) Question: Let P stand for the point in space at which the top vertices of the five outside triangles meet. When you set this figure down on its pentagonal base, P is directly above some point, Q, of the pentagon. Which point? Unfold your cutout and mark the location of Q. Then explain how you found that location.

   Solution: As before, Q must be the center of the circle that passes through the five vertices of the base because it must be equidistant from all those vertices. Q is the intersection point of the perpendicular bisectors of any two sides of the pentagon.

   (c) Question: When your figure is folded, what is the distance between P and Q? Explain how to calculate this distance without measuring it. Then measure it and see if your calculation agrees with your measurement within a reasonable margin of error.

   Solution: Again, this is the most difficult of the three questions. It depends on visualizing the same kind of triangle—APQ, where A is a vertex of the pentagon. If the students worked through the sequence of questions for question 1(v), they can follow the same line of reasoning here.

   (i) If B is another vertex of the base, what kind of triangle is AQB? How do you know? It is isosceles because AQ and BQ are radii of the same circle.

   (ii) What is the measure of $\angle AQB$? How do you know? 72°. The five radii from Q to the vertices subdivide the pentagon into five conguent triangles (by SSS), so the five angles at Q must be equal. Since their sum is 360°, each one must be a fifth of that.

(iii) What is the measure of $\angle QAB$ and $\angle QBA$? Why? They are equal and total 108°, (180° − 72°), so each must be 54°.

(iv) How long is AQ? Explain. The information gathered so far determines triangle AQB (by ASA), so AQ can be found somehow. If you had your students write out the steps of the argument for question 1(iv), then they can mimic them here.

Let r denote the length of AQ. Now, AB, the base of isosceles triangle AQB, is a side of the original base triangle, so it has length 6 cm. Drop a perpendicular from Q to AB; it will bisect AB at some point D. Thus, ADQ is a right triangle with a 54° angle at A, its hypotenuse has length r, and side AD is 3 cm long. This means that $r\cos 54° = 3$ cm, so $r \approx 5.1$ cm.

(v) How long is PQ? Explain. Triangle APQ is a right triangle and its hypotenuse, AP, is 6 cm long. Side AQ is 5.1 cm long (by the previous part), so

$$6^2 = 5.1^2 + (PQ)^2$$

That is, PQ $\approx \sqrt{9.99} \approx 3.2$ cm

3. A completed version of Display 5.5 appears in Display 5.3T.

Type of Base	No. of Sides	AQ (cm)	Height (cm)
triangle	3	3.46	4.9
square	4	4.24	4.24
pentagon	5	5.1	3.2
hexagon	?	?	?

Display 5.3T

Chapter 5

5.1 Another Dimension

Now let's see if there are any interesting trends or patterns. For pyramids with regular bases with 3, 4, or 5 sides, you now know two measurements,

- the distance from a vertex to the *center* of the base (the length of AQ);

- the height of the pyramid (the length of PQ).

3. Organize your data by copying and completing the table of Display 5.5, except for the question mark entries.

Type of Base	No. of Sides	AQ (cm)	Height (cm)
triangle	3		
square	4		
pentagon	5		
hexagon	?	?	?

Display 5.5

4. Describe any general trends you see in the data for Display 5.5.

5. What about the question marks in Display 5.5? Use the trends in your data to estimate what these entries should be.

6. Notice that the heights of the pyramids decrease as the number of base sides increase. Do you think that the number of base sides will ever get so large that the height will be 0? If so, how can you predict the number of sides for which that will happen? If not, can you explain why it will never happen?

7. Rewrite questions (a), (b), and (c) of part 1 to apply to a pyramid with a base that is a regular hexagon 6 cm on a side. Then answer them.

5.7

What goes wrong when you try to make a pyramid with a regular hexagon as its base and equilateral triangles as its sides? See if you can extend your observation to complete this general statement about polygonal shapes in space.

For polygons that fit together at a common vertex to form a *peak* (a three dimensional angle) there, . . .

Published by IT'S ABOUT TIME, Inc. © 2000 MATHconx, LLC

374

4. Students should at least observe that, as the number of base sides increases, the length of AQ increases, but the height of the pyramid decreases. It is worthwhile that they recognize these as trends, not just random occurrences. Also, they should see that these numerical trends conform to the visual sense of what happens as the size of the base increases.

5. The answer for the first question mark is easy; a hexagon has 6 sides.

   For the second question mark, the sequence 3.46, 4.24, 5.1 might lead one to estimate a value near 6, give or take a little. Some students might see something suspicious here—a value larger than 6 means that the pyramid has collapsed! Don't worry if no one observes this, however; it comes up again in this Exploration.

   For the third question mark, the sequence 4.9, 4.24, 3.2 invites next term estimates between 1 and 2.5, but *all* such estimates are wrong! This fact should emerge naturally in the discussion that follows.

6. Ballpark guessing from the trends of the table of Display 5.5 should lead students to conjecture that the 0-height case might happen sometime soon, perhaps for 7 or 8 base sides.

7. These questions are the same as (a), (b), and (c) of part 2 with hexagon in place of pentagon. The cutout should provide the first evidence that something is wrong; the pyramid comes out looking *very* flat!

   Mimicking the solution to 2(b), we soon get confirmation that something is amiss. The measure of ∠AQB is 60° because it is one of six congruent triangles that meet at Q. Since it is isosceles, each base angle is 60°, too, so it is *congruent* to the side triangle that is being folded over! This means that the folded over triangle will fit exactly on top of it. It also means that the length of AQ must be 6 cm, so the height of this pyramid is 0. This settles parts 5 and 6, as well.

5.7

This question can be done without having completed the previous Exploration, but it won't be as natural or as obvious that way. In any event, have this class discussion before moving on. It sets up the key principle of the section.

For polygons that fit together at a common vertex to form a peak (a three dimensional angle) there, the sum of the angles at this vertex must be less than 360°.

The principle at work in the previous discussion is a simple fact used by every tailor, seamstress, and sheetmetal worker.

To make a peak at some point of a flat sheet, cut a wedge shaped piece from that point outward and then sew, glue, rivet, or weld the edges together.

Display 5.6 shows how this idea is used to make a cone from a disk.

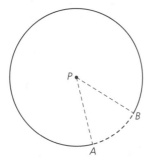
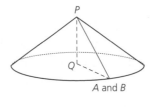

Display 5.6

In Display 5.6, a sector (a wedge-shaped piece) *APB* is cut out of the disk. Then the two edges, *PA* and *PB*, are joined together to form a cone. As you can see, the height, the radius (of the circular edge), and the *steepness* of the cone depend on how big a sector is cut out. The size of the sector is measured by its angle at *P*.

EXPLORATION 2

The Winter Wonder Woodstove Co. makes 10 inch radius galvanized steel disks that can be cut and shaped into stovepipe caps, as pictured in Display 5.6. They want to sell these disks with a chart telling do-it-yourself homeowners how to cut the disks to different sizes and slopes. Your job is to make this chart. The company also sells a special strip to close the seam, so you don't have to overlap the edges of the cut.

5.8

1. Cut four sample disks out of paper, using a scale of 1 cm = 1 inch. Cut a sector out of each disk – one of 30°, one of 60°, one of 90°, and one of your choice. Now make each one into a cone by taping the two edges together. Measure the height of the cone and the radius of the circular edge of its base. Copy the table of Display 5.7 and record your results in the first two columns.

375

It is essential that students see this fact as obvious, since we do not prove it in the text. The statement is important to us because it implies the famous result that there are only 5 regular solids, which is the final part of this section.

5.8 This Exploration can be done individually or in small groups, or it can be assigned as a homework project. The early parts are easy, but a little time consuming, and they involve cutting and taping. The later parts might require some discussion. Besides illustrating the process of making a cone, this Exploration reviews the ideas of similarity, slope and function.

The function questions (6−9) are worthwhile and quite challenging, but they are not essential to the rest of the section.

1. At the scale of 1 cm = 1 inch, these disks can be cut out of regular 8.5 × 11 inch paper. If you prefer, you can hand out precut paper disks to start this Exploration. Then, if necessary, students can remove the sectors by creasing and tearing, thereby avoiding the need for scissors. Please do not skip this hands-on step; it is an important stage in helping students to visualize 3-space shapes from 2-space diagrams. Also, the nuisance and imprecision of this cut and measure process may help students to see that computations and functions sometimes provide much easier, more efficient ways to do things.

A filled in version of the table of Display 5.7 appears as Display 5.4T. Student entries for the first two columns will likely vary a bit from ours because of measurement error. We have chosen 45° for the last row, so that you have other data for yourself, if needed. Students can pick any size they want. To measure the radius of the base, we placed the cone on a piece of graph paper and used the lines to guide us. The diameter can be seen that way, then halved for the radius. The height is very difficult to measure with any accuracy. If you don't tape the edges too close to the peak, you can slip a thin ruler into the seam.

Cutout Sector	Measured		Calculated		
	Radius	Height	Radius	Height	Slope
30°	9.1	4.1	9.17	3.99	0.44
60°	8.3	5.6	8.33	5.53	0.66
90°	7.6	6.7	7.5	6.61	0.88
45°	8.8	5.0	8.75	4.84	0.55

Display 5.4T

Chapter 5

Cutout Sector	Measured		Calculated		
	radius	height	radius	height	slope
30°					
60°					
90°					
(yours)					

Display 5.7

2. Was it awkward to measure the cones? Are you confident that your measurements are accurate? It might be better to calculate the radius and height; it certainly will be more precise. Here are some questions to guide you through the calculations for the 30° cutout cone. Round your answers to two decimal places.

   (a) What is the circumference of the base of this cone? How do you know?

   (b) If you know the circumference of a circle, how can you find its radius? Find the radius of the base of the cone in this way. Write your result in your copy of Display 5.7.

   (c) Can you think of a way to find the radius of the base of the cone without first finding its circumference? *Hint:* Use the fact that all circles are similar.

   (d) Look at the cone in Display 5.6. Which segment represents the height of the cone? Which segment represents the radius of its base? How long is *PA*? Use the Pythagorean Theorem to write an equation that relates these three lengths.

   (e) Find the height of the cone by solving the equation you just wrote. Write your result in your copy of Display 5.7.

3. Use the procedure of the previous part to find the radius and height of each of your other three cones. Write your results in your copy of Display 5.7 and compare them with your measurements. How close did you come?

4. What do you think we mean by the *slope* of the cone? Define it as precisely as you can, then use your definition to calculate the slopes of the four cones. Write your results in your copy of Display 5.7.

376

2. It should not be difficult to get students to admit that their measurements might not be very precise.

   (a) Approximately 18.33π. The fraction of the disk circumference left after the 30° sector has been cut out is $\dfrac{360 - 30}{360}\left(=\dfrac{11}{12}\right)$. The whole disk circumference is $2\pi \cdot 10 = 20\pi$. Multiply this by the fraction left.

   (b) Since $C = 2\pi \cdot r$, the circumference divided by 2π gives you the radius. In this case, $\dfrac{18.33\pi}{2\pi} \approx 9.17$.

   (c) The fact that all circles are similar means that linear measures are in proportion. We saw this in Chapter 4. Thus, the radius ratio of the cone edge to the disk edge is the same as the circumference ratio of these two circles. But that's the same as the ratio of the central angle remaining to 360°. Therefore, to find the radius of the edge of the cone, subtract the cutout central angle from 360° and divide by 360° to get the ratio; then multiply by 10, the disk radius.

   (d) PQ represents the height, h, of the cone. QA or QB represents the radius, r, of its edge. The length of PA is 10; the radius of the original disk.

   $$h^2 + r^2 = 10^2$$

   (e) In this case, $r = 9.17$, so

   $$h^2 + 9.17^2 = 10^2$$
   $$h^2 + 84.0889 = 100$$
   $$h^2 = 100 - 84.0889$$
   $$h = \sqrt{15.9111} \approx 3.99$$

3. See 1 of 5.8. of Display 5.4T the completed table of Display 5.7.

4. The slope of the cone is its height divided by the radius of its edge. If you think of it as a roof, this is its pitch, as discussed in Chapter 2. -179This ratio tells you how steep the slant of the cone is. The values appear in Display 5.4T.

Chapter 5

5. Do you see any interesting patterns in your results? If so, describe them. See if you can find any more examples to confirm or refute these patterns. If any pattern seems to hold, see if you can find an explanation for it.

Now think about this: When you found the radius, height and slope for the four cones, you went through the same three procedures four times. The only thing different each time was the angle size of the cutout sector, right? This means that each of these procedures is a *function* of the angle size of the sector!

"So what," we hear you say.

Well, if you can write a formula for each function, then you can get your calculator to do all the routine arithmetic for you. All you have to do is put in the angle size.

6. Let x stand for the angle size of the cutout sector. Write a formula for the radius of the base of the cone in terms of x. Put this formula into the function list of your graphing calculator. For the TI-82 (TI-83) calculators, this is the Y= list. Then use the calculator's value process to check the four radius values in your table.

7. Now let x stand for the radius of the base of the cone. Write a formula for the height of the cone in terms of x. Put this formula into the function list of your graphing calculator. Then use the calculator to check the four height values in your table.

8. Can you put the previous two functions together to get a function for the height of the cone in terms of the angle of its cutout sector? Try it. If you get one, see if its results agree with your results using the previous function.

Now let's put together some observations from these two explorations.
- To make a cone, we had to cut a sector out of the 360° disk. To make the point of the cone, we *had to* use less than the full 360° portion of the disk.

- Similarly, the fold up pyramids worked fine as long as the sum of the angles at the top vertex was less than 360°. The one that didn't work was the one for which the angle sum at its top vertex equaled 360°.

Published by IT'S ABOUT TIME, Inc. © 2000 MATHconx, LLC

377

5. This question is open-ended; students might see various things, depending in part on which angle they chose for their fourth cone. Nevertheless, there is one peculiar pattern. When the slopes are rounded to two digits, the 30°, 60° and 90° cases are 0.44, 0.66 and 0.88, respectively. Checking the angles halfway between these, we get slopes of 0.55 and 0.77 for the 45° and 75° cases, respectively! This pattern doesn't appear to have any interesting conceptual foundation; it's largely a result of the rounding process.

6. $\dfrac{360-x}{10^{\frac{360x}{36}}} \cdot 10$. There are many equivalent forms, the simplest of which is $10 - \dfrac{360x}{36}$.

7. $\sqrt{100-x^2}$

8. This is algebraically ugly, but it's a good example of function composition. The most difficult part for the students may be keeping straight the differing meanings of x. You might have them rewrite the height formula using r, instead of x, for the radius. Using the simple form of the radius formula, we get

$$\sqrt{100 - \left(10 - \frac{x}{36}\right)^2}$$

If you care to put your students through a little exercise in algebraic multiplication, etc., this can be simplified a little (but not much), to

$$\sqrt{\frac{720x - x^2}{1296}}$$

The key to both of these questions is the principle immediately before them.

NOTES

Chapter 5

These observations lead to a key principle about three dimensional shapes with polygonal sides.

For polygons that fit together at a common vertex to form a peak (a three dimensional angle), the sum of the angles at this vertex must be less than 360°.

5.9

1. Design a pyramid with a regular hexagon as its base. Draw the cutout pattern for it, then cut it out and fold it up.

2. Design a pyramid with a regular octagon as its base. Draw the cutout pattern for it, then cut it out and fold it up.

About Words

Poly- means many and *-hedron* means figure with flat sides, so a *polyhedron* is a figure with many flat sides. The plural of *polyhedron* is *polyhedra*.

A three dimensional shape formed from polygonal pieces joined at their edges is called a **polyhedron.** Each polygonal piece is called a **face** of the polyhedron. Using these words, the key principle just stated becomes:

A Fact to Know: The face angles at any vertex of a polyhedron must total less than 360°.

This important fact was known to the ancient Greeks. A proof of it appears in the last book of Euclid's famous *Elements,* a summary of all mathematics known by the Greeks in 300 B.C. The last part of Euclid's *Elements* applies this principle to a kind of polyhedron we have seen before. Here's another look at these special polyhedra, this time through the eyes of the ancient Greeks.

The Greeks were very fond of symmetry. You can see it everywhere in their art, their architecture, and their mathematics. As you know, the most symmetric polygons are the *regular* ones—polygons with all sides and all angles congruent. In three dimensional space, a polyhedron is regular if all its faces are congruent regular polygons. For example, a cube is a regular polyhedron; all its faces are squares of the same size. We began Exploration 1 in this section by looking at another regular polyhedron—the pyramid formed by folding up four congruent equilateral triangles. This figure is called a (regular) **tetrahedron** because it has four faces.

378

5.9

1. The sum of the six top angles must be less than 360°. If the pyramid is to be as symmetric as possible (which is not a necessity here), the six triangles must be isosceles with top angles less than 60° each. It is also possible to design such a pyramid with noncongruent faces, but the cutout pattern is more difficult to draw.

2. Similarly, the sum of the eight top angles must be less than 360°. If all faces are isosceles triangles, each top angle must be less than 45°. In both of these cases, the smaller the top angles are, the taller the pyramid will be.

The proof of this remarkable, famous fact of geometry is well within the

NOTES

Chapter 5

The Greeks knew that there are regular polygons with any number of sides. However, they could only find five different types of regular polyhedra, as shown in Display 5.8.

> **tetrahedron** — 4 faces (triangles)
> **hexahedron** — 6 faces (squares)
> **octahedron** — 8 faces (triangles)
> **dodecahedron** — 12 faces (pentagons)
> **icosahedron** — 20 faces (triangles)

At first, this may have seemed puzzling. But, as the very last theorem of his *Elements,* Euclid proved that there cannot be any other regular polyhedra; *these five are the only possible types!*

tetrahedron   hexahedron   octahedron   dodecahedron   icosahedron

The regular polyhedra

Display 5.8

Do what Euclid did. Give a convincing argument to prove that there are only these five different types of regular polyhedra. Use the fact that the face angles at any vertex of a polyhedron must total less than 360°.

5.10

A regular polyhedron can be inscribed in a sphere; that is, it can be placed inside a spherical shell in such a way that *all* its vertices touch (are points of) the sphere. This fact was of great significance to the Pythagoreans, who regarded the sphere as the most perfect three dimensional figure. Pythagoras may have learned of the tetrahedron, the cube, and the octahedron from the Egyptians. In studying these figures, the Pythagoreans discovered the regular icosahedron. These four figures were used to represent the four elements of the physical world.

> **fire** — tetrahedron
> **earth** — hexahedron (cube)
> **air** — octahedron
> **water** — icosahedron

Later the Pythagoreans discovered the regular dodecahedron, which they used to represent the universe. Plato (429–348 B.C.) used this representation of the universe and its elements as the foundation for an elaborate theory of matter in which

Published by IT'S ABOUT TIME, Inc. © 2000 MATHconx, LLC

379

5.10

capabilities of most classes. A class discussion of it is a good way to pull together the main ideas of this section and to give students a bit of history at the same time.

Outline of argument. If your students have trouble finding a place to start, suggest that they think about how they would make one of these shapes. A problem at the end of this section asks them to construct a regular octahedron.

- At least three polygonal faces must meet at any vertex (peak) of the polyhedron.

- Since the polyhedron is regular, the situation at any vertex is the same as at any other. We only have to consider what happens at a typical vertex.

- The sum of all the face angles at the vertex must be less than 360°.

- Since all the faces are congruent, their angle sum at a vertex must be divided equally among the faces.

- Now let's look at the possible kinds of regular polygons.
  - Triangles. Each angle of an equilateral triangle measures 60°. How many could you have together and still total less than 360°? 3 (180°), 4 (240°), or 5 (300°). As we saw at the end of Exploration 1, 6 is too many! These three cases give you the tetrahedron, octahedron, and icosahedron.

  - Squares. Each angle measures 90°. Three of them totaling 270° could meet at a vertex of a cube, but 4 is too many.

  - Pentagons. Each angle of a regular pentagon measures 108°, $((3 \cdot 180) \div 5)$. Three of them totaling 324° could meet at a vertex of a dodecahedron, but 4 is too many.

  - Hexagons. Each angle of a regular hexagon measures 120°, $((4 \cdot 180) \div 6)$. Three of them total 360°, which is too much! Thus, there is no regular polyhedron with hexagonal faces.

  - Other regular polygons. Each angle of a regular polygon with more than 6 sides must measure more than 120°. Three of them coming together at a point would total more than 360°. Therefore, there is *no* regular polyhedron with faces of any other kind.

Chapter 5

everything is composed of right triangles. Because of this theory, these five polyhedra are known as the **Platonic Solids.**

Plato's triangular theory of matter has not stood the test of time, but the Platonic Solids can still be found in the Earth's elements.

Plato

- The crystalline structures of lead ore and rock salt are cubic;
- Fluorite forms octahedral crystals;
- Garnet forms dodecahedral crystals;
- Iron pyrite crystals come in all three of these forms; and
- The basic crystalline form of the silicates which form about 95% of the rocks in the Earth's crust is the smallest of the regular triangular solids, the tetrahedron.

5.11

Look up the Platonic Solids in an encyclopedia. Sometimes they are called the Platonic Bodies. Find out something interesting about them that hasn't been mentioned in this book. Describe what you find. Write about one page.

Problem Set: 5.1

1. A company that makes window blinds wants you to make a cutout pattern for a mailing carton. The carton should be 36 in. long, with right triangular ends that measure $3 \times 4 \times 5$ in.

   (a) Make a careful diagram of this carton cutout using a scale of $\frac{1}{4}$ in. per inch. Don't include any glue flaps; it will be taped at the seams.
   (b) Cut out and fold your diagram to make sure that it fits together the way you planned it.
   (c) Is there more than one way to design the cutout pattern for these measurements? If you can think of any others, which one do you think would be the best to use? Why?

2. Fussy Frank has just visited the Winter Wonder Woodstove Co. and bought a disk to make his stovepipe cap. (See Exploration 2.) He wants to make a cap that is exactly 5 inches high at its peak.

   (a) What size sector should he cut out of it? Estimate your answer by looking at your copy of Display 5.7. Then calculate the size to two decimal places.
   (b) If he does that, will the cap be big enough to cover

Published by IT'S ABOUT TIME, Inc. © 2000 MATHconx, LLC

5.11

Some or all of these writeups could be circulated around the class to see if (a) they are understandable, and (b) they are interesting. It might be advisable to protect the anonymity of the writers, unless you have a particularly mature, nonjudgmental group of students. Note that a video and workbook on The Platonic Solids are available from Key Curriculum Press.

Problem Set: 5.1

1. One cutout pattern (Problem 1(a)) for this carton appears in Display 5.5T. There are several other designs, depending on which sides of the triangles are folded and which are cut. There is not a unique answer for "best"; arguments can be made for several different choices. The one shown in Display 5.5T is easier to draw than one in which the right angles of the triangles are on the outside, but folding along the hypotenuse, instead of along a leg, would make the tape seams a little shorter. Of course, it is possible to have long, thin patterns with more than one folded edge on the triangles, but these cutout cartons would be very awkward to assemble and tape.

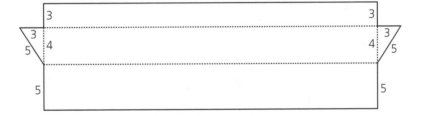

Display 5.5T

2. This is an exercise in working backwards.
    (a) A reasonable estimate would be somewhere around 45°. To calculate it, work backwards using the height function to get the radius.

    $$5 = \sqrt{100 - r^2}$$
    $$5^2 + r^2 = 100$$
    $$r = \sqrt{100 - 5^2} \approx 8.66 \text{ cm}$$

    There should be $\frac{8.66}{10}$ of 360° left, so Frank needs to cut out a sector of $\frac{1.34}{10} \cdot 360° = 48.24°$.

    (b) With an edge radius of 8.66 in., there will be a 4.66 in. overhang all around; the stovepipe radius is only 4 inches.

his 8 inch diameter stovepipe with at least 2 inches overhang all around?

(c) What will be the circumference of the edge of the cap?

3. Nifty Novelties, Inc. is trying out packaging patterns for a miniature fir tree. Two possible patterns for this clear plastic packaging are shown in Display 5.9.

- Package A has a 6 cm square base with right triangles on two opposite sides. The longer leg of each right triangle measures 10 cm. The other two triangles are shaped so that all four triangles will meet at a point at the top of the package when it is folded.

- Package B has an equilateral triangle as its base and congruent isosceles triangles as the sides. Each side of the base is 9 cm long, and all the other sides of the triangles are 11 cm long. All three side triangles will meet at a point at the top of the package when it is folded.

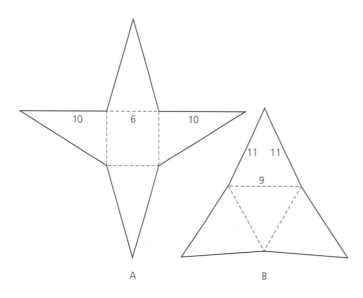

Display 5.9

If both packages are at least 9 cm high, the company probably will choose the one with the smaller surface area because it will be less expensive. Calculate the answers to these questions, accurate to one millimeter. To help you

(c) The circumference is $2\pi \cdot 8.66 = 17.32\pi \approx 54.41$ inches.

3. These questions review some facts and techniques from Chapters 2 and 3 in a three dimensional setting. No piece of this problem is particularly difficult, but the many computations required may place some strain on your students' patience and organizational skills, especially in part (b). You can require the students to make the cutout models or not, as you see fit. The answers can be found without making the models.

(a) The height of Package A is the altitude of the isosceles triangle at the top of the diagram. The two equal sides of this triangle are 10 cm long. Thus, the measure of each base angle is $\cos^{-1}\left(\frac{3}{10}\right) \approx 72.5°$, so its altitude is $10 \sin 72.5° \approx 9.5$ cm.

The height of Package B is more difficult to find; it requires visualizing the package folded. The reasoning is like that used in Exploration 1. If the top point of the package is P, the point on the base directly below it is Q, and one of the base vertices is A, then PQA is a right triangle with hypotenuse 11. As in Exploration 1, to find the length of PQ, you must first find the length of AQ. Following the reasoning of Exploration 1, AQ is the hypotenuse of a right triangle with angles of 30° and 60°. If we let r be the length of AQ, then $4.5 = r \cos 30°$, so $r = \frac{4.5}{\cos 30°} \approx 5.2$ cm. Now we use this length and Pythagorean Theorem to find the length of PQ.

$$(AQ)^2 + (PQ)^2 = 11^2$$

$$(5.2)^2 + (PQ)^2 = 121$$

$$PQ = \sqrt{121 - 27} \approx 9.7 \text{ cm}$$

(b) For each package, this is just a matter of computing the areas of the various triangles and the square separately and then summing them.

A. The area of the square is 36 sq. cm. The area of each right triangle is $\frac{1}{2} \cdot 6 \cdot 10 = 30$ sq. cm. The altitude of the top triangle was already found to be 9.5 cm, so its area is $\frac{1}{2} \cdot 6 \cdot 9.5 = 28.5$ sq. cm. The bottom triangle is the nuisance. To find its area, you must first find its altitude. To find that, you must first find the length of its longer sides, which is the same as the hypotenuse of the side right triangles.

$$\sqrt{6^2 + 10^2} \approx 11.66 \text{ cm}$$

visualize these packages, make a scale drawing of each one, cut it out, and fold it.

(a) What is the height of each package when it is folded?
(b) Which package do you think has the smaller total surface area? What is the total surface area of each package?

A third design has just been proposed—a cone made by cutting a sector out of a disk with radius 10 cm.

(c) If the diameter of the base must be 6 cm, what size sector should be cut out?
(d) If that size sector is cut out, how tall will the cone be?
(e) What will the surface area of this package be? Don't forget the base disk.
(f) Do you think this package is better than the pyramid packages? Why or why not?

4. Construct a regular octahedron by drawing a pattern on a piece of paper, cutting it out, and folding it.

5. Display 5.10 shows one pattern for an icosahedron.

   (a) Make a scale drawing of this pattern with each triangle 4 cm on a side. Remember: All the triangles are *equilateral*.
   (b) Number all the outside edges of the pattern (the solid lines) so that edges to be put together have the same number.
   (c) Cut out your pattern and fold it to form an icosahedron. Did your numbered edges come out right?

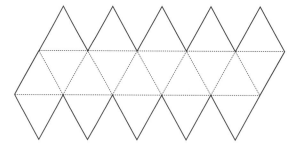

Display 5.10

The altitude, a, must satisfy $3^2 + a^2 = (11.66)^2$, so $a \approx 11.27$ cm. Thus, the area of this triangle is approximately $\frac{1}{2} \cdot 6 \cdot 11.27 = 33.81$ cm. Adding up the areas of the five regions, we get approximately 158.4 sq. cm.

B. Use the Pythagorean Theorem to find the altitudes of the triangles. For each isosceles triangle face, the altitude a must satisfy $4.5^2 + a^2 = 11^2$, so $a \approx 10.0$ cm. The area of each face, then, is approximately $\frac{1}{2} \cdot 9 \cdot 10 = 45$ sq. cm. Similarly, the altitude of the base is $\sqrt{9^2 - 4.5^2} \approx 7.8$ cm, so its area is $\frac{1}{2} \cdot 9 \cdot 7.8 \approx 35.1$ sq. cm. The sum of the four triangular areas yields approximately 170.6 sq. cm.

(c) The circumference of the base must be 6π. The circumference of the original disk is 20π, so 14π can be cut out. That is, a $\frac{14}{20} \cdot 360°$ sector can be cut out. This leaves a sector of $108°$ for the cone.

(d) As in Exploration 2, let P be the peak of the cone and let Q be the center of its base disk. Then, by the Pythagorean Theorem, $(PQ)^2 + 3^2 = 10^2$, so the length of PQ is approximately 9.5 cm.

(e) The surface area of the cone is $\frac{108}{360} \cdot \pi \cdot 10^2 = 30\pi$ sq. cm. The area of the base is 9π, so the total area of the package is $39\pi \approx 122.5$ sq. cm.

(f) There might be some creative opinions for this answer, which should not be discouraged if they are defended reasonably. Under the conditions of the problem, it seems that the conical package is best. It satisfies the minimum height requirement and its total surface area is less than either of the other designs. Moreover, it would have fewer seams, making it more attractive and probably easier to assemble.

Chapter 5

4. One possible pattern (an octahedron) is shown in Display 5.6T.

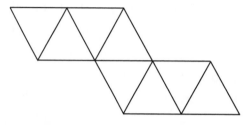

Display 5.6T

5. If the triangles in Display 5.10 are 4 cm on a side, this pattern will fit on a regular 8.5 × 11 in. piece of paper. The sides are numbered in Display 5.7T.

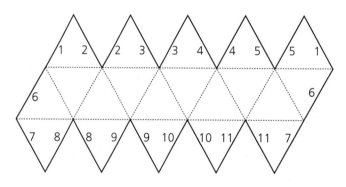

Display 5.7T

NOTES

5.2 Three-Space in Layers

In this section we look at another way of using flat figures to think about three dimensional shapes. This approach views three dimensional shapes as solid objects, instead of just dealing with them as hollow shells as we did in Section 5.1. We begin with an example from mapmaking.

In order to display mountains, valleys, ridges, and the like on a flat map, surveyors and mapmakers devised a way of using (curved) lines to show differences in height. Here is a condensed version of the way the U.S. Geological Survey describes that method.[1] As you read it, use Display 5.11 to help you visualize their description.

A **contour** is an imaginary line on the ground, every part of which is at the same altitude above sea level. Such a line could be drawn at any altitude, but only the contours at certain regular intervals are shown. Zero altitude of the Geological Survey maps is mean sea level. Thus, the 20 foot contour line would be the shore line if the sea were to rise 20 feet above mean sea level.

Contour lines show the shape and height of the hills, mountains, and valleys. Successive contour lines that are far apart on the map indicate a gentle slope, lines that are close together indicate a steep slope, and lines that run together indicate a cliff.

The **contour interval**, the vertical distance between one contour and the next, is stated (in feet) at the bottom of each map. This interval varies according to the topography of the area mapped: in a flat country it may be as small as 1 foot; in a mountainous region it may be as great as 250 feet. In order that the contours may be read more easily, every fourth or fifth contour line is made heavier than the others and its altitude is printed with it. The altitudes of many points to the nearest foot are also printed on the maps.

Learning Outcomes

After studying this section you will be able to:

Interpret the contour lines on topographical maps;

Use contour lines and cross sections to describe three dimensional objects in layers;

Describe the shapes of cylinders and prisms using cross sections;

Find the side surface area of a prism.

About Words

The word *contour* comes from the Italian word *contorno* meaning an outline of a curving or irregular figure or shape. In agriculture, farmers use contour plowings on sloping land in furrows that follow the level contour lines, to slow down erosion from rainwater runoff.

---

[1]This is a shortened, paraphrased version of the description of contour lines that appears on the back of topographical maps published by the U.S. Geological Survey.

Published by IT'S ABOUT TIME, Inc. © 2000 MATHconx, LLC

383

5.2 Three-Space in Layers

Continuing the theme of looking at two dimensional representations of three dimensional objects, this section examines 3-space figures in *layers*, slices parallel to some base plane. It begins by exploring topographical maps. The ideas of contour lines and contour intervals taken from the topographical map example are used to represent simple solid shapes in layers. Contour lines lead to horizontal cross sections (planar regions bounded by contour lines), which lead, in turn, to vertical cross sections. Most of the material in this section discusses how horizontal and vertical cross sections are used to represent three dimensional figures. The section ends with a writing exercise about an important application of this layer approach to 3-space—CAT (Computer Assisted Tomography) scans.

Note that this section and the next should be considered as a unit. Much of what is done here sets up the approach to volume via Cavalieri's Principle, which is carried out in Section 5.3.

Chapter 5

Additional Support Materials:

Assessments	Qty
Form (A)	1
Form (B)	1

Blackline Masters	Qty
Student pp. 390-392	1

Extensions	Qty

Supplements	Qty
Three-Space in Layers	2

Illustration taken from U.S. Geological Survey map descriptions.

Display 5.11

Maps of this kind are called **contour maps** or **topographical maps**. Display 5.11 shows how they work. The top frame is a sketch of a river valley between two hills, with an ocean bay and hooked sandbar in the foreground. The bottom frame is a contour map of this region viewed from above. The contour interval for this map is 20 feet; that is, each contour line represents the base outline of the land at a level 20 feet higher than the previous one. Display 5.12 is another example of a contour map.

5.12

These questions refer to Display 5.12, a contour map of an area in east-central Maine. The top of the map is North. The contour interval for that map is 20 feet, so answer all height questions in terms of 20 foot intervals.

1. Where is the highest place in this area?

2. How high is the hill just north of Upper Pistol Lake?

3. How high is the hill in the northwest corner of the map?

Published by IT'S ABOUT TIME, Inc. © 2000 MATHconx, LLC

5.12

These questions are designed to get students comfortable with interpreting contour lines.

1. The highest point is the top of Duck Mountain. Note that we did not ask how high it is because of an ambiguity in the map here. The point marked 1169 should be the highest point, but the small contour line to the left of the D suggests a higher spot. However, that small oval could actually indicate a lower spot, a depression at the top.

2. At least 560 feet and less than 580 feet.

3. 600–620 feet. This one is a little more difficult because the 500 foot line of that hill has to be inferred from the nearby 500 foot mark.

NOTES

Chapter 5

←North   Contour interval 20 feet   Scale $\frac{1}{51250}$

From U. S. Geological Survey map of
Nicatous Lake quadrangle, Maine

Display 5.12

385

NOTES

4. If we had not told you that the contour interval for this map is 20 feet, how could you tell that from the map?

5. The scale of this map is $\frac{1}{51250}$. What distance does 1 inch represent? Express your answer in inches, in feet, and in miles. Round your answers to one decimal place.

6. The dashed lines are trails. If you hiked along the trail south from Upper Pistol Lake to where it joins the east-west trail, about how high would you have climbed? Explain your answer. About how long is this part of the trail?

None of the hills mapped in Display 5.12 are very steep. Display 5.13 is a simplified contour map of a steeper, but smaller hill, Mole Mountain. We'll use this small map to illustrate how two dimensional contours tell us about three dimensional shapes.

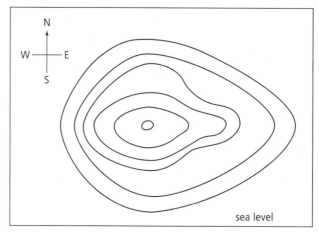

Contour interval 10 feet   Scale: 1 inch = 80 feet
Mole Mountain
Display 5.13

Suppose you were standing far to the south of Mole Mountain, looking north. What would the silhouette (outline) of its side view look like? Can you draw that outline from the information contained in the topogaphical map? If so, how?

1. How would you start to draw the silhouette of Mole Mountain? Describe the steps you would take to translate the information of the contour lines into a side view of this hill.

5.13

4. The darker, 100 foot contour lines are marked with their heights and there are five contour intervals between successive darker lines. Therefore, each contour interval must be 20 feet.

5. This scale is approximate, adjusted for the enlargement used in printing this book. It is expressed as a fraction because that is the way the scale is printed on the original map. One inch represents 51,250 inches, 4270.8 feet, or 0.8 miles.

6. The altitude of the lake is 421 feet. The trails meet very close to the 620 foot contour line. Therefore, this is about a 200 foot climb in roughly a $\frac{3}{4}$ mile hike.

5.13

This problem is very important for helping students to visualize how contour lines are translated into information about height. It sets up the rest of the section. The text following these questions describes how to draw a rough side view, but you might want students to try it on their own before reading ahead. The analogous question, drawing an east to west side view, is left entirely for the students in 5.14.

1. A brief class discussion of this question should tell you which students, if any, might need help with understanding the whole idea of contour lines, etc. The focus should be, "How would you start?"

Chapter 5

NOTES

_____

....

....

....

....

....

....

....

....

2. About how high is Mole Mountain? How do you know?

3. Why do you think the text says "*far* to the south," instead of just "to the south"?

Look again at the idea of a contour line. If the hill were sliced horizontally at the level of a particular contour line, the contour line would be a boundary of the flat surface made by the slice. The flat surface is called a (horizontal) **cross section**. Display 5.14 shows the horizontal cross sections of Mole Mountain that are bounded by the top three contour lines. You might think of them as very thin slices of the hill. The frames represent the plane of each slice.

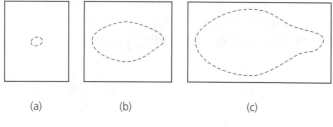

(a)   (b)   (c)

Display 5.14

Of course, looking at them from the side, you would only see the edges of these very thin slices; they would look like line segments. Display 5.15 shows the three cross sections tilted over (along with the imaginary frames around them). Display 5.16 shows the edges of these three cross sections viewed from far to the south. The lengths of these horizontal segments are the maximum east-west diameters of the cross sections.

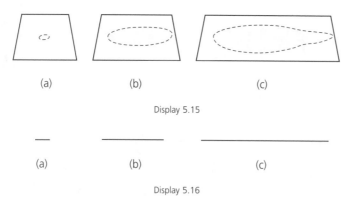

(a)   (b)   (c)

Display 5.15

(a)   (b)   (c)

Display 5.16

Published by IT'S ABOUT TIME, Inc. © 2000 MATHconx, LLC

387

2. It is at least 60, but less than 70, feet high above sea level. There are six contour lines, and the contour interval is 10 feet.

3. If you were standing just a little way to the south, your side view would be distorted by the angles of your sight lines to various parts of the hill, up to the top, off to one side or the other. Standing far away minimizes this distortion; those angles can be ignored for the purposes of making a rough sketch of the silhouette. In formal terms, we could regard the silhouette as a south-north parallel projection.

NOTES

To get a rough outline of this view of Mole Mountain, we can stack up the edge views of the contour line cross sections. We just have to measure the largest east-west diameters carefully, and then stack the resulting line segments in their proper positions. But what are their proper positions?

Fortunately, the largest east-west diameters of all the contour regions lie along the same straight line, a line through the top of the hill. (See Display 5.17.) By marking the center of the hilltop and measuring each diameter with reference to that point, you can see how the segments should be placed in the stack.

Display 5.18 shows the edges of the contour regions stacked up in their proper positions. It is drawn to the same scale as the original map: 1 inch = 80 feet. Since the contour interval is 10 feet, the lines are spaced $\frac{1}{8}$ inch apart. The dashed line represents the center of the hilltop.

In Display 5.18, we have *not* completed the outline by connecting the ends of the segments. That's because the contour map does not specify the shape of each 10 foot step between contour levels. The hillside could be smooth from top to bottom, or it could look like a jagged staircase. To get a better picture of it from a contour map, we would need a smaller contour interval. A 5 foot contour interval would give us twice as many lines and a better picture. A 1 foot contour interval would give us ten times as many lines and a picture that would make the outline quite accurate, but the contour map would be almost unreadable!

5.14

Suppose you were standing far to the east of Mole Mountain, looking west. What would the silhouette of its side view look like? Using the preceding discussion and solution to guide you, draw an approximate outline of this view of Mole Mountain. When you get the stack of horizontal segments, make a second copy of them. Then fill in the outline in two different ways, one that would make Mole Mountain easy to climb and one that would make it very difficult.

If you were to slice Mole Mountain down the middle along the east-west line shown in Display 5.17, the face of the slice would be the same two dimensional shape as the silhouette approximated by Display 5.18. The face of a vertical slice is called a (vertical) **cross section**. We designed Mole Mountain so that its silhouette would match a vertical cross section. This is not always the case. (See problem 2.)

388

5.14
This question is labeled as Discuss this, rather than a Do this now, because it might make a good small group activity. It need not be done at all if you think your students have a good enough grasp of the Mole Mountain problem and its solution. It is a reinforcement question. Having students actually do it themselves brings home the difficulty of finding the proper diameters and aligning the segments properly.

The "rigging" of the Mole Mountain map also applies to this problem: All the maximum north south diameters lie along the same line through the hilltop. Your students should end up with a drawing that looks like Display 5.8T which does not appear in the student text.

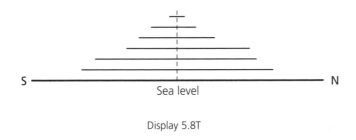

Display 5.8T

Chapter 5

NOTES

We arranged things this way because a silhouette is a good way to visualize a cross section.

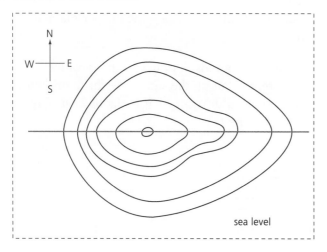

Contour interval 10 feet   Scale: 1 inch = 80 feet
Mole Mountain with imaginary reference line

Display 5.17

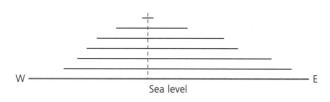

Display 5.18

Cross sections are two dimensional shapes. Horizontal and vertical cross sections together can be used to describe three dimensional shapes. Moreover, thinking of three dimensional shapes in layers can be a powerful tool for finding their volumes, as we shall see in the next section. In the rest of this section we look at ways of describing common three dimensional shapes by means of contour lines and cross sections, beginning with a cone.

These questions refer to Display 5.19, the contour map of a cone with actual base radius of 3.5 cm. The largest circle represents the base.

5.15

1. **What is the scale of this contour map? You'll have to measure something to find out.**

Published by IT'S ABOUT TIME, Inc. © 2000 MATHconx, LLC

389

5.15

Questions 5 and 6 are challenging. A class discussion might be appropriate for some students. Address the fact that Display 5.19 could be the contour map of *any* cone. These questions are foreshadowed by asking about scale in question 1 and varying the contour intervals in questions 2 and 3. Even if you know the scale and the contour interval, a vertical cross section is needed to specify that it is a cone, not just a stack of cylinders or other shapes that happen to have circular contours at just the right levels. See question 4.

1. The radius of the base in Display 5.19 (the largest circle) should measure 2.1 cm. If the actual base is 3.5, the scale is $\frac{2.1}{3.5}$, or $\frac{3}{5}$.

2. The dot in the center is intended to represent the peak, right at a contour interval. You might have to clarify this for the students. There are 7 contour steps from base to peak, so the height of the cone is 7 cm. See Display 5.9T for a full-size vertical cross section.

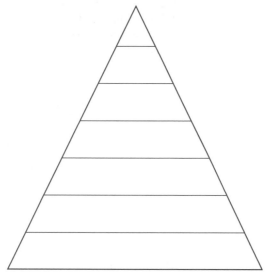

The contour lines, included for reference, are not a necessary feature.

Display 5.9T

Chapter 5

2. If the contour interval is 1 cm, how tall is this cone? Draw a full size vertical cross section that goes through its peak.

3. If the contour interval is $\frac{1}{4}$ inch, how tall is this cone? Draw a full size vertical cross section that goes through its peak.

4. Draw vertical cross sections of two different shapes that are not cones, but fit the contour map in Display 5.19.

5. Could Display 5.19 be the contour map of a cone with base radius of 1 foot and height 5 feet? If so, what are the appropriate scale and contour interval? If not, why not?

6. Could Display 5.19 be the contour map of a cone with base radius r and height h? If so, what are the appropriate scale and contour interval? If not, why not?

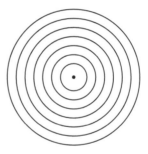

Display 5.19

EXPLORATION

5.16 SETUP

For this Exploration you will need some small square tiles about $\frac{1}{8}$ inch thick. You can cut them out of cardboard. Corrugated cardboard from an old box will do fine. Cut one square of each of these side lengths:

1 in. 1.5 in. 2 in. 2.5 in. 3 in. 3.5 in. 4 in.

Display 5.20 contains four contour maps, showing four different ways to stack your seven tiles. In each one, the outermost square represents both the base and the first contour. Each contour is a square, and the contour interval is whatever thickness your tiles are. The squares are always stacked in size

390

3. Its height is $1\frac{3}{4}$ inches. See Display 5.10T for a full-size vertical cross section.

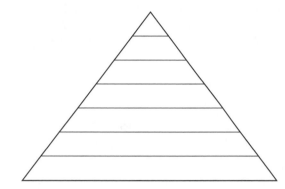

The contour lines, included for reference, are not a necessary feature.

Display 5.10T

4. See Display 5.11T for two such cross sections, not drawn to full size. There are many others.

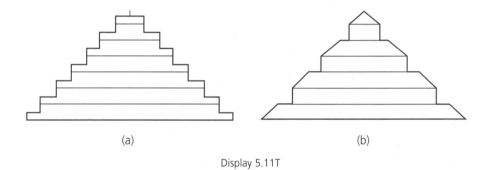

(a)   (b)

Display 5.11T

5. Yes. To find the scale, divide 1 ft. by 2.1 cm (the radius of the base as it is drawn). The contour interval would be $\frac{5}{7}$ feet.

6. Yes. The scale would be $\frac{r}{2.1\text{ cm}}$. The contour interval would be $\frac{h}{7}$ of whatever unit is used to measure h.

5.16

This is a hands-on activity with simple manipulatives to get students to see how contour diagrams and cross sections are related to actual three dimensional figures. It can be done individually or in small groups. The use of relatively thin layers paves the way for finding volumes by Cavalieri's Principle, to be discussed in the next section.

Chapter 5

order, from largest on the bottom to smallest on the top. The dashed lines are not contours; they are explained in Activity 2. The scale of these maps is 1:2.

ACTIVITIES AND QUESTIONS

1. Stack your square tiles in each of the four ways shown in Display 5.20.

2. The dashed lines show the locations of some vertical cross sections of these stacks. Make full size drawings of these six cross sections.

3. Two of these six cross sections are the same as two others. Which are they? Pair them up.

4. Stack your seven tiles in an arrangement that is different than any of the ones in Display 5.20. Then draw a contour map of your arrangement.

5. Draw a dashed horizontal line across your (new) contour map so that it passes through the smallest square. Then draw the cross section determined by your dashed line.

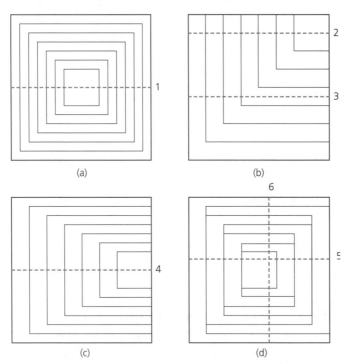

(a)   (b)

(c)   (d)

Display 5.20

1. Students' work here may help you see who understands the contour map idea and who has trouble visualizing three dimensions from planar diagrams.

2. These six cross sections are shown half-size in Display 5.12T.

3. 2 and 4 are the same; 1 and 6 are the same.

4. The results here depend entirely on student choice.

5. The same is true here. The requirement that the line go through the smallest square is just a way to avoid the ambiguity of having a line that matches one of the edges of a square.

6. This last question could be very difficult, depending on the student's stack arrangement. If you choose to do it, the explanation might best be handled via group discussion.

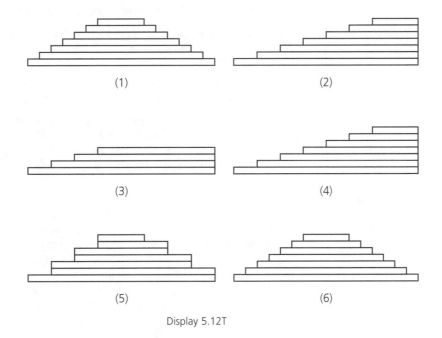

Display 5.12T

5.2 Three-Space in Layers

6. Does *any* cross section of your new stack match *any* cross section of the four stacks of Display 5.20? Don't limit yourself just to the cross sections that are marked. If so, show where the matching cross sections are. If not, explain how you know that no matches are possible.

5.17

Imagine a pyramid with a square base 4 inches on a side and sides that are isosceles triangles. The peak of the pyramid is 4 inches above the center of the base.

1. Draw a full size contour map of this pyramid, using a contour interval of $\frac{1}{2}$ inch.

2. In what way(s) is this contour map similar to Display 5.20(a)? In what ways is it different?

3. Imagine cutting this pyramid directly down through its peak along a line parallel to one side of its base, that is perpendicular to the plane of the base. Draw the cross section that you would get in this way.

4. Describe at least two ways in which this cross section differs from cross section 1 of Display 5.20(a).

5.18

1. Can you represent a cube by a contour map? If so, do it. If not, describe what goes wrong when you try.

2. Two cylindrical stovepipes, 8 inches in diameter, are standing side by side. One is 12 inches tall; the other is 20 inches tall. Represent this pair of shapes by a single contour map, if you can. If you cannot, describe what goes wrong when you try.

Contour maps are useful for representing shapes that get smaller from bottom to top—cones, pyramids, domes, mountains, etc. They are not at all useful for representing some of the most basic building-block shapes of our 3-D world, such as cubes and cylinders. If all the horizontal cross sections of a figure are congruent and are stacked up straight with no offset, there is no way to tell one contour line from another.

For instance, think of a cube sitting on a table. Its base (the side facing down) is a square. So is its top. So is every one of its horizontal cross sections. And all these squares are exactly the same size. No matter what size contour interval you use, every contour is exactly the same size and shape as every other one, so there is no way to tell them apart on a map.

Published by IT'S ABOUT TIME, Inc. © 2000 MATHconx, LLC

This can be done individually or in groups. It is, in part, a reading exercise. It asks students to translate a verbal description of a three dimensional shape into a mental image, then to represent that image by horizontal and vertical cross sections. Since a main objective here is word to picture translation, please do not be too quick to help out with this part of the exercise by supplying the picture for your students. Rather, if they are having difficulty, help them to take apart the word description piece by piece until they see the picture for themselves.

1. Apart from size, the only real difference is the addition of a smaller square and a center point to represent the peak. A subtle difference in interpretation is that the largest square in this case represents only the base, whereas in Display 5.20(a) the largest square represents both the base and the first contour interval above it.

2. See Display 5.13T(a), which is drawn to half-size.

3. See Display 5.13T(b).

4. (a) This cross section is much taller than that of Display 5.20(a).
   (b) This cross section has straight (oblique) sides, not staircase sides.
   (c) This cross section comes to a peak at the top; the other one is flat at the top.

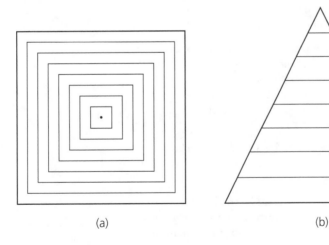

(a)   (b)

Display 5.13T

This discussion should be done right away. The main point of both questions is the same: Some figures are too nicely shaped to be represented by contour maps. In particular, if all the horizontal cross sections of a figure are congruent and are stacked up straight with no offset, it is impossible to distinguish contour lines at different levels. Cubes and cylinders are examples of such figures. This observation appears in the text directly after these questions, but it would be better for students to discover it first for themselves.

Chapter 5

Nevertheless, it can be very useful to think of such a figure as a stack of congruent cross sections. You can think of a cube, for example, as a stack of copies of its square base, one at every level from bottom to top. (See Display 5.21.) If each side of the cube is 2 inches long, say, then there is a copy of the 2 by 2 inch base square for each of the (infinitely many) numbers between its 0 inch base height and its 2 inch top height. In this way, the three dimensional shape (the cube) is made by repeating infinitely many copies of the two dimensional base figure (a square).

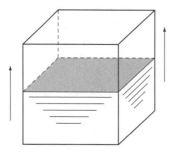

Display 5.21

Other three dimensional shapes can be visualized in the same way.

- A cylinder has congruent circular cross sections all along its length. You can think of a cylinder standing upright on one end as an infinite stack of circles.

- If the walls of the cylinder are thick (as in a piece of pipe, for instance), each cross section is an annulus (like a washer). Thus, you can think of the upright pipe as a stack of infinitely many very thin washers. Since a cross section is a two dimensional thing, these imaginary washers actually have no thickness at all.

- A solid cylindrical shape can be imagined as a stack of very thin dimes or quarters, provided that you realize that the "coins" have no thickness.[†]

- Copies of any polygonal region can be stacked up to form a prism, which is a shape formed by two congruent polygons in parallel planes and parallelograms on all the other sides.

[†]If coins got thinner as their buying power decreased, these cross section coins would represent the ultimate inflation.

Published by IT'S ABOUT TIME, Inc. © 2000 MATHconx, LLC

NOTES

If the stacking of the cross sections is perpendicular to the base plane, with no offset or slant, then the cylinders and prisms are called **right cylinders** and **right prisms**, respectively. The word *right* signifies that the walls of the figures make right angles with the plane of the base. (See Display 5.22.) Prisms and cylinders that are stacked on a slant are said to be **oblique**. (See Display 5.23.)

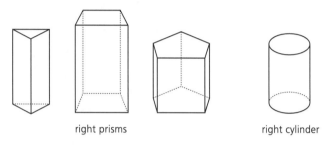

right prisms   right cylinder

Display 5.22

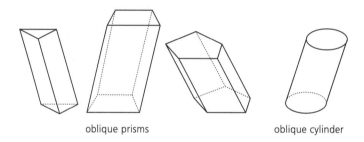

oblique prisms   oblique cylinder

Display 5.23

5.19

1. All upright walls of all right prisms have the same shape. What is that shape?

2. Find at least three different objects in your classroom that are shaped like prisms.

3. Suppose a right prism is 6 cm high and has a triangular base with side lengths 3, 4, and 5 cm. Describe its upright sides; then find its total surface area.

Published by IT'S ABOUT TIME, Inc. © 2000 MATHconx, LLC

5.19

This exercise reinforces the students' visualization of right prisms. It also provides review practice in computing area and recognizing that a 3-4-5 triangle is a right triangle.

1. They are rectangles.

2. Any rectangular box is a prism. Triangular, square, and hexagonal shaped objects are common in everyday life.

3. The three rectangular faces have dimensions 3×6, 4×6, and 5×6 cm. Thus, the total area of the faces is $18 + 24 + 30 = 72$ sq. cm. The base and top are congruent right triangles with legs 3 and 4 cm, so their combined area is 12 sq. cm. The total surface area of the prism is 84 sq. cm.

NOTES

Chapter 5

We often describe cylinders and prisms as standing upright, but they can appear in many different orientations. A cereal box is a cereal box, whether it is standing up or lying down. A length of pipe is shaped like a cylinder, whether it is standing upright, lying on the floor, or stuck into the ground at an angle. A right prism with triangular ends can be lying on one of its rectangular sides; it's still a right prism. In any orientation, a **cross section** is a planar slice that cuts directly across (is perpendicular to) some specified axis (direction line) for that figure. This is true whether the shape is symmetric (like a cone or a prism) or not (like a mountain or a person). In fact, one of the most important tools in modern medical technology depends on looking at cross sections of a person. It's called a **CAT scan,** a process of taking cross sectional X rays of a person and then using a computer to assemble these images into a three dimensional picture of that person's inner organs.

About Words

In English, something is standing upright if it is perpendicular to some horizontal floor or base plane.

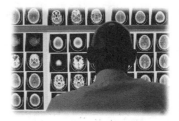

Write a one page paper explaining CAT scans in terms that your family or other members of your class would understand. Begin by explaining what the acronym CAT stands for. Relate your explanation to the ideas of cross sections that were covered in this section.

5.20

Suggestions. Start by looking in some encyclopedias, searching the Internet, or asking a librarian to help you locate what you need. Then, if you need more information, try the back issues of magazines such as *Popular Science* or *Scientific America.*

Problem Set: 5.2

1. Before answering the following questions, which refer to Display 5.12, you want to review your answers about that display.

   (a) Duck Mountain is actually a cluster of four small peaks. The heights of two of them are marked. What are the approximate heights of the other two?

   (b) In the rectangle that includes Duck Lake, how many hilltops are shown? Don't count Duck Mountain. Which of these hilltops is the highest? About how high is it?

   (c) If you hiked the trail east from No. 2 Camp, would you be going downhill or uphill in the first half mile? What if you hiked the trail west? What if you hiked the trail south?

5.20

This is a research exercise with direct relevance to the material of this section. CAT stands for Computer Assisted Tomography (some might say Computer *Assembled* Tomography). Tomography is an X ray technique for taking cross sectional pictures of internal organs. The computer helps to put these pictures together in a way that allows doctors to see (virtual) three dimensional representations of the organs and thereby to detect abnormalities more easily. You might consider linking this assignment with your students' English composition and/or biology courses.

Problem Set: 5.2

1. (a) The peak just above the n in Mtn is 900–920 feet high. The peak farthest to the northwest (just beyond the 1099 foot peak) is 1000–1020 feet high.
    (b) There are five or six small peaks (depending on how you count the double-top at the bottom edge of the map). The highest of these is 840–860 feet high.
    (c) The trail east goes downhill; the trail west goes uphill; the trail south is very nearly level for the first half mile, then dips down slightly to a stream, after which it begins to climb.

NOTES

Chapter 5

2. There are many ways to do this. A C shaped ridge or an S shaped ridge will work. A volcanic crater (a mountain with a hollow cone in its top) will also work, but the contour lines become confusing.

3. This problem is simple to comprehend conceptually, but fairly difficult to solve computationally. Besides reinforcing contour map and cross section ideas, it reviews some ideas from Section 5.1.

   (a) The base of this pyramid is an equilateral triangle 10 cm on a side. Its height is approx. 8.2 cm. Computing the height is difficult. It requires the same argument as in question 3 of Exploration 1,
   Section 5.1. If your students seem to have difficulty, you might help them here as you did there.

   (b) See Display 5.14T(a), which is half-size. Calculating the exact positions of the contour line vertices is tedious and unnecessary. It is far better for students to have the general insight that these contour lines are evenly spaced, nested triangles. A drawing that captures this idea should suffice here.

   (c) Display 5.14T, parts (b) and (c), show half-size approximations of these cross sections. (b) is the cross section parallel to a base edge. It is an isosceles triangle with base length approx. 6.8 cm and height approximately 8.2 cm. (c) is the cross section perpendicular to a base edge. It is also an isosceles triangle, but one of its two equal sides lies on the base of the pyramid. The two equal sides are approximately 8.66 cm, ($= 10 \cdot \sin 60°$) long; the third side is 10 cm long (it is an edge of the tetrahedron).

   (d) Many different answers are possible here. To begin with, both cross sections are isosceles triangles, but they differ in size and orientation, relative to the base of the pyramid.

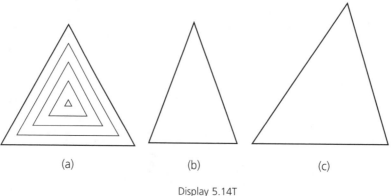

(a)   (b)   (c)

Display 5.14T

NOTES

Chapter 5

2. In the Mole Mountain problem, you saw that the east-west and north-south silhouettes (outlines) of this hill were the same as its vertical cross sections. Draw a contour map of an imaginary hill for which some silhouette is not the same as any of its vertical cross sections. Explain how your drawing satisifes this property.

3. This question is about a regular tetrahedron that measures 10 cm along each edge.

   (a) Describe this shape as a kind of pyramid. What shape is its base? How tall is it? *Hint:* Section 5.1 contains some helpful ideas about tetrahedra and the heights of pyramids, particularly in Exploration 1.
   (b) Draw a full-size contour map of this figure, using a contour interval of 2 cm.
   (c) Draw two full size vertical cross sections of this figure, both passing through the top vertex. Make one of them parallel to a base edge and make the other one perpendicular to a base edge.
   (d) Compare the two cross sections you drew. How are they similar? How are they different?

4. A hemisphere (half a sphere) of radius 10 cm is placed flat side down; that is, its base is a circle of radius 10 cm. Next to it is an upward pointing cone of base radius 10 cm and height 10 cm. The goal of this problem is to make and compare the contour maps of these two shapes.

   If you use 1 cm as the contour interval, then each map consists of ten concentric[†] circles. The common center of these circles represents the topmost point of the shape. To make the two maps, all you need to know is the radius of each circle.

   (a) Copy the table of Display 5.24 and fill it in with the radii of the circles for each map. In each case, contour 1 is the largest of the ten circles and contour 10 is the smallest. Round your answers to two decimal places. *Hint:* Try drawing vertical cross sections to guide your computations.
   (b) Draw full size contour maps of the cone and the hemisphere.

---

†Coplanar circles that have the same center are said to be **concentric**.

Published by IT'S ABOUT TIME, Inc. © 2000 MATHconx, LLC

4. (a) See Display 5.15T for the completed table of Display 5.24. Note that Contour 1 is at zero cm level. The most difficult part of this problem is filling in the column of contour radii for the hemisphere. It can be done using a vertical cross section in either of two ways, depending on the level of computational rigor you want to require of your students. One way is to draw to scale the cross section, which is a semicircle, and then measure the chord length at each centimeter level. This is good enough for rough drawing purposes, but the second decimal place of accuracy will be meaningless.

The other way is to use the Pythagorean Theorem, as illustrated by Display 5.16T. To find the radius r of the contour at level c cm, solve the equation

$$r^2 + c^2 = 10^2$$

for r. Thus, for example, the radius at the 6 cm level is

$$r = \sqrt{10^2 - 6^2} = 8$$

Contour	Contour Radius	
	Hemisphere	Cone
1	10	10
2	9.95	9
3	9.80	8
4	9.54	7
5	9.17	6
6	8.66	5
7	8	4
8	7.14	3
9	6	2
10	4.36	1

Display 5.15T

Published by IT'S ABOUT TIME, Inc. © 2000 MATHconx, LLC

Chapter 5

(c) What is the main difference between these two contour maps? Which map has a more regular pattern? Can you relate that regularity to the slope of some line? If so, which line? Look at a vertical cross section.

5. (a) How many vertices does a prism with a triangular base have? What if the base is a quadrilateral? What if it is a pentagon? What if it is an *n*-gon?

Contour	Contour Radius	
	Hemisphere	Cone
1	10	10
2		
3		
4		
5		
6		
7		
8		
9		
10		

Display 5.24

(b) A right prism is 7 cm high; its base is an equilateral triangle 5 cm on a side. It is open at the top and the bottom. What is the total area of its sides? Will the side area be different if the prism is not a right prism? Explain.

(c) A prism is 7 cm high, open at top and bottom, and its base is a square 5 cm on a side. What is the total area of its sides?

(d) A prism is 7 cm high, open at top and bottom, and its base is a regular pentagon 5 cm on a side. What is the total area of its sides?

(e) A prism is 7 cm high, open at top and bottom, and its base is a regular *n*-gon 5 cm on a side. What is the total area of its sides? Write your answer as a function of *n*.

Published by IT'S ABOUT TIME, Inc. © 2000 MATHconx, LLC

397

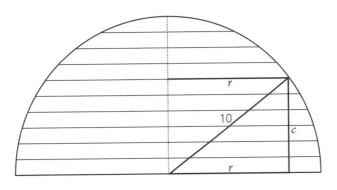

Display 5.16T

(b) See Display 5.17T for reduced size versions of these contour maps.

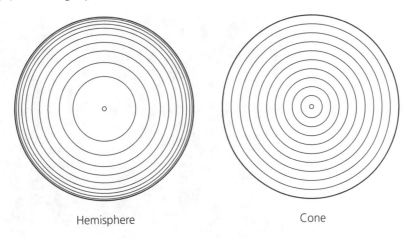

Hemisphere

Cone

Display 5.17T

(c) The main difference is in the spacing between the circles, of course. The cone map has the more regular pattern; the difference in radius between any two successive circles is 1 cm. That is, for each centimeter of height, there is a 1 centimeter change in radius. This is represented by the slopes of the two oblique lines of the cross section, which have slopes of 1 and -1.

5. (a) 6; 8; 10; $2n$
   (b) $3 \cdot (7 \cdot 5) = 105$ sq. cm. The side area will be the same if the prism is not a right prism, provided that height truly is vertical height. A parallelogram with base b and height h has area bh, regardless of the measure of its angles measure. Therefore, the remaining parts of this problem are not ambiguous.
   (c) $4 \cdot (7 \cdot 5) = 140$ sq. cm
   (d) $5 \cdot (7 \cdot 5) = 175$ sq. cm
   (e) $n \cdot (7 \cdot 5) = 35n$ sq. cm

(f) A prism is h cm high, open at top and bottom, and its base is a regular n-gon of side length s. Express the total area of its sides as a fomula in three variables, h, n, and s.

(g) How would you generalize the results of this problem to the case of a prism with a base polygon that is *not* regular?

6. One dictionary defines *topography* as "the configuration of a surface including its relief and the position of its natural and man-made features." Explain what the words *configuration* and *relief* mean in this context. Use this definition to explain why *topographical map* is a good name for a contour map.

398

(f) $n \cdot (h \cdot s) = nhs$ sq. cm

(g) To find the total side area of such a prism, first find the perimeter of the base polygon, then multiply it by the vertical height of the prism.

6. This is an exercise in understanding English words. Student answers probably will be quite varied. Look for basic understanding.

NOTES

5.3 Finding Volumes by Stacking

In Chapter 1 you saw that volume is measured in cubic units. Whatever unit of measure you choose—inches, centimeters, feet, Wobbits, etc.—you can imagine the volume of an object as if it were filled with little cubes of that side length. Of course, the shape of an object might be too irregular to fit cubes in it exactly. In such cases, you can think of the cubic unit as a sort of measuring cup; the volume of the three dimensional shape is the number of cubic unit cupfuls of sand or water that it would take to fill up that shape. Keep this in mind as a way of generalizing from actual cubes to cubic measure as we look at some simple examples.

Learning Outcomes

After studying this section you will be able to:

Find the volumes of prisms and cylinders;

Explain Cavalieri's Principle for volumes.

You have a rectangular cake pan with a base area of 117 sq. in. and sides that come up at right angles to the base.

5.21

1. A batch of brownies baked in this pan is 1 inch thick. What is the total volume of the brownies?

2. A single layer cake baked in this pan is 2 inches thick. What is the total volume of the cake?

Now generalize the idea of questions 1 and 2.

3. If you have a rectangular box with a base area of 117 sq. in. and a height of h in., what is its volume?

4. If you have a right prism with a base area of 117 sq. in. and a height of h in., what is its volume?

5. If you have a right prism with a base area of B sq. in. and a height of h in., what is its volume?

6. If you have a right cylinder with a base area of B sq. in. and a height of h in., what is its volume?

Published by IT'S ABOUT TIME, Inc. © 2000 MATHconx, LLC

399

5.3 Finding Volumes by Stacking

The theme of this section is the development of Cavalieri's Principle of Indivisibles. The approach is to get students thinking of volume first in terms of layers one unit thick, then in terms of thinner and thinner layers. Similarity (and sometimes congruence) of the layer cross sections leads to Cavalieri's Principle, and to easy volume formulas for prisms and cylinders. In the next section, Cavalieri's Principle is used to find formulas for the volumes of pyramids, cones, and spheres.

5.21

These questions should be routine and immediate. Their purpose is to get students thinking of volume in terms of base area and layers of unit thickness. (1) 117 cu. in. (2) 234 cu. in. (3) $117h$ cu. in. (4) The point of this question (and question 6) is that the shape of the base doesn't matter. $117h$ cu. in. (5) and (6) Bh cu. in.

Chapter 5

Additional Support Materials:

Assessments	Qty
Form (A)	1
Form (B)	1

Blackline Masters	Qty

Extensions	Qty

Supplements	Qty
Finding Volumes by Stacking	2

These questions lead to a simple, powerful idea.

If you imagine the volume of a shape in layers of the same thickness, then it becomes easier to calculate the volumes of many three dimensional shapes.

This statement is not very precise, but it contains the germ of the key idea. The following questions help to turn that idea into a useful mathematical principle.

5.22

These questions are about a rectangular box with a base of 2 feet by 3 feet and a height of 2 feet. They are all easy, but the *pattern* of their answers is important. The pattern will be easier to see if you copy the table in Display 5.25 and fill it in as you go along.

Number of Layers	Thickness of Each Layer (ft.)	Volume of Each Layer (cu. ft.)	Volume of Box (cu. ft.)
2	1		
4	$\frac{1}{2}$		
n			

Display 5.25

1. **If you think of the volume of this box as being composed of 1 foot cubes, how many cubes are there? (See Display 5.26(a).)**

2. **Now think of these unit cubes in two layers, each layer 1 foot thick. (See Display 5.26(a) again.) What is the volume of each layer? What is the total volume of the box? Fill in the first row of the table.**

3. **Split each layer into two by cutting it in half horizontally. Now the thickness of each layer is $\frac{1}{2}$ foot. What is the volume of each layer? What is the total volume of the box? Fill in the second row of the table. (See Display 5.26(b).)**

4. **Repeat the splitting process; cutting each layer in half horizontally again. How many layers are there now? What is the thickness of each layer? What is**

Published by IT'S ABOUT TIME, Inc. © 2000 MATHconx, LLC

5.22

Besides being about layers and volumes, these questions provide practice in visualizing three dimensional shapes in a context where the results are helpful to the students. This exercise is particularly suitable for helping students gain confidence and insight.

There are 12 unit cubes in the box. The rest of the answers appear in Display 5.18T. The questions for part 5 are: "Repeat the splitting process; cutting each layer in half horizontally again. How many layers are there now? What is the thickness of each layer? What is the volume of each layer? What is the total volume of the box? Fill in the fourth row of the table."

If your students are doing spreadsheet work from time to time, the answers in the last row of the table suggest a way of extending this table indefinitely. It might be an enlightening homework exercise for some students to do that.

Number of Layers	Thickness of Each Layer (ft.)	Volume of Each Layer (cu. ft.)	Volume of Box (cu. ft.)
2	1	6	12
4	$\frac{1}{2}$	3	12
8	$\frac{1}{4}$	1.5	12
16	$\frac{1}{8}$	0.75	12
n	$\frac{2}{n}$	$\frac{12}{n}$	$n \cdot \frac{12}{n}$

Display 5.18T

Chapter 5

the volume of each layer? What is the total volume of the box? Fill in the third row of the table. (See Display 5.26(c).)

5. Write the questions for this part, following the pattern of the previous parts. Then answer them.

6. Suppose you have subdivided the box into n horizontal layers of equal height. What is the thickness of each layer? What is the volume of each layer? Write these two answers as functions of n. Express the total volume of the box as a combination of these two functions. Put your answers in the last row of the table.

Could you have filled in the last column of Display 5.25 without doing the other parts? Of course! If you ask a bakery to slice a loaf of bread, you get the same amount of bread whether the slices are thick or thin. So it is with the box; its volume is not affected by changing the thickness of the slices. Display 5.27 illustrates this fact.

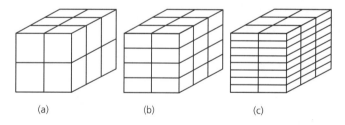

(a)   (b)   (c)

Display 5.26

Display 5.27

NOTES

Chapter 5

5.23

Cut some 2 by 2 inch squares—at least 8 of them, but more if you'd like—out of corrugated cardboard. Think of them as solid tiles and stack them to form a right rectangular prism.

1. What is the total volume of the corrugated cardboard in your prism? How did you get your answer?

2. Now stack your tiles off center, as if you were making small stairs. What is the total volume of the corrugated cardboard in this stack? How do you know?

3. Now twist your stack into a spiral. What is the total volume of the corrugated cardboard in this stack? How do you know?

By putting the results of these questions together with a 400 year old idea, we have the main principle of this section. At the very end of the 16th century, scientists such as Galileo and Kepler began to think about 3-space as if it were an infinite stack of parallel planes that had no thickness at all. They imagined it to be something like a huge stack of very thin paper. Since these planes had no thickness, they couldn't be divided into thinner "sheets," so they were called *indivisibles*. These imaginary layers of space provided a way of working on 3-space questions with 2-space tools.

Cavalieri

Perhaps the best of these tools was formulated by Bonaventura Cavalieri, an Italian priest and mathematics professor who had been a pupil of Galileo. In the 1620's Cavalieri was studying three dimensional shapes, often called **solid figures** or just **solids**, trying to determine their volumes and surface areas. Thinking of each solid figure as an infinite stack of parallel planar regions, he proposed the following principle.

Cavalieri's Principle: Two solids have the same volume if their cross sections at equal heights always have equal areas.

Important: Notice that the cross sections don't have to be congruent shapes; they just must have equal areas!

5.3 Finding Volumes by Stacking

5.23

It is not necessary to have exactly 8 tiles; almost any number of them will do. However, the visual/kinesthetic effect will not be as striking if there are only a few squares. Once your students have the square tiles, answering the questions should take very little time.

1. Answers will vary, depending on thickness of cardboard and number of tiles. The usual way to find the volume is to multiply the base area, $2 \times 2 = 4$ sq. in., by the height of the stack. When we did it, our stack of 8 cardboard squares measured 1.25 in. high, so the total volume was $4 \times 1.25 = 5$ cu. in.

2. The stack is made up of the same cardboard tiles, so the volume has to be the same as in question 1.

3. Same answer as 2.

The observation here is simple, but powerful: The *configuration* of the stack makes no difference; since all the tiles have the same base area, the height of the stack determines its volume. This leads right into Cavalieri's Principle.

Note that Cavalieri's Principle is also known as the Principle of Indivisibles. The principle itself is a sound mathematical insight, but the idea of indivisibles is not. Modern mathematics deals with Cavalieri's Principle as part of integral calculus. To avoid reinforcing the erroneous underpinnings of this useful idea, we refer to it by using Cavalieri's name.

NOTES

Chapter 5

To see why Cavalieri and we should think this is reasonable, look at the rectangular prism and the oblique prism in Display 5.28. Both of these solids are the same vertical height and they have equal-area cross sections at every height. In fact, in this case the cross sections actually are congruent. Cavalieri's Principle claims that they should have the same volume. Does that seem reasonable to you? Can you be sure from the picture?

Maybe not. But think about your stack of cardboard tiles and the questions about them. When you slanted the stack to one side, the volume didn't change, did it?

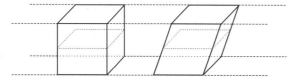

Two prisms with the same cross sections at every height

Display 5.28

"No, but that was different," we hear you say. "The slanted stack wasn't a prism then, because the staircase effect made two of its sides have ridges in them."

We agree. It wasn't a prism then, but it was pretty close to being one. If you used a stack of thinner cardboards of the same total height—pieces of file cards, say—the ridges would be less noticeable. And if you used a stack of thin paper squares, the ridges would be almost invisible. At each of these stages, your layers would be thinner and you would need more of them to make up a stack of the same height. But the volume would always be the same because you would be starting with a rectangular stack of the same size each time.

Display 5.29 shows the first few stages of this process. If you approximate the prism with a stack of 4 rectangular layers, as in (b), the ridges (steps) are quite large and the slanted stack doesn't look much like a prism. Doubling the number of layers, as in (c), makes the ridges smaller, and doubling the number of layers again makes the stack shown in (d) look much more like

Published by IT'S ABOUT TIME, Inc. © 2000 MATHconx, LLC

NOTES

a prism than the others. And that stack only has 16 layers in it. Imagine how close to a prism a stack of 100 or 1000 (much thinner) layers would be!

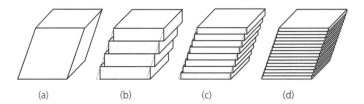

(a)   (b)   (c)   (d)

Display 5.29

Cavalieri just pushed this idea to its limit, thinking of "infinitely many, infinitely thin" layers. Infinitely thin layers don't exist, of course, *but limits do.* Cavalieri's basic understanding of what was going on was correct, even though his arguments were shaky. His use of indivisible layers of 0 thickness was incorrect, but his idea of seeing where the trend of thinner and thinner sheets was heading, the *limit* of the process, was right on target. This idea foreshadowed integral calculus, the logical treatment of limits that would justify his insight. But integral calculus was a thing of the future for Cavalieri (as it may be for you). He had to be content with a less formal justification. And so will you.

5.24

Round these answers to two decimal places, if you wish.

1. Find the volume of a right rectangular prism with a base rectangle measuring 10 cm by 15 cm and a vertical height of 20 cm.

2. Find the volume of a right prism with an equilateral triangular base measuring 10 cm on a side and a vertical height of 20 cm.

3. Find the volume of a right cylinder with a base radius of 5 cm and a vertical height of 20 cm.

4. Suppose you wanted a right cylinder to have the same height and the same volume as the rectangular box of question 1. What would be the radius of its base?

Published by IT'S ABOUT TIME, Inc. © 2000 MATHconx, LLC

5.24

The first three of these questions are routine volume computations. They emphasize that the volume of any right cylinder or right prism can be found by multiplying its base area by its height. The last two questions ask students to work the volume computations "backwards," providing some practice in solving simple equations that involve square roots.

1. $10 \times 15 \times 20 = 3000$ cc

2. This reviews earlier work with triangles. The area of the base equilateral triangle is
$$\frac{1}{2} \cdot 10 \cdot (10 \sin 60°) = 43.30 \text{ sq. cm (approximately)}.$$
The volume, then, is $43.30 \cdot 20 = 866$ cc.

3. $\pi \cdot 5^2 \cdot 20 = 500\pi = 1570.80$ cc (approximately).

4. Solve $\pi \cdot r^2 \cdot 20 = 3000$ for r. Since the heights are the same, you can simplify the initial equation to looking at equality of base area $\pi \cdot r^2 = 150$. Thus,
$$r = \sqrt{\frac{150}{\pi}} = 6.91 \text{ cm (approximately)}.$$

NOTES

Chapter 5

5. Suppose you wanted a right prism with an equilateral triangular base to have the same height and the same volume as the rectangular box of question 1. What would be the side length of its base?

1. Did you find it awkward that we kept repeating the word "right" when describing the prisms and cylinders in the previous questions? Would it have mattered if we had just left out that word? Explain.

5.25

2. You are in a large room upstairs that measures 10 feet from floor to ceiling.

   (a) An 8 inch diameter cylindrical stovepipe runs straight up from floor to ceiling. What volume does that stovepipe enclose?

   (b) Another 8 inch diameter cylindrical stovepipe runs from floor to ceiling at an angle of 45°. About how long is that pipe? Answer to the nearest foot.

   (c) Which of the two pipes encloses the larger volume of space, or do they enclose the same amount?

   (d) What is the apparent conflict between common sense and Cavalieri's Principle in this situation? How can you resolve it?

As the preceding questions point out, an oblique cylinder is not a right cylinder that has been tipped a little! (It's more like a Slinky® toy, a coil of thin, flat metal that has been stacked at an angle.) Similarly, an oblique prism with a square base is not the same as a tipped right prism with a square base. A prism is oblique if its side edges intersect the base plane at some angle other than 90°. In such cases, the length of an edge is not the same as the vertical height of the prism. The edge length of an oblique prism is called its slant height.

Published by IT'S ABOUT TIME, Inc. © 2000 MATHconx, LLC

405

5. Again, looking at equality of base area, solve

$$\frac{1}{2} \cdot r \cdot (r \sin 60°) = 150$$

That is,

$$r^2 \cdot \sin 60° = 300$$

so

$$r = \sqrt{\frac{300}{\sin 60°}} = 18.61 \text{ cm (approximately)}.$$

5.25

The main point of the first question is straightforward, but it leads to the slightly trickier second discussion question.

1. The word "right" *could* have been left out without altering either the results or the approach, *provided that* the height still is understood to be vertical height. Cavalieri's Principle guarantees that prisms and cylinders with the same base areas and same vertical heights must have the same volume. However, there is a possibility here for a misunderstanding about cross sections. Question 2 addresses this difficulty.

2. (a) The radius is 4 inches, so the volume is

$$\pi \cdot 4^2 \cdot 120 = 1920\pi \approx 6032 \text{ cu. in., about 3.5 cu. ft.}$$

   (b) The approximate length of the second pipe is that of the hypotenuse of an isosceles right triangle with leg length 10 ft. By the Pythagorean Theorem, this is $\sqrt{200} = 14$ ft. (approximately).

   (c) Since both pipes have the same diameter, the longer one encloses more volume.

   (d) This part is the point of the entire question. Guide your students' discussion so that they see both the apparent conflict and its resolution. You might help them by drawing a rough sketch on the board.

   The conflict arises from the temptation to regard the second stovepipe as an oblique (circular) cylinder. It is not. If it were, then since both cylinders have the same base radius and the same height, they would have to enclose the same volume, which is not possible. The reason the second pipe is not an oblique cylinder is that its *horizontal* cross section is not circular; it's an oval. Have students think about the hole they would have to cut in the ceiling to put the pipe through at a 45° angle. As the following text says, an oblique cylinder is not just a right cylinder that has been tipped a little. Thus, Cavalieri's Principle does not apply in this situation.

Chapter 5

 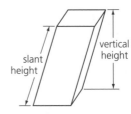

Tipped right prism   Oblique prism

Display 5.30

5.26

1. How would you define the slant height of an oblique cylinder?

2. How would you determine the angle that an oblique cylinder makes with its base plane?

Cavalieri's Principle lets us describe the volumes of all prisms and cylinders, whether they are right or oblique, by a single formula:

A Fact to Know: The volume of any prism or cylinder with base area B (sq. units) and vertical height h (units) is Bh (cubic units). In symbols,

$$V = Bh$$

Problem Set: 5.3

Round all your numerical answers to one decimal place.

1. Each of these parts describes a prism with a square base 6 inches on a side. Display 5.30 should help you to visualize each one.

   (a) This is a right prism with vertical height 15 inches. What is its volume?

   (b) This is an oblique prism that makes a 75° angle with its base plane and has vertical height 15 inches. What is its volume?

   (c) This is an oblique prism that makes a 75° angle with its base plane and has slant height 15 inches. What is its volume?

   (d) This is an oblique prism that makes a 60° angle with its base plane and has volume 400 cubic inches. What is its vertical height?

5.26

1. The easiest way is to look at the length of its center line, the line connecting the centers of its base and top circles.

2. Some students might need additional help for this somewhat more difficult question. It is a good instance of how to make a concept "well defined" (unambiguous). One would probably want to measure the angle that the center line makes with the base plane, but there is no obvious way to determine the direction in which this angle should be viewed. One way is to define this angle as the smallest angle that the center line makes with the base plane. However, this doesn't tell one how to find such an angle. In a practical situation, one might find this angle by hanging a plumb line from the top of the cylinder and taking the complement of the angle that this line makes with the center line, or with a line along the cylinder that is parallel to the center line.

Problem Set: 5.3

Note that "equality" in each of these answers is rounded to one decimal place.

1. (a) $V = 6^2 \cdot 15 = 540$ cu. in.
    (b) By Cavalieri's Principle, this is the same as the answer to part (a): 540 cu. in. The angle with the base plane is irrelevant information in this part; it is included to set up the contrast with part (c).
    (c) Look at Display 5.30 to see why the vertical height, h, equals the slant height times the sine of the angle with the base plane.
    $$h = 15 \cdot \sin 75° = 14.5$$
    Thus,
    $$V = 6^2 \cdot 14.5 = 521.6 \text{ cu. in.}$$
    (d) $400 = 6^2 \cdot h$, so $h = \frac{400}{36} = 11.1$ inches

Chapter 5

(e) This is an oblique prism that makes a 60° angle with its base plane and has volume 400 cubic inches. What is its slant height?

2. (a) A dairy is considering replacing its standard large milk carton with a plastic cylinder. The current carton is a rectangular box with a 4 inch square base and a height of 8 inches. The peaked, reclosable spout top takes up another 2 inches of height. (See Display 5.31.) To fit in the same packing crates, the cylinders must have a base diameter of 4 inches. The tops of these plastic cylinders don't require any extra space for opening and closing. They use a flat, snap-out plug. In order to hold the same volume of milk as the cartons, how tall must the cylinders be? Round your answer *up* to the nearest tenth of an inch.

(b) Assuming that the two kinds of containers cost the same amount to produce, which do you think would be better? Why?

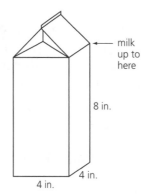

milk
up to
here

8 in.

4 in.

4 in.

Display 5.31

3. (a) Find the volume of a right cylinder *A* that has base radius 7 cm and height 25 cm.

(b) An oblique cylinder *B* makes a 45° angle with its base plane and has base radius 7 cm and vertical height 25 cm. Is its volume more, less, or the same as that of cylinder *A*? Justify your answer.

(c) An oblique cylinder *C* makes a 45° angle with its base plane and has base radius 7 cm and slant height 25 cm. Is its volume more, less, or the same as that of cylinder *B*? Calculate the volume of cylinder *C*.

Published by IT'S ABOUT TIME, Inc. © 2000 MATHconx, LLC

407

(e) By part (d), the vertical height of this prism is 11.1 inches. If s is its slant height, then

$$11.1 = s \cdot \sin 60°$$

so,

$$s = \frac{11.1}{\sin 60°} = 12.8 \text{ inches}$$

2. (a) The usable volume of the rectangular carton is $4 \times 4 \times 8 = 128$ cu. in. The radius of the cylinder is 2 in., so its base area is $\pi \cdot 2^2 = 4\pi$ sq. in. To find the necessary height, solve $4\pi h = 128$ to get 10.2 inches, rounded to the nearest tenth of an inch.

   (b) This question has no well defined answer. Its purpose is to get the students to focus on the problem's context. There probably are some good reasons not to change carton style here, but most of them have very little to do with the mathematical result. The only reason related to the computation is that the cylinders have to be slightly taller (0.2 inches), which might create storage problems for large quantities of them. However, the difference is not big enough for this to be conclusive.

3. Taken together, these parts illustrate Cavalieri's Principle and the distinction between slant height and vertical height. The comparison questions encourage students to use what they know to check on the reasonableness of their computations.

   (a) $\pi \cdot 7^2 \cdot 25 = 1225\pi = 3848.5$ cu. cm

   (b) By Cavalieri's Principle, this answer is the same as for part (a).

   (c) This requires finding the vertical height. The slant height is the hypotenuse of an isosceles right triangle for which the vertical height is the length of one leg. Encourage your students to make a sketch to help them visualize this. Since the vertical height will be less than the slant height, the volume of this cylinder should be less than that of cylinder B. The vertical height is $25 \sin 45° = 17.68$ (approximately), so the volume is

$$\pi \cdot 7^2 \cdot 17.68 = 2721.6 \text{ cu. cm}$$

Chapter 5

(d) An oblique cylinder D makes a 60° angle with its base plane and has base radius 7 cm and slant height 25 cm. Is its volume more, less, or the same as that of cylinder C? Calculate the volume of cylinder D.

4. The design of a chute for dropping sand from one level of an industrial plant to another level 10 feet below is shown in Display 5.32. The plan calls for a 2 foot square opening at each level and an angle of 76°.

(a) The volume of sand that this chute will hold between the two levels is the maximum amount that the structure will support. What is that volume?

(b) Tests show that the 76° angle is too steep; the sand drops too fast inside the chute. An angle of 64° is proposed, instead. However, the total volume of sand cannot be increased. What size square opening will allow for exactly the same maximum amount of sand between the two levels? Explain your answer.

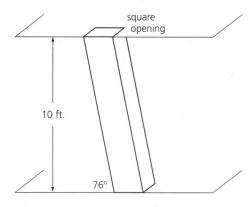

Display 5.32

5. (a) Find the volume of a right prism with height 10 cm and a base that is a parallelogram with sides of lengths 4 cm and 6 cm and an obtuse angle of 150°.

(b) An oblique prism with the same base as the prism of part (a) makes a 50° angle with its base plane and has the same volume. What is the slant height of this prism?

6. Cavalieri's Principle, as given in this section, is a statement about volumes of three dimensional shapes. It has a two dimensional analogue; what do you think it is? Explain your idea.

408

Published by IT'S ABOUT TIME, Inc. © 2000 MATHconx, LLC

(d) This is similar to part (c). Without doing any computations, students should be able to see that the larger angle in this case (60° vs. 45°) means that the vertical height of this cylinder will be larger than the vertical height of cylinder C, making the volume greater. The vertical height is 25 sin60° = 21.65 (approx.), so the volume is

$$\pi \cdot 7^2 \cdot 21.65 = 3332.8 \text{ cu. cm}$$

A common error here is to use *cos* instead of *sin*. The estimated comparison with cylinder C should allow students to correct this error on their own.

4. (a) $V = 2^2 \cdot 10 = 400$ cu. ft. Notice that the angle plays no role in this computation.

(b) The same size opening. If the cross section (parallel to the floors) is the same and the vertical height is the same, then Cavalieri's Principle says that the volume must be the same.

5. Part (a) is a straightforward computation. Part (b) applies Cavalieri's Principle in reverse.

(a) Using the longer side of the parallelogram as its base, its altitude must be 4 sin30° = 2 cm. Thus, the area of the base of the prism is 6 · 2 = 12 sq. cm, so its volume is 12 · 10 = 120 cu. cm.

(b) The reasoning in this part is more difficult than the computation. The two prisms have the same base area, 12 sq. cm, and the same volume, 120 cu. cm. By Cavalieri's Principle, the vertical height of the second prism must be 10 cm. Solve 120 = 12h. If s is the slant height of this prism, then 10 = s · sin50°, so

$$s = \frac{10}{\sin 50°} = 13.1 \text{ cm}$$

6. Cavalieri's Principle for 2-space: Two planar regions have the same area if their cross sections at equal heights always have the same length. In this case, cross sections are line segments, and the heights have to be measured at right angles to some base line. A good illustration of the 2-space principle at work is the formula for finding the area of a parallelogram.

Chapter 5

5.4 The Volumes of Spheres, Cones, and Pyramids

If Cavalieri's Principle were useful only for finding volumes of prisms and cylinders, it would not deserve to be famous. However, it is much more powerful and flexible than that. In this section we use some of that power and flexibility to solve a difficult problem of elementary solid geometry— finding the volume inside a *sphere*.

(5.27)

As you probably know, a **sphere** is the set of all points in space that are a fixed distance from a particular point, its **center**. In everyday terms, it is the hollow shell of a round ball. (See Display 5.33.) The sphere is as important in three dimensional geometry as its two dimensional analogue is in plane geometry. What is the two dimensional analogue of a sphere? Although the idea of this shape is simple, finding its volume is quite difficult. Volume is measured by unit cubes, and cubes don't fit very well inside a ball!

A sphere

Display 5.33

In ancient Greece, Archimedes unraveled the mysteries of the volume and surface area of a sphere. Because of his work and that of others (such as Cavalieri) during the following 2000 years, it will be much easier for us. Still, this is not an easy question to answer. We'll need many of the things you have studied this year, including similarity and scaling factors, limiting arguments, and some facts about circles and disks. Along the way, we'll pick up some other useful information, such as formulas for the volumes of pyramids and cones. To help you see the "big ideas" without getting lost in the details, we are going to move pretty fast, so hang on tight! If you feel yourself getting lost, *ask for help*.

Published by IT'S ABOUT TIME, Inc. © 2000 MATHconx, LLC

5.4 The Volumes of Spheres, Cones, and Pyramids

This section applies Cavalieri's Principle to the problem of finding the volumes of spheres, cones and pyramids. Big Steps 1–5 illustrate the process of successive generalization that is so common throughout mathematics. Seeing and understanding how this process works is a valuable lesson, far more valuable than memorizing volume formulas! Even a correctly remembered formula for the volume of a cone will not tell a student that the formula applies to "generalized cones" with irregular shaped bases. Once students see how the volumes of cylinders, cones, and spheres relate to each other and to the idea of measuring base area with squares, the formulas (which usually can be looked up, if needed, later in life) become meaningful, flexible ideas, rather than magic incantations.

(5.27) The two dimensional analogue of a sphere is a circle and not a disk. This is as much a question about analogy as it is about circles and spheres.

Chapter 5

Additional Support Materials:

Assessments	Qty
Form (A)	1
Form (B)	1

Blackline Masters	Qty

Extensions	Qty

Supplements	Qty
The Volumes of Spheres, Cones, and Pyramids	2

Big Step 1

We begin by finding the volume of a particularly "nice" pyramid—a pyramid with a square base, a peak directly above the center of the base, and vertical height exactly half of the side length of the base. The trick to finding the area of this pyramid is realizing that six congruent copies of it can be put together to form a cube. Display 5.34 shows how this works. The original pyramid is *PABCD*; its peak is at *P* and its vertical height (its altitude) is labeled *a*. Can you see the other five congruent pyramids in this cube? The peaks of all of them are at *P*.

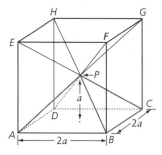

Display 5.34

5.28

All these questions refer to Display 5.34.

1. Identify the other five pyramids that are congruent to pyramid *PABCD* by listing their vertices.

2. What is the total volume of the six pyramid cube? Express your answer in terms of *a*, the height of the original pyramid.

3. In terms of *a*, what is the volume of pyramid *PABCD*?

4. In terms of *a*, what is the area of the base of this pyramid?

5. Let *B* stand for the area of the base. Rewrite the expression for the volume of the pyramid in terms of *a* and *B*.

6. If this pyramid has a base that is 12 meters on a side, what is its base area? What is its volume?

410

5.28

1. *PABFE, PBCGF, PFGHE, PADHE,* and *PCGHD*

2. $(2a)^3 = 8a^3$

3. $\frac{1}{6} \cdot 8a^3 = \frac{4}{3}a^3$

4. $2a \cdot 2a = 4a^2$

5. $\frac{4}{3}a^3 = 4a^2 \cdot \frac{1}{3}a = \frac{1}{3}Ba$

6. Base area = 144 sq. m; volume = 576 cu. m

NOTES

Chapter 5

Big Step 2

Now we start to generalize. Suppose next that you have a pyramid with a square base, a peak directly above the center of the base, and vertical height *h*, as in Display 5.35.

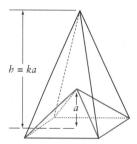

Display 5.35

We know from Big Step 1 that a pyramid with a square base and height *a* equal to half of a base side has volume $\frac{1}{3}aB$, where *B* is the area of the base. The only difference between such a pyramid and the pyramid of Display 5.35 is the height. The key to this part is that we can express *h* as a multiple of *a*, say *ka*. Before going any further, answer the following questions to make sure you understand what we have done so far.

> These questions relate the notation that we have defined in Big Step 2 to a pyramid with a square base that is 8 cm on a side, a peak directly above the center of the base, and vertical height of 25 cm.

5.29

1. What is *h* in this case?

2. What is *B* in this case?

3. What is *a* in this case?

4. What is *k* in this case? How did you find it?

When you studied the effect of scaling on volume in Chapter 2, you saw that a solid figure scaled by a factor *k* has its volume changed by k^3. The sides of the *small cubes* by which volume is measured change by a factor of *k* in all three dimensional directions, so their volume changes by $k \cdot k \cdot k$. But if a solid figure is scaled by *k only in one direction*, as is the case here, then the small cubes will be stretched or shrunk by the scaling factor in one dimension. This means that the

Thinking Tip

Use appropriate notation. Notice how the algebra expresses what's going on!

Published by IT'S ABOUT TIME, Inc. © 2000 MATHconx, LLC

411

5.`29

Go through these questions quickly in class, just to make sure that everyone understands the notation. If you spend too much time on them, the bigger picture may fade.

1. 25 cm

2. $8^2 = 64$ sq. cm

3. 4 cm

4. $\frac{25}{4}$, or 6.25 (*not* cm, just a constant). This is the most important of the questions because it is the least obvious. It is important for students to see that k is always the quotient $h \div a$.

NOTES

Chapter 5

total volume will change only by a factor of k. In this case then, the volume of the pyramid with height h is k times the volume of the pyramid with height a. In symbols,

$$\text{Volume of } h \text{ height pyramid} = k \cdot (\tfrac{1}{3}aB) = \tfrac{1}{3}kaB$$

But $ka = h$, so–writing V for the volume of this pyramid—we have

$$V = \tfrac{1}{3}hB$$

where B is the area of the base.

5.30

1. **Find the volume of the pyramid described in the previous questions. It has a square base that is 8 cm on a side, a peak directly above the center of the base, and vertical height of 25 cm.**

2. **Choose any positive numbers you want for the vertical height and the base side length of a pyramid with a square base and its peak directly above the center of the base. Then compute the volume of that pyramid.**

Big Step 3

Does it really matter if the peak of the pyramid is directly over the center of the base? What if the pyramid were pushed off center, so to speak? In Cavalieri's terms, what if the stack of horizontal cross sections were slid over horizontally in some direction, as in Display 5.36? By Cavalieri's Principle, the volume is the same. Thus, we can say that the volume, V, of *any* pyramid with a square base and height h is given by the formula

$$V = \tfrac{1}{3}hB$$

where B is the area of the base.

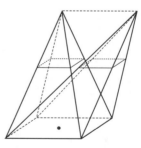

Display 5.36

5.30

1. This is an easy, straightforward question. $B = 64$ sq. cm,
    so $V = \frac{1}{3} \cdot 64 \cdot 25 = \frac{1600}{3} \approx 533.33$ cu. cm

2. This free choice question emphasizes that the formula works for *any* choice of numbers. Students should see that they don't have to construct the special pyramid of Big Step 1; they can just compute the base area and then plug the numbers into the formula.

NOTES

Chapter 5

A pyramid with a square base 10 cm on a side has its peak 15 cm directly above its front left corner. What is its volume?

5.31

Big Step 4

Now we are ready to tackle the general case, the volume of a pyramid with *any* shape of base. We'll get a bonus, too. This argument applies just as well to the volume of a cone.

In Chapter 1 you saw that the area of any region in the plane can be approximated as closely as we please by tiling it with small squares. Think of a cone or a pyramid with peak P any shape of base you choose. Approximate its base by tiling it with small squares. Then think of each square as a pyramid with the same peak point, P. Since all these square based pyramids lie within your pyramid, the sum of their volumes is an approximation of the volume of your pyramid.

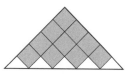
Triangular base

Circular base

Base area approximated by squares

Display 5.37 (a)

5.37(a) illustrates this process. Its top part shows the triangular base of a pyramid and the circular base of a cone, each approximately tiled with little squares. In this picture there are ten squares inside the triangular base. We'll use this small case as a typical example. Write the areas of these ten squares as $B_1, B_2, ..., B_{10}$. Then

$$B \text{ approximately equals } B_1 + B_2 + ... + B_{10}$$

Published by IT'S ABOUT TIME, Inc. © 2000 MATHconx, LLC

5.31 This question should be almost instantaneous! $V = \frac{1}{3} \cdot 10^2 \cdot 15 = 500$ cu. cm

NOTES

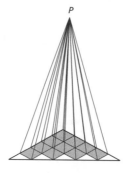

Pyramid with triangular base   Cone with circular base

Volume approximated by square base pyramids

Display 5.37 (b)

Now, since all the squares are inside the base of the big pyramid, all the square based pyramids with peak P must be inside the big pyramid. Thus, the sum of their volumes must approximate the volume of the big pyramid. But we know how to find the volumes of pyramids with square bases! If h is the vertical height of all these pyramids, then the volumes of the ten small pyramids are

$$\frac{1}{3}B_1h, \ \frac{1}{3}B_2h, \ ..., \ \frac{1}{3}B_{10}h$$

Adding them up, we get

$$\frac{1}{3}B_1h + \frac{1}{3}B_2h + ... + \frac{1}{3}B_{10}h = \frac{1}{3}(B_1 + B_2 + ... + B_{10})h$$

5.32 What law of algebra allows us to "pull out" the common factors $\frac{1}{3}$ and h? By the statement that B approximately equals $B_1 + B_2 + ... + B_{10}$, this is approximately $\frac{1}{3}Bh$.

Of course, ten squares inside a triangle is not a very close area approximation, but with more squares we can get as close as we please to the area, B, inside any polygon or circle or other region. If we used n squares (where n could be 1000 or 1,000,000 or whatever), then

$$B_1 + B_2 + ... + B_n$$

would be *very* close to B, so

$$\frac{1}{3}(B_1 + B_2 + ... + B_n)h$$

would be very close to $\frac{1}{3}Bh$.

This value that is approached closer and closer as the number of squares increases is called the *limit* of the approximation process. Calculus provides ways of verifying that this limit is the correct value for this approximation process. Thus, the actual volume of any pyramid or cone is given by the formula

$$V = \frac{1}{3}Bh$$

Published by IT'S ABOUT TIME, Inc. © 2000 MATHconx, LLC

414

(5.32)   The Distributive Law for multiplication over addition.

NOTES

where *B* is the area of the base (of any shape) and *h* is the vertical height.

These questions refer to the cone pictured in Display 5.37(b), which has a base radius of 9 cm and a vertical height of 21 cm. Each of the 12 squares within its base is 4 cm on a side.

5.33

1. Find the volume of one of the square base pyramids.

2. What is the total volume of the 12 square base pyramids?

3. What is the actual volume of the cone?

4. Compute the difference between the volume of the cone and its approximation by these 12 pyramids. Express this difference as a percentage of the volume of the cone to the nearest whole percent.

5. Compute the difference in base area between the disk and the sum of the 12 squares. Express this difference as a percentage of the disk area to the nearest whole percent. What do you notice?

6. To get a better approximation of the base by squares, you could insert four smaller squares, 2 cm on a side. Do you see where you can put them? Find the volume of one of the smaller square base pyramids.

7. Using all 16 square base pyramids, what is the approximate total volume now? What is the difference between this approximation and the actual volume? Express this difference as a percentage of the volume of the cone to the nearest whole percent.

8. Compute the difference in base area between the disk and the sum of the 16 squares. Express this difference as a percentage of the disk area to the nearest whole percent. What do you notice?

Some people call a shape that is like a cone but does not have a circular base a *generalized cone*.

5.34

1. How would you *define* a generalized cone?

2. Sketch an example of a generalized cone that is not a pyramid.

3. Does the volume formula $V = \frac{1}{3} Bh$ apply to generalized cones? Why or why not?

Published by IT'S ABOUT TIME, Inc. © 2000 MATHconx, LLC

5.33

These questions lead students through the actual computations of two successive approximations of the volume of a cone. It also provides practice in computational techniques from earlier material.

1. $\frac{1}{3} \cdot 4^2 \cdot 21 = 112$ cu. cm

2. $12 \cdot 112 = 1344$ cu. cm

3. $\frac{1}{3} \cdot (\pi \cdot 9^2) \cdot 21 = 567\pi = 1781.28$ cu. cm rounded to two places.

4. 437.28 cu. cm; 25%

5. $\pi \cdot 9^2 - 12 \cdot 4^2 = 81\pi - 192 = 254.47 - 192 = 62.47$ sq. cm rounded to two decimal places; 25%. The percentage of difference in area exactly matches that of difference in volume.

6. $\frac{1}{3} \cdot 2^2 \cdot 21 = 28$ cu. cm

7. New approximation: $4 \cdot 28 + 1344 = 1456$ cu. cm
   New difference: 325.28 cu. cm; 18%

8. $\pi \cdot 9^2 - 12 \cdot 4^2 - 4 \cdot 2^2 = 46.47$ sq. cm rounded to two places; 18%. Again, the percentage of difference in area exactly matches that of difference in volume. These percentage questions emphasize the fact that an arbitrarily close approximation to base area will result in an equally close approximation to volume.

5.34

Knowing how this volume formula was derived also tells you that it applies to other figures besides pyramids and cones. These questions make that point.

1. The figure formed by connecting each point of a closed curve in a plane with some particular point — the peak — not on that plane.

2. This part will depend on students' graphic capabilities. In theory, it is limited only by their imagination in creating a closed curve base. If they seem stuck for an idea, you might suggest a semicircular or crescent shaped base, or a star, or a footprint, or ....

3. Yes. Volume approximation by pyramids with square bases, which leads to this volume formula, depends only on the fact that the base can be approximated by squares. This is true of any planar region, so the formula applies to any generalized cone.

Chapter 5

Step 5, the Last and Biggest Step

Now for the sphere. Actually, we'll find the volume inside a hemisphere (half a sphere) and then double it. Unlike most of what we have done up to now, the approach here is not at all obvious or natural. But it's slick! We have to

- Ask the right question.
- Draw the right picture.
- Translate the question into algebra.
- Interpret the algebraic answer.
- Apply Cavalieri's Principle.

In this case, the algebra makes a surprising connection between two geometric shapes that do not look alike at all. The two shapes appear in Display 5.38. One of them is a hemisphere; the other is a right cylinder with a cone cut out of it. The height and radius of the cylinder are the same as the radius, r, of the hemisphere. The right question is

> How are the horizontal cross sections of these two shapes related?

Specifically, if we take horizontal cross sections of the two shapes at the same height, how are their areas related? Display 5.38 shows a cross section of each figure at a particular height, a. One is a disk; the other is an annulus. We want to know how cross sections at every height are related, so we use a letter, instead of a number, to stand for *any* height. It actually makes our work easier!

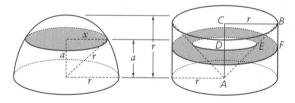

Display 5.38

To compare these areas we need to express them in terms of the known numbers, r and a. Answer the following questions to see how this is done.

Published by IT'S ABOUT TIME, Inc. © 2000 MATHconx, LLC

NOTES

Chapter 5

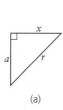
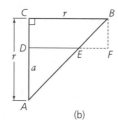

(a)   (b)

Display 5.39

Both the hemisphere and the cylinder in Display 5.38 have radius r; the cylinder has height r. The cross section of each figure is height a above its base. These questions lead you to express the areas of these two cross sections in terms of a and r. Each step asks you first to work out the example $r = 6$ cm and $a = 4$ cm, and then to write the answer algebraically with the letters a and r.

5.35

1. The area of the disk depends on its radius, marked x in Display 5.38. This radius x is one leg of a right triangle whose other two sides are a and r. (See Display 5.39(a).)

   (a) If $a = 4$ cm and $r = 6$ cm, what is the area of this disk?

   (b) In terms of a and r, what is the area of this disk?

2. The area of the annulus is the difference of two disk areas.

   (a) If $a = 4$ cm and $r = 6$ cm, what is the radius of the larger disk? What is its area?

   (b) In terms of a and r, what is the radius of the larger disk? What is its area?

   (c) The radius of the smaller disk is the length of DE. (See Displays 5.38 and 5.39(b).) Justify these statements: $\triangle ABC$ is an isosceles right triangle, and it is similar to $\triangle AED$. Therefore, the length of DE is determined.

   (d) If $a = 4$ cm and $r = 6$ cm, what is the length of DE? What is the area of the smaller disk?

   (e) In terms of a and r, what is the length of DE? What is the area of the smaller disk?

   (f) If $a = 4$ cm and $r = 6$ cm, what is the area of the annulus?

   (g) In terms of a and r, what is the area of the annulus?

3. How are the algebraic expressions for the areas of the two cross sections related? How do you know?

Published by IT'S ABOUT TIME, Inc. © 2000 MATHconx, LLC

417

5.35

Work through these questions with your students quickly; almost all of them are easy. It is important that your students do not lose track of the big picture as they handle these routine details.

1. To find the area of the disk, you don't need x itself; you need x^2.
   (a) The Pythagorean Theorem tells us that $4^2 + x^2 = 6^2$, so $x^2 = 36 - 16 = 20$. Thus, the area of the disk is 20π.
   (b) As in part (a), the Pythagorean Theorem tells us that $x^2 = r^2 - a^2$, so the area of the disk is $\pi(r^2 - a^2)$.

2. (a) The larger disk has radius 6 and area 36π.
   (b) The larger disk has radius r and area πr^2.
   (c) $\triangle ABC$ is a right triangle because AC is the vertical center line of the cylinder and BC is on the horizontal top plane; it is isosceles because r is both the height and the radius of the cylinder. $\triangle ABC$ is similar to $\triangle AED$ by AAA because DE is on a horizontal cross section plane and hence is parallel to BC. Therefore, $\triangle AED$ is also an isosceles right triangle, so $DE = AD$.
   (d) In this case, the length of DE is 4 cm, so the area of the smaller disk is 16π. At this point, some of your students might see where we're going.
   (e) The length of DE is a, so the area of the smaller disk is πa^2.

   (f) The area of the annulus is the difference in the two disk areas

   $$36\pi - 16\pi = 20\pi$$

   (g) The area of the annulus is $\pi r^2 - \pi a^2$.

3. The algebraic expressions for the two cross sections are equal! By the Distributive Law for multiplication over subtraction,

   $$\pi(r^2 - a^2) = \pi r^2 - \pi a^2$$

Now all the hard work is done. You did most of it! We have just shown that the cross sections of the two shapes of Display 5.38 have the same area at every height *a*, so Cavalieri's Principle applies. That is, both shapes have the same volume!

a

5.36

Finish the work of this section by answering these questions about the shapes shown in Display 5.38. Give a reason to justify each answer.

1. What is the volume of the cylinder?

2. What is the volume of the cone that has been taken out of the cylinder?

3. What is the volume of the shape that is left after the cone is taken out?

4. What is the volume of a hemisphere of radius *r*?

5. What is the volume of a sphere of radius *r*?

b

5.37

The radius of the Earth is approximately 4000 miles. If the Earth were perfectly spherical, what would be its volume? Round your answer to the nearest 1000 cubic miles.

5.38

The Statewide Basketball Association is designing a trophy for the best amateur team in the state. It features a silver basketball 20 cm in diameter. To keep down both cost and weight, this basketball is to be a hollow shell 1 cm thick. The trophy company says that the silver they use weighs 10 grams per cm³ and costs $6 per ounce. How much will this silver basketball cost? Round your answer to the nearest dollar.

Useful Facts
1 kilogram (1000 grams) = 2.2 pounds (approx.)
1 pound = 16 ounces

Do you think you could improve on the design of this trophy basketball? If so, explain how your design change would be an improvement.

A review of important formulas

- The volume of any pyramid or cone with base area *B* (sq. units) and vertical height *h* (units) is $\frac{1}{3}Bh$ (cubic units). That is,

$$V = \frac{1}{3}Bh$$

418

Published by IT'S ABOUT TIME, Inc. © 2000 MATHconx, LLC

5.36

These questions put all the pieces together. Work through them carefully to be sure that your students understand them.

1. By the formula given in this section, the volume of the cylinder is

$$\pi r^2 \cdot r = \pi r^3$$

2. By the formula derived in this section, the volume of the cone is

$$\frac{1}{3}\left(\pi r^2\right) \cdot r = \frac{1}{3}\pi r^3$$

3. $\pi r^3 - \frac{1}{3}\pi r^3 = \frac{2}{3}\pi r^3$

4. By Cavalieri's Principle, it is the same as the previous answer; $\frac{2}{3}\pi r^3$

5. $2 \cdot \frac{2}{3}\pi r^3 = \frac{4}{3}\pi r^3$

5.37

Your students probably know that the Earth is almost spherical, but not quite. This is a simple application of the formula just derived.

$$V = \frac{4}{3}\pi \cdot 4000^3 = 268,082,573,000 \text{ cu. mi.}$$

5.38

This is mostly a series of routine computations, once the problem solving strategy is chosen. Early discussion should focus on finding a good strategy. A good approach here would be analogous to the one used in the derivation of the sphere volume formula: find the volume of a desired shape by subtracting the volume of one known shape from that of another. In this case, the two shapes are the outer sphere and the inner sphere which bound the silver shell. Once students see this as a productive approach, the rest is routine.

- The radius of the outer shell is 10 cm, so its volume is

$$\frac{4}{3}\pi \cdot 10^3 \approx 4188.79 \text{ cu. cm}$$

- The radius of the inner shell is 9 cm, so its volume is

$$\frac{4}{3}\pi \cdot 9^3 \approx 3053.63 \text{ cu. cm}$$

- The volume of silver in the ball is approximately

$$4188.79 - 3053.63 = 1135.16 \text{ cu. cm}$$

- The weight of silver in the ball is

$$10 \cdot 1136.16 = 11,361.6 \text{ g} = 11.3616 \text{ kg} \approx 25 \text{ pounds} = 400 \text{ oz.}$$

- This silver basketball would cost $2,400. (!!!)

Design improvement should focus on reducing cost and weight. Two obvious ways to do this are (1) reduce the size of the ball, and/or (2) reduce the thickness of the shell. There are common sense limitations on each of these approaches, of course. If the class is interested, this problem can be extended to have students evaluate the weight and cost of various design suggestions.

- The volume of a sphere with radius r (units) is $\frac{4}{3}\pi r^3$ (cubic units). That is,

$$V = \frac{4}{3}\pi r^3$$

Problem Set: 5.4

Round all numerical answers to one decimal place, unless instructed otherwise.

1. Find the volume of

   (a) a pyramid with a 3 by 4 by 5 inch triangular base and a vertical height of 8 inches;

   (b) a cone with its peak directly above the center of its base, a base diameter of 12 feet, and a slant height of 20 feet.

2. Find the volume of

   (a) a sphere of radius 3 meters;

   (b) a hemispherical dome 700 yards in diameter.

3. (a) A square based pyramid has *all eight* of its edges the same length: 50 m. What is its volume?

   (b) Find a general formula to solve *all* problems like part (a). That is, if a pyramid with a square base has all eight of its edges the same length, s, find a formula for its volume, V, expressed as a function of s. Use your function to solve part (a). Do you get the same answer as before?

4. Display 5.40 shows a cutout pattern for a pyramid with a square base that has its peak, P, directly above one base corner, A. The base of this pyramid measures 4 cm on a side and its vertical height (the length of AP) is 7 cm.

   (a) Calculate the lengths of all the other edges. Do not round off your answers.

   (b) What is the volume of this pyramid?

   (c) Make two full size cutouts of this pattern from heavy paper or light cardboard. Fold them into pyramids. Tape them so that they stay folded.

   (d) The volume of each copy of the pyramid described is $\frac{1}{3}$ of the volume of the rectangular box with base $ABCD$ and height 7 cm, but three copies of this pyramid cannot be arranged to form this box. Arrange your two

Published by IT'S ABOUT TIME, Inc. © 2000 MATHconx, LLC

419

Problem Set: 5.4

Note that all statements of numerical equality are rounded to one decimal place.

1. These are routine applications of the first volume formula at the end of the section. They require an extra step or two, drawing on ideas from earlier chapters.

   (a) A 3-4-5 triangle is a right triangle, so the area of the base is $\frac{1}{2} \cdot 3 \cdot 4 = 6$ sq. in. Thus, its volume is $\frac{1}{3} \cdot 6 \cdot 8 = 16$ cu. in.

   (b) The only wrinkle here is finding the vertical height. Consider the right triangle formed by a radius of the base, the vertical core line of the cone, and a hypotenuse line on the surface of the cone. The base radius is 6 ft. and the hypotenuse is 20 ft., so the vertical height, h, must satisfy $6^2 + h^2 = 20^2$. Solve this to get $h = 19.08$. Thus, the volume is $\frac{1}{3} \cdot \pi \cdot 6^2 \cdot 19.08 = 719.3$ cu. ft.

2. These are routine applications of the second volume formula at the end of the section.

   (a) $\frac{4}{3}\pi \cdot 3^3 = 36\pi = 113.0$ cu. m

   (b) The radius of the dome is 350 yds; its volume is

   $$\frac{2}{3}\pi \cdot 350^3 = 89{,}797{,}190.0 \text{ cu. yds.}$$

3. The purpose of part (b) is to connect this geometric work with algebra, and particularly with functions.

   (a) The only tricky part is calculating the vertical height. Since all edges are the same length, the peak must be directly over the center of the square base. Use the Pythagorean Theorem to get the length of a diagonal of the base, $50\sqrt{2}$ m. Half of that is the distance from a base corner to the base center. Now, using the Pythagorean Theorem again, solve

   $$h^2 + (25\sqrt{2})^2 = 50^2$$

   to get the height, $h = \sqrt{1250}$ m. Then, by the formula for the volume of a pyramid,

   $$V = \frac{1}{3} \cdot 50^2 \cdot \sqrt{1250} = 29{,}462.8 \text{ cu. m}$$

   (b) $V(s) = \frac{\sqrt{2}}{6}s^3$

4. This exercise relates the cutout patterns of Section 5.1 to the work of this section. It also strengthens students' ability to visualize objects in 3-space and to think through the implications of diagrammed relationships. You might use it as an in-class group activity.

copies on a table so that one side of the box is formed by matching two right triangle sides along their hypotenuses. The point of each pyramid will touch a base corner of the other. Can you visualize the shape of the remaining $\frac{1}{3}$ of the box?

(e) The missing part of the rectangular box is easier to visualize if you think of it in two congruent pieces, but even then it can be difficult. Each piece is a tetrahedron. Can you make a cutout pattern for this tetrahedron? Try it. A tetrahedron has four triangular sides, but they may not all be congruent. Reason your way, step by step: What must be the measurements of one triangle? How is it connected to another one? What must be the measurements of that triangle? And so on. When you think you have all the measurements, draw a cutout pattern for this tetrahedron and make two copies of it. Cut them out, fold them up, and see if they complete the rectangular box.

(f) What is the volume of each tetrahedron? How do you know?

5. The pyramid of Display 5.40 has a square base 4 cm on a side, and its peak is directly above one base corner. Its vertical height is 7 cm. Let us compare this pyramid with a pyramid of the same vertical height and base, but with its peak directly over the center of the base.

(a) What are the volumes of these two pyramids? Which pyramid has the larger volume, or are they equal?

(b) Draw a cutout pattern for the pyramid with its peak centered over the base. Mark the length of each edge.

(c) What are the surface areas of these two pyramids? Which pyramid has the larger surface area, or are they equal?

(d) If you had a pyramid of the same vertical height and base, but with its peak 5 meters to the left of the center of the square, how would its volume and surface area compare to those of the other two pyramids? Don't calculate these quantities; just answer *larger*, *smaller*, or *equal* in each case.

(e) What pattern (if any) do you see here? Do you think you could find a pyramid with the same vertical height and base, but with surface area smaller than any of

Published by IT'S ABOUT TIME, Inc. © 2000 MATHconx, LLC

(a) By the Pythagorean Theorem, $BP = DP = \sqrt{65}$ cm. If you leave this measurement in radical form, it works very nicely in the next computation. By the Pythagorean Theorem again, $CP = 9$.

(b) The volume is $\frac{1}{3} \cdot 4^2 \cdot 7 = 37\frac{1}{3}$ cu. cm

(c) The full-size pattern will fit on a standard $8.5 \times$ by 11 inch sheet. Old file folders are ideal for this, but even fairly heavy copier paper will work.

(d) When properly placed, the matched triangles on the bottom form a 4 by 7 cm rectangle and the squares are vertically upright at the ends. The missing third of this rectangular box is *very difficult* to visualize!

(e) This part requires a little patience and ingenuity, but not much sophistication. A cutout pattern appears in Display 5.19T. The measurements marked are in cm. Again, this full-size pattern will fit on a standard 8.5×11 inch sheet. Students' patterns may differ in the arrangement of the triangles, but they should all fold into the same shape.

(f) The volume of each tetrahedron is half the volume of the pyramid because together they make up the remaining $\frac{1}{3}$ of the box: $18\frac{2}{3}$ cu. cm

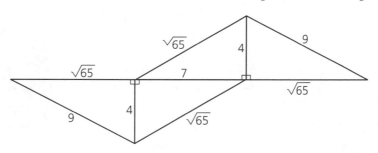

Display 5.19T

5. (a) The volumes are equal: $\frac{1}{3} \cdot 4^2 \cdot 7 = 37\frac{1}{3}$ cu. cm

(b) The sides are four congruent isosceles triangles, one on each edge of the base. The edge of each one, from base corner to peak, is the hypotenuse of a 7 by $2\sqrt{2}$ cm right triangle; it is $\sqrt{57}$ cm long.

(c) This part takes some calculation. The pattern in Display 5.40 shows that the triangular sides of that pyramid come in congruent pairs, so each pair can form a rectangle. This makes the calculations less tedious. If students did not already do problem 4, they will have to calculate the length of BP (and DP), which is $\sqrt{65}$. The surface area of this pyramid, then, is

$$4 \cdot 7 + 4 \cdot \sqrt{65} + 4^2 = 76.3 \text{ sq. cm}$$

The area of each isosceles side of the other pyramid is $2\sqrt{53}$ sq. cm. The total area of the 4 sides and the base is

$$4 \cdot 2\sqrt{53} + 4^2 = 74.2 \text{ sq. cm}$$

these three other pyramids? Explain.

(f) Do you think you could find a pyramid with the same vertical height and base and with 100 square kilometers of surface area? Explain. If you could, what would its volume be?

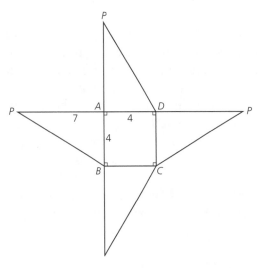

Display 5.40

6. Explorers in a remote part of northern Africa have found a metal sphere in an ancient king's tomb. It is 12 cm in diameter and appears to be pure gold.

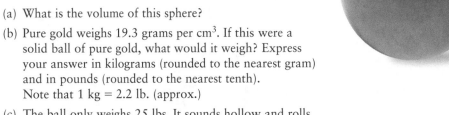

(a) What is the volume of this sphere?

(b) Pure gold weighs 19.3 grams per cm^3. If this were a solid ball of pure gold, what would it weigh? Express your answer in kilograms (rounded to the nearest gram) and in pounds (rounded to the nearest tenth). Note that 1 kg = 2.2 lb. (approx.)

(c) The ball only weighs 25 lbs. It sounds hollow and rolls as if there is no unevenness in the width of its shell. The explorers are convinced that it is pure gold. Assuming they are right, how thick is the gold shell of this sphere?

7. If you were offered your choice of

(a) a solid gold ball 2 inches in diameter, or

(b) a hollow gold ball 4 inches in diameter, but only 0.1 inch thick, which would you take? Why? Justify your answer.

421

Thus, the pyramid with the centered peak has the smaller surface area.

(d) Its volume would be the same, but its surface area would be *much* larger.

(e) It appears that moving the peak horizontally away from the vertical center line of the pyramid increases the surface area. If that's the pattern, then the pyramid with the centered peak must have the smallest surface area.

(f) Theoretically, yes. Moving the peak horizontally increases the surface area and there is no theoretical limit to the amount of that increase. Practically, no. The peak of the pyramid would have to be offset many thousands of kilometers to describe 100 sq. km of surface area. If you could make such a pyramid, its volume would be $37\frac{1}{3}$ cu. cm, just like the others.

6. Parts (a) and (b) of this exercise are completely straightforward computational exercises. Some strategic thought is required in setting up part (c), making this problem suitable for small group work. Part (c) is a good example of reasoning backwards on a known process and of the problem solving interplay between algebra and geometry.

(a) Its volume is $\frac{4}{3}\pi \cdot 6^3 = 288\pi = 904.8$ cu. cm

(b) 17.463 kg; 38.4 lbs.

(c) Assuming it is pure gold, the first step is to find the volume of the gold in cubic centimeters.

$$\frac{25}{2.2} = 11.364 \text{ kg} = 11,364 \text{ g}$$

$$\frac{11,364}{19.3} = 588.8 \text{ cu. cm}$$

Assuming the ball is hollow and of even thickness, this 588.8 cu. cm of gold is contained between two concentric spheres. The outer one has a diameter of 12 cm, so its radius is 6 cm. The radius, r, of the inner sphere is unknown. If we knew it, then $6 - r$ would give us the thickness of the shell.

Chapter 5

To find r, observe that the volume of the shell is the difference in the volumes of the spheres. Algebraically,

$$904.8 - \frac{4}{3}\pi \cdot r^3 = 588.8$$

$$\frac{4}{3}\pi \cdot r^3 = 316$$

$$\pi \cdot r^3 = 237$$

$$r^3 = 75.4$$

How your students proceed from here depends on what they know about cubes and cube roots. The direct way is to use a calculator to find the cube root. If they don't know what a cube root is, however, this will not be meaningful. They can estimate it on the calculator (or on a spreadsheet) by plugging various values for r into the function r^3. A little such experimentation should easily produce an answer correct to two or three decimal places: 4.225 cm. Thus, the thickness of the gold shell is $6 - 4.225 = 1.775$ cm.

7. (2), because it contains more gold. The volume of ball (1) is $\frac{4}{3}\pi$ cu. in. The volume of gold in ball (2) is given by

$$\frac{4}{3}\pi \cdot 2^3 - \frac{4}{3}\pi \cdot 1.9^3$$

which is

$$\frac{4}{3}\pi \cdot (8 - 6.859)$$

Since $8 - 6.859 = 1.141 > 1$, ball (2) contains more gold.

NOTES

Chapter 5

5.5 Turning Around an Axis

Learning Outcomes

After studying this section you will be able to:

Describe how to form a solid of revolution from a planar region;

Find the volumes of various solids of revolution;

Find the center of gravity of a planar region bounded by a regular polygon, rectangle, triangle, circle, or semicircle.

Another way to turn two dimensional shapes into three dimensional ones is by spinning them around. For instance, tape a square of paper to the top of a pencil, like a flag. When you spin the flag around by rubbing the pencil between your hands, the square sweeps out a cylinder in space. (See Display 5.41.) Try it!

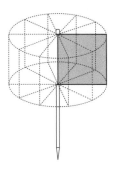

Display 5.41

 5.39

What three dimensional shape(s) would you get if you used a right triangular flag? Would it depend on which edge you tape to the pencil?

The pencil flag is a crude example of a useful idea. When a two dimensional figure (the flag) revolves around a straight line in its plane (the pencil), it passes through a part of 3-space. The shape of the region it passes through is called a **solid of revolution**. The straight line is called the **axis of revolution**, or simply the **axis**, of the figure.

Many everyday objects are shaped like solids of revolution. Some are even made that way. A baseball bat or a table leg shaped on a lathe is made as a solid of revolution. A bowl or a vase shaped on a potter's wheel is made as a solid of revolution. In such cases, the two dimensional figure does not exist at the start except, perhaps, in the woodworker's mind or the potter's hands. The outline is shaped slowly as the wood or clay turns on its axis. Nevertheless, the outline that takes shape really is a two dimensional thing. The woodworker's cutting tools stay (more or less) on a plane with the axis (as in Display 5.42), and so do the potter's hands.

5.5 Turning Around an Axis

This section is about solids of revolution. It begins by looking at some shapes that we have already developed in other ways (cones, cylinders, spheres) and then looks at common shapes that we can't describe conveniently in those other ways (e.g., doughnuts, footballs). The theoretical focus of the section is the Pappus-Guldin Theorem, which says that the volume of a solid of revolution is the product of the area of the revolved planar region times the distance traveled by the centroid (center of gravity) of that region. This leads to some basic facts about centroids of various planar shapes.

In developing these ideas, we extend two techniques from earlier sections.
• Approximating volumes by seeing what happens to "little squares of area" (from Section 5.4);

• "Nip's way vs. Tuck's way" of computing the area of an annulus from Chapter 4.

A major pedagogical objective of this section is to connect commonsense ways of thinking about three dimensional shapes with some fundamental concepts from plane geometry.

This is a fairly long section, but there is much for students to do along the way. In preparation for the Exploration about the centroid, you should get some scrap cardboard that can be cut into various planar shapes. Corrugated cardboard does *not* work well for this because the distribution of its weight over its area tends not to be very uniform. The back cardboard from writing tablets (or something like it) is much better for this.

5.39

These preliminary visualization exercises are not intended to take much time or effort. Their purpose is simply to get students to imagine in some concrete way the portion of space that a planar region passes through when it is rotated around an axis.

Yes, it depends on which edge is taped to the pencil. If either leg is taped along the pencil, the resulting figure is a cone—one pointed up, the other pointed down. If the hypotenuse is taped along the pencil, the figure is like two cones with their wide ends joined.

Assessments   Blackline Masters   Extensions   Supplements

For Additional Support Materials see page T-299

From *Cabinetmaking and Millwork*, 2nd ed., by John L. Feirer.
Copyright © 1982 by Glencoe/McGraw Hill. Reprinted by permission.

Display 5.42

A simple example of a solid of revolution and its two dimensional pattern appears in Display 5.43. The shape of the baseball bat is determined by the two dimensional region and its axis of revolution which are shown just below it. Sketch a two dimensional revolution pattern for each of the solid shapes in Display 5.44.

5.40

(a) a rolling pin
(b) a football (ignore the laces)
(c) a lightbulb (ignore the threading at its base)
(d) a spool (without thread on it)

Identify at least three other examples of solids of revolution from your everyday life. In each case, describe the axis of revolution and the planar figure that is being revolved.

5.41

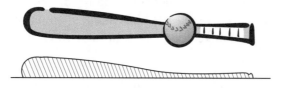

Display 5.43

(a) rolling pin   (b) football   (c) lightbulb   (d) spool

Display 5.44

5.40

There's no need to be fussy about careful drawing here, as long as students see the basic pattern of revolution. It is important that each pattern lie entirely on one side of its axis. The patterns appear in Display 5.20T.

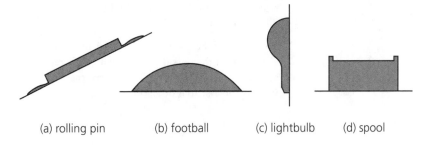

(a) rolling pin   (b) football   (c) lightbulb   (d) spool

Display 5.20T

5.41

This exercise speaks for itself. It is an (optional) extension of Do this now 5.40. If your students need some help getting started, you might suggest a styrofoam cup or a soft drink bottle. The axis in each of these cases is the line directly down through the center, and the region is half of the cross section outline of its styrofoam or glass or plastic shell.

Chapter 5

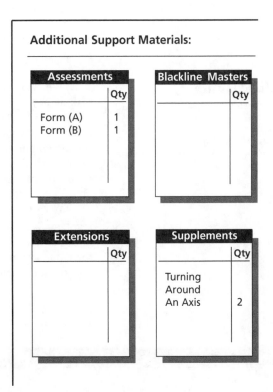

Additional Support Materials:

Assessments	Qty
Form (A)	1
Form (B)	1

Blackline Masters	Qty

Extensions	Qty

Supplements	Qty
Turning Around An Axis	2

5.42

1. Describe the solid of revolution formed by each region and axis in Display 5.45. Use a ruler and the scale given with the figure to find the approximate dimensions of each object.

2. Two pairs of these solids look identical when viewed from the outside, but they are different in an essential way. Which two pairs are they, and what is the essential difference?

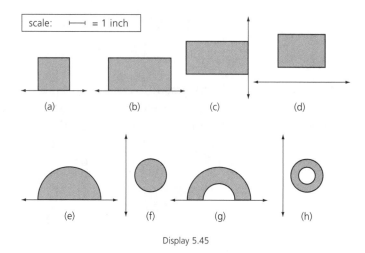

Display 5.45

As you can see, many different shapes can be made as solids of revolution. *Any* region of a plane and *any* line in that plane can be used to form one. This approach gives us a way of forming 3-space shapes that we cannot make easily by folding or stacking.

But there is something more important here than just making new figures. Thinking of figures in different ways helps us discover hidden similarities among shapes. These patterns and analogies sometimes lead to powerful, general methods for finding volumes, surface areas, and other basic features of three-dimensional shapes. The rest of this section is an example of this process at work.

We'll focus on the doughnut shaped figure formed by revolving a circle around a line that it does not intersect. If we revolve a disk as in Display 5.45(f), we get a solid three dimensional shape. If we revolve only the circle, we get just its outer shell (the glazed doughnut's sugar coating). This

Thinking Tip

Remember these thinking habits?
Look for a pattern.
Argue by analogy.
Generalize what you did.

424

5.42

This is an *essential* exercise. Several of the figures and ideas examined here suggest the theoretical basis for the rest of the section. It is particularly suitable for small group work.

1. Exact measurements are not critical; the measurement part of the question is just another way of checking to see that students are visualizing the revolution process properly. Note that unless a distortion has occurred in the printing process, the scale here is 5 mm = 1 in.

   (a) A right cylinder, lying on its side. Its base radius is 2 inches and its height is 2 inches.

   (b) A longer right cylinder, lying on its side. Its base radius is 2 inches and its height is 4 inches.

   (c) A short right cylinder, standing on its base. Its base radius is 4 inches and its height is 2 inches.

   (d) A right cylinder, lying on its side, with a hollow core. Its outer base radius is 3 inches, the radius of the core hole is 1 inch, and its height is 3 inches.

   (e) A solid (round) ball of radius 2 inches.

   (f) A doughnut (solid torus). The term *torus* will be formally introduced shortly; "doughnut" will do for now. The diameter of the cross section shown is 2 inches; the overall diameter of the doughnut is 6 inches.

   (g) A (relatively thick) round ball with a hollow center. The radius of the ball is 2 inches; the radius of its hollow core is 1 inch.

   (h) A doughnut with a hollow core. The outer dimensions of this doughnut are the same as those of (f); its wall is a half-inch thick; the diameter of its hollow core is 1 inch.

2. This question assumes that the solids are made of some kind of opaque material. The two balls, (e) and (g), look the same from the outside, but (g) has a hollow center. Similarly, the doughnuts (f) and (h) look the same, but (h) has a hollow core.

Chapter 5

shape is called a **torus**. We'll use the word for both the hollow and the solid cases. Display 5.46 is a sketch of a torus.

Display 5.46

A Go-Kart Track (Part IV)

Nip and Tuck, whom you met in Chapter 4, want to make a three dimensional sign for their Go-Kart track, a HUGE sign that can be seen from the nearby highway. They want it to look like a big, soft tire. Its shape is a torus. They have decided that the outside diameter of the tire should be 24 feet, and the diameter of the hole in the middle should be 10 feet.

The sign needs to be sturdy enough to withstand all kinds of weather and bumps from birds and small flying objects, but light enough to be supported easily. Nip and Tuck are thinking of making it out of a tough plastic "skin" filled with recycled styrofoam or fiberglass or shredded wastepaper. In order to compare the costs and weights of their options, they need to know the volume of this sign.

"I know an easy way to find the volume," says Tuck. "Pretend the tire is a big rubber hose bent around on itself. If you cut it apart and straighten it out, it's just a cylinder! Finding the volume of a cylinder is easy."

"That sounds easy, but it probably won't work," says Nip. "When you straighten the tire out, its inner ring will have to stretch, so all the measurements will change."

"But the outside ring will shrink as the inside one stretches," says Tuck. "I bet they just about balance out."

"'Just about' isn't good enough to risk money on," objects Nip. "I won't trust your easy way unless you come up with a better argument than that."

Published by IT'S ABOUT TIME, Inc. © 2000 MATHconx, LLC

425

NOTES

5.43

1. Do you understand Tuck's idea? If so, explain it in your own words. If not, ask a question about what puzzles you.

2. Do you understand Nip's objection? If so, explain it in your own words. If not, ask a question about what puzzles you.

3. Which of the twins do you think is correct, Nip or Tuck? What argument can you give to someone who disagrees with you?

Can we find a better argument for Tuck, or is Nip right to be doubtful? Let's see which one we should believe by looking at a simpler, but analogous, figure. Instead of revolving a circle around an axis to form the torus, we'll revolve a square around an axis, as shown in Display 5.47. The measurements in this figure are given in feet.

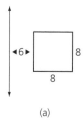
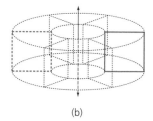

(a)   (b)

Display 5.47

These questions refer to the solid of revolution in Display 5.47(b).

5.44

1. Describe its shape. What are its dimensions?

2. How do you think Nip would find its volume? Do it that way.

3. How do you think Tuck would find its volume? How is Display 5.48 related to Tuck's way? What is the length marked d? Find the volume that way.

Published by IT'S ABOUT TIME, Inc. © 2000 MATHconx, LLC

5.43

This is a brief class discussion to clarify the question and to set up the need for the argument that follows. Its main purpose is to let you, the teacher, help the students visualize clearly the cutting and straightening out process that was described in the loose language of the dialogue.

5.44

Be prepared to spend some time on these questions. They are an important step toward the main result of the section.

1. It is a right cylinder with a hollow core. Its base is an annulus with inner radius 6 feet and outer radius 14 feet. It is 8 feet high.

2. Nip would not distort the figure. By Cavalieri's Principle, its volume is the area of the base annulus times the height. The area of the base annulus is
$$\pi \cdot 14^2 - \pi \cdot 6^2 = \pi(196 - 36) = 160\pi \text{ sq. ft.}$$
Thus, the volume is $8 \cdot 160\pi = 1280\pi$ cu. ft.

3. Tuck would turn the cylinder into a rectangular box by cutting it and unbending it, as shown in Display 5.48. The only real problem with this is deciding how long the box is. By analogy with Tuck's approach to the area of an annulus, it makes sense to use the middle line of the annular base as the length, assuming that the stretching on one side of it will be matched by the shrinking on the other. That center line is a circle of radius 10 feet, so its length, d, is 20π. Thus, the volume of the box is

$$8 \cdot 8 \cdot 20\pi = 1280\pi \text{ cu. ft.}$$

4. Since both approaches yield the same answer for this figure, it is tempting to believe Tuck's approach. Nevertheless, the difference between a square and a circle might be significant here, so this example does not guarantee that Tuck's approach works for the torus. That is, (a) is more likely, but not certain; strictly speaking, (c) is the best answer.

Chapter 5

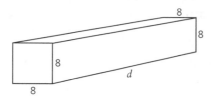

Display 5.48

4. Based on your answers to questions 2 and 3, choose one of these.
   (a) Tuck's approach would work for the torus.
   (b) Nip's objection is correct.
   (c) There is enough difference between this figure and the torus that we can't be sure.

   Justify your choice.

Putting Algebra to Work

By using the language of algebra, we can show that Nip's way and Tuck's way come out the same for *any* square and *any* axis. That is,

(*)   When a square is revolved around a line parallel to one of its sides, the volume of the solid formed equals the area of the square multiplied by the distance traveled by its center.

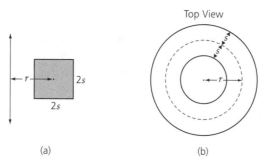

(a)   (b)

Display 5.49

A picture of the situation described by Statement (*) appears in Display 5.49. It is labeled with letters, rather than with specific numbers. Why? Use these letters and the language of algebra to answer the following questions about the solid of revolution described by this figure.

5.45

Published by IT'S ABOUT TIME, Inc. © 2000 MATHconx, LLC

5.45

This is an exercise in some elementary algebraic techniques, in the context of proving a geometric statement. Both aspects are important. Students should be gaining some facility in expressing ideas algebraically and in using routine algebraic procedures routinely. This kind of exercise helps them to see that the manipulative procedures of algebra are not ends in themselves, but are convenient ways of identifying and confirming patterns, relationships, and the like.

Letters are used instead of numbers to guarantee that the process works for *any* numbers and doesn't depend on some particularly lucky (or unlucky) choice in an example.

1. Nip's way
   (a) $r - s$
   (b) $r + s$
   (c) $\pi(r + s)^2 - \pi(r - s)^2$
   (d) This part provides much of the algebra practice. Help your students to work through these steps.
   $$\text{Area} = \pi(r + s)^2 - \pi(r - s)^2 \quad \text{(difference of disks)}$$
   $$= \pi((r + s)^2 - (r - s)^2) \quad \text{(Distributive Law)}$$
   $$= \pi((r^2 + 2rs + s^2) - (r^2 - 2rs + s^2)) \quad \text{(Distributive Law, used four times)}$$
   $$\text{Area} = \pi(4rs) \quad \text{(distribute, collect like terms, etc.)}$$
   (e) $4\pi rs \cdot 2s = 8\pi rs^2$

2. Tuck's way
   (a) $2s$. It was chosen this way (i.e., rather than just s) to make it easier to deal with the midpoint of the square. This way, there is no need to deal with halves and related fractions.
   (b) $(2s)^2 = 4s^2$
   (c) r
   (d) $2\pi r$
   (e) Multiply them together: $4s^2 \cdot 2\pi r = 8\pi rs^2$

3. Students who do the algebra correctly might end up with terms in different orders, for instance. Commutativity of multiplication fixes this immediately. Other discrepancies suggest that they recheck their work.

Chapter 5

1. Nip's way.
   (a) What is the radius of the inner circle of revolution?
   (b) What is the radius of the outer circle of revolution?
   (c) What is the area of the base annulus?
   (d) The area of the base annulus can be written as the product of four numbers or single letters. $\_ \cdot \_ \cdot \_ \cdot \_$. Can *you* do that? If you can, explain how you got your answer.
   (e) Using your answer to (d), write a simple expression for the volume of the figure.

2. Tuck's way.
   (a) What is the side length of the square cross section? Why do you think that the notation was chosen this way?
   (b) What is the area of the square cross section?
   (c) How far is the center of the square cross section from the axis of revolution?
   (d) How far does that center point travel when it is revolved?
   (e) How would Tuck use your answers to (b) and (d) to find the volume of the figure? Do it.

3. Compare your results from parts 1 and 2. Are they the same? They should be. If not, can you make them the same without breaking any algebraic rules?

Tuck's idea seems OK so far, but squares are not circles. To connect what we've done here to Nip and Tuck's problem, we need to generalize Statement (*) a bit.

(5.46) Suppose you revolve a 3 by 1 (any unit of measure) rectangle around an axis, as shown in Display 5.50(a). What shape do you get? Think of this process as the revolution around the same axis of three unit (1 by 1) squares, as shown in Display 5.50(b).

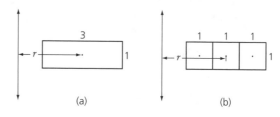

(a)   (b)

Display 5.50

5.46

This part sets up the idea of the centroid of a planar figure, an essential ingredient in the Pappus-Guldin Theorem. Even if you do not do the algebraic steps in detail, make sure your students see the basic idea that there is a balance point for all the centers of all the squares that make up the planar region. Intuitively, this balance point must be a point of symmetry with respect to the centers of pairs of squares. Its distance from the axis is the mean distance of the centers of all the same sized squares that make up the region.

The shape of the solid described by Display 5.50 is a short cylinder with a hollow core. Its base is an annulus. Its wall is 3 units thick; its height is 1 unit.

In regard to $r - 1$ and $r + 1$: Help students to see that, if the distance from the axis to the center of the middle square is r, then the distance to the nearer square is $r - 1$ and the distance to the farther square is $r + 1$.

You might want to point out to your students that this entire section is being developed along the lines of the Thinking Tip earlier in this section: We are *generalizing*, step by step, by looking at *analogous* earlier work and *patterns* in the results that emerge.

NOTES

Chapter 5

Using Statement (*), we can get the volume of this hollow-core cylinder by adding the results of revolving each unit square around the axis. The area of each unit square is 1, so the total volume is

$$V = 2\pi(r - 1) + 2\pi r + 2\pi(r + 1)$$

Where do $r - 1$ and $r + 1$ come from? Factoring out the common term 2π, we have

$$\begin{aligned} V &= 2\pi((r - 1) + r + (r + 1)) \\ &= 2\pi \cdot 3r \\ &= 2\pi r \cdot 3 \end{aligned}$$

That is, the result is the same as multiplying the area of the *entire* rectangle 3 by the distance traveled by its center $2\pi r$.

But you knew that, right? At least, you could have figured it out easily in another way. So, why did we bother with this argument? The answer is *because it generalizes to shapes that we can't handle easily in other ways*. The key to this generalization is the idea of a balance point for a two dimensional region. The balance point of the 3 by 1 rectangle in Display 5.50 (a) is the center of the middle square. If the rectangle were a 3 inch by 1 inch piece of cardboard, it would balance on a pencil point at that spot.

Words to Know: The balance point of a planar region is called its **center of gravity**, particularly in physical science. In mathematics, it is also called the **centroid** of the region.

In earlier sections, we used the fact that area can be approximated as closely as we please by using smaller and smaller squares. That fact is useful here, too. We can think of any region as being made up of tiny squares in such a way that each square is *balanced* by another square. The centers of the two balanced squares are on the same straight line through the center of gravity, equal distances away, on opposite sides.

Published by IT'S ABOUT TIME, Inc. © 2000 MATHconx, LLC

429

NOTES

Chapter 5

a

5.47

1. Where is the center of gravity of a disk? Of a square region? Of a rectangular region? Of a region bounded by a regular polygon?

2. Display 5.51 shows a rough approximation of a disk by squares. The center of gravity is marked with a dot. Pair up each numbered square with the square that balances it; list all the pairings. Can you describe the pairings by a numerical pattern? If so, how?

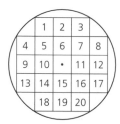

Display 5.51

All our work so far in this section leads to a very important fact.

A Fact to Know: The volume of a solid of revolution is the product of the area of the revolved planar region times the distance traveled by the centroid (center of gravity) of that region.

This fact is known as the **Pappus-Guldin Theorem.** It is named after two people who discovered it independently of each other – Pappus, a Greek mathematican of the 3rd century A.D., and Paul Guldin, a Swiss scholar of the 16th century. A rigorous proof of this theorem is best done with calculus, but our work with approximations by small squares should give you a good idea of why it is true.

b

5.48

Tuck was right.

1. Use the Pappus-Guldin Theorem to find the volume of Nip and Tuck's big tire sign.

2. This approach to finding the volume of a torus is analogous to Tuck's way for finding the area of an annulus described in Chapter 4. Explain this analogy.

430

5.47

The first question is just a way of matching intuition with vocabulary. The second question checks to see if students are envisioning the balancing picture correctly.

1. The center of gravity of a disk is its center. The center of gravity of a square or any region bounded by a regular polygon of the center of its circumscribing circle. This is also true for any rectangle, but it probably is easier to think of the centroid of a rectangle as the intersection of its two main diagonals.

2. The pairings are (1, 20), (2, 19), (3, 18), (4, 17), (5, 16), (6, 15), (7, 14), (8, 13), (9, 12), (10, 11). Each pair adds up to 21.

5.48

The first question is for everyone. The second question (challenging for many students) is optional, but is strongly recommended for some students. It serves to sharpen the students' sense of analogy and to develop their abstract thinking skills.

1. The tire sign is a torus, which is made by revolving a disk. By the measurements given earlier in this section, the diameter of the disk must be 7 feet, so its radius is 3.5 feet. You might suggest that your students make a sketch and label it with the various measurements as they go along. Thus, the area of the disk is $\pi \cdot 3.5^2 = 12.25\pi$ sq. ft. The center of this disk (its center of gravity) is $5 + 3.5 = 8.5$ feet away from the axis of revolution, so it travels $2\pi \cdot 8.5 = 17\pi$ feet in one revolution. Thus, the volume of the tire sign is

$$12.25\pi \cdot 17\pi = 208.25\pi^2 \approx 2055.3 \text{ cu. ft.}$$

2. The annular region is the two dimensional analogue of the (solid) torus. It can be formed by revolving a line segment about a point not on the segment, but on the same line. In this case, the midpoint of the line segment is its center of gravity, and the length of the line segment is analogous to the area of the disk.

Chapter 5

The Pappus-Guldin Theorem works even when the region used to form the solid of revolution is not symmetric in any way. If you can find the center of gravity of a region, you can compute the volume of any solid of revolution made from it.

We could spend lots of time working on finding the centers of gravity of various planar regions, but that would distract us from the purpose of this chapter, the discussion of three dimensional shapes. Instead, we shall look only at two special cases: triangular regions and semidisks. It's time for an Exploration.

EXPLORATION

For this Exploration you'll need a piece of fairly heavy plain cardboard, a long pencil with an eraser, a ruler, a compass and a pair of scissors.

(5.49)

1. Use your compass and ruler to outline on the cardboard a semidisk of radius 10 cm. Position it at one end of your cardboard sheet, so that a large part of the cardboard is left for the rest of the experiment. Cut out the semidisk carefully and put it aside for later.

2. On the remaining piece of cardboard, draw a large triangle. It can be any shape you want. Cut it out carefully.

3. Experiment until you find the point where your triangular region balances on the eraser of your pencil. Mark that point.

4. Hold your pencil or ruler horizontal. Balance the triangle on your pencil so that one vertex is right on top of the pencil or ruler. Draw the line that best approximates the balance line through that vertex.

5. Repeat the previous step for the other two vertices. What do you think we are trying to get you to notice? Why do you think it makes sense?

6. Measure the distances from the point where each balance line intersects a side to the vertices at the ends of that side. Write down your measurements. Do you notice anything special about the numbers you get? If you do, can you turn it into a conjecture about the property of balance lines?

7. Measure the distances from the center of gravity to each end of each balance line. Write down your measurements in pairs, one pair for each balance line. Do you notice

Published by IT'S ABOUT TIME, Inc. © 2000 MATHconx, LLC

431

This is a hands-on experience with centers of gravity of planar regions. It can be done in small groups, rather than individually, if you prefer. Each student or group doing the experiment will need a piece of stiff cardboard about the size of a standard $8\frac{1}{2} \times 11$ inch sheet of paper.

1. Placing the compass point in the middle of one end of the $8\frac{1}{2} \times 11$ inch sheet will work fine, with a little margin for error. The semicircle will be used *after* this Experiment, as part of the closing problem of the section. We suggest that it be cut here just for efficiency's sake.

2. A large triangle behaves better when balanced. If it is too large, however, students might need to use a ruler, instead of a pencil, for the balancing in parts 4 and 5. No specific dimensions are needed.

3. This is self explanatory. In preparation for future steps, advise your students always to mark their triangles on the same side, so that they can see how the marks relate to each other.

4. For large triangles, a ruler might be necessary here, but it is not as forgiving when students are trying to find the balance line. You might remind the too conscientious ones that all we want is an approximation.

5. We want them to notice that the balance point (center of gravity) is on all three balance lines, and hence the three balance lines all intersect at the center of gravity. The underlying intuition here is that the balance point for the region must be on *every* balance line for that region.

6. The measurements from the intersection point to each endpoint of the side should be very close to being equal—perhaps off by 1 or 2 millimeters, depending on the students' care in balancing and drawing. The general conjecture is that the balance lines bisect the sides.

7. This time the general property is less obvious. The measurements should approximate the fact that the ratio of the centroid vertex distance to the centroid midpoint distance is 2 to 1. That is, on each median the centroid of a triangle is $\frac{2}{3}$ of the way from the vertex to the midpoint of the opposite side.

Chapter 5

anything special about the numbers you get? If you do, can you turn it into a conjecture about property of balance lines? If you don't, divide the smaller number of each pair into the larger one, and look again.

Let's summarize our observations. First, here's a word that will make things a little simpler to write down.

A Word to Know: A line segment between a vertex of a triangle and the midpoint of its opposite side is called a **median** of the triangle.

The word median makes it easy to state three important facts suggested by the Exploration.

Facts to Know:
- Each median of a triangle is a balance line for that triangular region.
- All three medians meet at the the center of gravity of the triangular region (the centroid of the triangle).
- On each median, the centroid is $\frac{2}{3}$ of the way from the vertex to the midpoint of the opposite side.

If a solid of revolution is made from a triangular region, then its volume can always be found by the Pappus-Guldin Theorem. Sometimes the computations are annoyingly fussy, but they are never impossible. If you know enough about the triangle to determine all its sides and angles, then you know enough to determine its centroid and its area. These two facts, along with the location of the axis of revolution, are enough to determine the volume and the surface area. An example appears as problem 6 in the Problem Set.

Are you wondering why we asked you to cut out a cardboard semidisk? It's to help with this problem:

Iris Ironhands is an artist who specalizes in large metal sculpture. She is working on a piece for the main hall of the Steelworkers' Union International headquarters. The piece features a huge stainless steel semidisk, perfectly flat and of uniform thickness, which is to be hung from a single steel

Published by IT'S ABOUT TIME, Inc. © 2000 MATHconx, LLC

NOTES

Chapter 5

cable held by a large crane. The semidisk has a radius of 10 feet and weighs more than a ton. In order to hang parallel to the ground, it must be fastened to the cable precisely at its center of gravity. Since it is so heavy, there is no easy way to find this point by trial and error. Where should Iris place the cable hook?

Use your cardboard semidisk to find an approximate answer for Iris by answering these questions

a

5.50

1. Your cardboard semidisk is a scale model of the stainless steel semidisk. What is the scale?

2. A line of symmetry will be a balance line for the semidisk. Find one and mark it on your semidisk.

3. Try balancing your semidisk on the eraser end of your pencil until you find the center of gravity. Then mark that point. What is the distance from the center of gravity to the midpoint of the semidisk's straight edge?

4. Where should Iris place the cable hook on the real semidisk? Do you think your answer is close enough so that the semidisk won't tilt?

"Answers found by marking and measuring a cardboard model aren't close enough," says Iris. "Can't you use your mathematics to get me a more reliable answer?"

Can you? If we tell you that the trick is to think about the Pappus-Guldin Theorem backwards and then use a little algebra, can you see how to get an exact answer for Iris? Here are some questions to help you along.

b

5.51

1. If you revolve a semidisk around its straight edge, what solid shape do you get?

2. If the semidisk has a radius of 10 feet, what is the volume of the solid shape? *Hint:* Look back at something you learned in Section 5.4.

3. How is the measurement that Iris needs related to this solid of revolution?

4. What is the area of the semidisk?

433

5.50

These questions should be done in class before going on to the end of the section.

1. 1 cm = 1 foot

2. The only line of symmetry is the perpendicular bisector of the straight edge.

3. Students should recognize right away that the balance point will lie along the line of symmetry. Reinforce that recognition. The measurement should come out somewhere between 4 and 4.5 cm along the line of symmetry.

4. According to the previous answer and the scaling factor, Iris should place it somewhere between 4 and 4.5 feet from the midpoint of the straight edge, along the line of symmetry. The inaccuracies of cutting cardboard, balancing, measuring and marking make it highly unlikely that any particular student answer will be accurate enough. Reinforce this by comparing the answers from different students or groups. You should find a variability of nearly half a centimeter among them, which translates into almost 6 inches of uncertainty in the real situation.
This uncertainty leads to the final punch line of the section.

5.51

This is a very nice use of the Pappus-Guldin Theorem to get a result that can't easily be found in other elementary ways. Be prepared to spend a little time with this. It provides a striking example of the power of algebra in a geometric setting.

1. You get a solid sphere.

2. By the formula for the volume of a sphere, $V = \frac{4}{3} \pi \cdot 10^3 = \frac{4000}{3} \pi$ cu. ft.

3. The vertical distance from the midpoint of the straight edge to the center of gravity is the radius of revolution for this sphere, considered as a solid of revolution. Call it r, as usual.

4. The area of the semidisk is $\frac{1}{2} \cdot \pi \cdot 10^2 = 50\pi$ sq. ft.

5. Use the Pappus-Guldin Theorem to set up an equation in which this measurement is the only variable.

6. Solve the equation. Get an exact answer, and then get an approximate answer that is accurate to 0.001 feet.

Problem Set: 5.5

1. Match each planar region in Display 5.52 with the three dimensional object in Display 5.53 that most closely resembles the solid of revolution generated by it. The planar shapes are not necessarily drawn to the same scale or orientation.

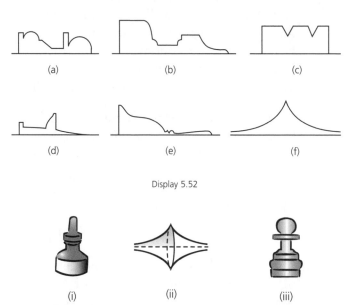

(a)   (b)   (c)

(d)   (e)   (f)

Display 5.52

(i)   (ii)   (iii)

(iv)   (v)   (vi)

Display 5.53

434

5. By the Pappus-Guldin Theorem, $V = 50\pi \cdot 2\pi r$, so $\frac{4000}{3}\pi = 50\pi \cdot 2\pi r$

6. $\frac{4000}{3}\pi = 100\pi^2 r$

$$\frac{40}{3} = \pi r$$

$$\frac{40}{3\pi} = r$$

Using the calculator, we get $r = 4.244$ feet (approximately).

Problem Set: 5.5

1. (a) and (iii); (b) and (i); (c) and (iv); (d) and (vi); (e) and (v); (f) and (ii).

2. Unless a distortion has occurred in the printing process, the scale here is 5 mm = 1". Exact measurements are not critical. The figures were designed to be described in terms of the nearest half-inch.

   (a) The region represents a 1.5" square. The solid of revolution is a cylinder, 1.5" high, but lying on its side. It has a diameter of 4" and a 1" diameter hole at its core.
   (b) This is a hollow cylinder 4" long (like a piece of thick walled pipe). Its wall is 1" thick and its outside diameter is 4". The diameter of its hollow core is 2".
   (c) This is another short, fat cylinder with a hollow core. It is 2" high, its outside diameter is 8", and the diameter of its hollow core is 2".
   (d) This is another cylinder with a hollow core. Notice that the rectangle that generates it has exactly the same dimensions as the one in (c). This cylinder is 3" high, but is lying on its side. Its outside diameter is 7" and the diameter of its hollow core is 3".
   (e) This is a torus, but its center hole is closed up like a doughnut that swelled too much when it was cooked. The diameter of the disk is 3"; the outside diameter of the torus is 6".
   (f) This is also a torus. It is formed from the same size disk as in (e), but is larger. Its outside diameter is 8"; the diameter of the doughnut hole is 2".
   (g) This is the inner half of the doughnut of (f), except, of course, that it's not really a half. That fact is the subject of a later problem.
   (h) This is the outer half of the doughnut of (f).

3. This problem is an exercise in observation. The obvious candidate pairs (because the areas are equal) are (c) and (d), (e) and (f), and (g) and (h). Of these, only the centroids of (c) and (d) are equally distant from the axis of revolution, so that's the correct pair.

2. Describe the solid of revolution formed by each region and axis in Display 5.54. Use a ruler and the scale given with the figure to find the approximate dimensions of each object.

3. Without calculating any of the volumes, pick out which two of the solids of Display 5.54 have exactly the same volume. Justify your choice. *Hint:* Think about the Pappus-Guldin Theorem.

4. Using the Pappus-Guldin method, calculate the volumes of the solids of revolution formed from parts (a)–(f) of Display 5.54. In each case, begin by identifying the center of gravity (the centroid) of the planar region. Then specify its distance from the axis of rotation. State your answer exactly (in terms of π) and then round to the nearest tenth of an inch.

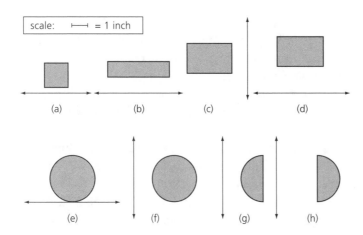

Display 5.54

5. Ginny Gemstone, a jeweler, has been asked to make a doughnut shaped gold pendant for the president of Dunk'em Donuts. It is to be made of pure gold, which is very heavy and very expensive. Pure gold weighs 19.3 grams per cubic centimeter and costs Ginny $14 per gram. Her customer wants it to be about the size of a real Dunk'em Donut: 12 cm across its outside diameter and 4 cm thick. In order to make the pendant light enough to wear comfortably, Ginny will have to make it as a hollow shell. However, pure gold is soft, so a thin shell will be easy to damage.

(a) What is the volume of a solid doughnut of this size?

(b) If Ginny makes the shell of the pendant 5 mm thick,

Published by IT'S ABOUT TIME, Inc. © 2000 MATHconx, LLC

435

4. This problem presumes that students know the dimensions of these planar regions, from problem 2. If they don't, they will have to measure them here, using the given scale. In (a)−(d), the centroid is the intersection of the diagonals; in (e) and (f), it is the center of the disk.

(a) The centroid of this 1.5″ square is 1.25″ from the axis.
The volume is $1.5^2 \cdot 2\pi \cdot 1.25 = 5.625\pi \approx 17.7$ cu. in.

(b) The centroid of this 4″ × 1″ rectangle is 1.5″ from the axis.
The volume is $4 \cdot 2\pi \cdot 1.5 = 12\pi \approx 37.7$ cu. in.

(c) The centroid of this 3″ × 2″ rectangle is 2.5″ from the axis.
The volume is $6 \cdot 2\pi \cdot 2.5 = 30\pi \approx 94.2$ cu. in.

(d) The centroid of this 3″ × 2″ rectangle is 2.5″ from the axis.
The volume is $6 \cdot 2\pi \cdot 2.5 = 30\pi \approx 94.2$ cu. in.

(e) The centroid of this 1.5″ radius disk is 1.5″ from the axis.
The volume is $\pi \cdot 1.5^2 \cdot 2\pi \cdot 1.5 = 6.75\pi^2 \approx 66.6$ cu in.

(f) The centroid of this 1.5″ radius disk is 2.5″ from the axis.
The volume is $\pi \cdot 1.5^2 \cdot 2\pi \cdot 2.5 = 11.25\pi^2 \approx 110.0$ cu in.

5. The volume of the torus shaped shell can be found by subtracting the volume of the hollow torus shaped inside from the volume of the whole doughnut. It can also be found by revolving an annulus around the axis.

(a) The larger torus (the whole donut) is formed by revolving a disk of radius 2 cm in such a way that the outermost diameter is 12 cm. You might suggest that students make a sketch to help them visualize this. This places the center of the disk 4 cm from the axis. Thus, the volume is $4\pi \cdot 8\pi = 32\pi^2$ cu. cm.

(b) The disk that is rotated to form the hollow core is concentric with the larger disk, so its center is also 4 cm from the axis. Its radius is smaller by 0.5 cm, the thickness of the shell. Its volume, then, is $1.5^2\pi \cdot 8\pi = 18\pi^2$ cu. cm. Subtract this from the answer to part (a) to get the volume of gold in the shell — $14\pi^2$ cu. cm ≈ 138.17 cu. cm. That's about 2667 grams of gold, which would cost Ginny $37,338. Moreover, since a kilogram is a little more than 2.2 pounds, this gold doughnut would weigh almost 6 pounds! This seems to be a *terrible* idea for a pendant.
A more efficient way to get the volume of the shell is to think of it as formed by an annulus with area $2\pi \cdot 1.75 \cdot 0.5 = 1.75\pi$ sq. cm. The centroid is the same (the center of the annulus), so the volume is $1.75\pi \cdot 8\pi = 14\pi^2$ cu. cm.

(c) This part mimics part (b) to emphasize what changes and what stays the same when the shell thickness is altered. The width of the annulus is 0.2 cm in this case, so its area is $2\pi \cdot 1.9 \cdot 0.2 = 0.76\pi$ sq. cm. The volume of the shell is $0.76\pi \cdot 8\pi = 6.08\pi^2$ cu. cm ≈ 60 cu. cm. That's about 1158 grams of gold, which would cost Ginny $16,212 and would weigh about 2.5 pounds.

Chapter 5

how much gold will she use? How much will the pendant weigh? How much will the gold cost her? Is this thickness a good idea?

(c) If Ginny makes the shell of the pendant 2 mm thick, how much gold will she use? How much will the pendant weigh? How much will the gold cost her? Is this thickness a good idea?

(d) The customer is willing to pay $10,000 for this pendant, provided it is at least 1 mm thick. Ginny wants to make at least $1,500 for her work and other materials. Should she take the job? Why or why not?

6. Toontown Trucking Co. is installing a large fuel tank in a level field next to its main terminal. The tank is a right cylinder 30 feet in diameter, mostly buried underground. The top 10 feet of the cylinder are above ground. Local safety regulations require that the part above ground be surrounded by a collar of earth that slopes down from the top of the tank to the level of the field. The diagram in Display 5.55 is a vertical cross section through the center of the tank and the earthwork around it. The 30° angle shown is the steepest angle that the bulldozer operator will allow. Your job is to estimate the number of cubic yards of earth that will be needed for this earthen collar.

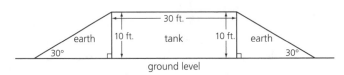

Display 5.55

(a) The earth needed is the volume formed by revolving a triangular region around the center line of the cylinder. What is the area of the triangle? Do you have enough information to find it? Is the triangle determined? Explain your answer.

(b) The radius of revolution depends on finding the center of gravity of the triangle. We have not shown you a way to calculate this, so estimate it by cutting out a scale model cardboard triangle and finding its balance point. Then use it to find the approximate radius of revolution. Estimate your answer to one decimal place.

436

(d) It might be tempting for some students to say that, since $8,500 is more than half of $16,212, a shell only half as thick can be made for $8,106. However, this is not quite right. For a 1 mm shell, the volume of the shell is $2\pi \cdot 1.95 \cdot 0.1 \cdot 8\pi = 3.12\pi^2$ cu. cm ≈ 30.8 cu. cm. That's about 594.5 grams of gold, which would cost Ginny $8,323. Yes, she should take the job; she'll make $1,677 for her work and other materials.

6. This problem reaches back to use ideas about triangles that were developed in Chapters 2 and 3. It is fairly demanding, in that it requires students to use a variety of techniques in one setting. However, no single part is particularly difficult.

(a) This is a right triangular region with another angle and one side specified, so it is determined. The trigonometric functions on a graphing calculator make it fairly easy to find the lengths of the hypotenuse and the horizontal side, and then the area: $\sin 30° = 0.5$, so 10 feet is half of the hypotenuse, which must be 20 feet.
The length of the horizontal side, then, is $20 \cos 30° \approx 17.32$ ft.
The area is approximately $\frac{1}{2} \cdot 17.32 \cdot 10 = 86.6$ sq. ft.

(b) The center of gravity of the triangle occurs at a point that is predictable (the intersection of the medians) but is not conveniently measurable, anyway. This hands-on activity, if done carefully, should provide a reasonably good estimate of its location. A scale of 1 cm = 1 ft. works well with ordinary cardboard. The center of gravity is about 6.0 ft. in from the vertical side. Thus, the approximate radius of revolution is $15 + 6.0 = 21.0$ ft.

(c) The approximate volume is $86.6 \cdot 2\pi \cdot 21 \approx 11,427$ cu. ft. One cubic yard is 27 cu. ft., so the approximate volume of earth needed is $\frac{11,427}{27} \approx 423$ cu. yds.

(d) This part just repeats the calculations of part (c) for two different radii of revolution, determined by adjusting the distance between the center of gravity of the triangle and its vertical side by 0.3 feet one way and then the other.

$$86.6 \cdot 2\pi \cdot 20.7 \approx 11,263 \text{ cu. ft.} \approx 417 \text{ cu. yds.}$$
$$86.6 \cdot 2\pi \cdot 21.3 \approx 11,590 \text{ cu. ft.} \approx 429 \text{ cu. yds.}$$

Thus, an error of 0.3 ft. in locating the center of gravity leads to a discrepancy of 6 cu. yds. of volume.

7. This exercise is a straightforward generalization of Do this now 5.51, the final result of the section. By working through the steps of Do this now 5.51 with s substituted for 10 feet, students should find this result fairly easily. The algebraic steps provide some practice in simplifying equations.

Chapter 5

(c) Use your results from parts (a) and (b) to calculate the approximate volume of earth, in cubic feet. Then convert your answer to cubic yards because that's the unit of measure commonly used by suppliers and truckers of earth fill. Start by calculating the number of cubic feet in one cubic yard.

(d) Your estimate in part (b) came from balancing a piece of cardboard, so it might not be very accurate. Assuming that it might be off by as much as 0.3 feet one way or the other, calculate lower and upper estimates of the amount of earth required (in cubic yards).

7. Construct a formula for finding the centroid of a semidisk of radius s. Then use your formula to locate the centroid of a semidisk of radius 7 cm. To check your work, cut out a cardboard semidisk of this size and see if it balances at the point predicted by your formula. *Hint:* Look back at the Iris Ironhands problem.

8. The solids of revolution described by parts (g) and (h) of Display 5.55 are the inner and outer "halves" of the torus described by part (f), except that they are not really halves. That is, neither figure has a volume that is exactly 50% of the volume of the whole torus. Find the volumes of the two solids generated by these semidisks. Round your answers to one decimal place. Then express each volume as a percentage of the volume of the whole torus.

Published by IT'S ABOUT TIME, Inc. © 2000 MATHconx, LLC

The volume of the solid sphere (of revolution) is $\frac{4}{3}\pi s^3$ and the area of the semidisk is $\frac{1}{2}\pi s^2$. If r is the (unknown) radius of revolution, then

$$\frac{4}{3}\pi s^3 = \frac{1}{2}\pi s^2 \cdot 2\pi r$$

This needs to be solved for r. Several things cancel easily, leading to $r = \frac{4s}{3\pi}$.

Students must understand that r is the distance from the center along the perpendicular bisector of the straight edge of the semidisk. Checking this understanding is the purpose of the cut-out-and-balance question. A semidisk of radius 7 cm should balance at the point

$$\frac{4 \cdot 7}{3\pi} = \frac{28}{3\pi} \approx 2.97 \approx 3 \text{ cm}$$

from the center along that perpendicular bisector.

8. This problem may be too demanding for some students, unless it is done with group discussion and/or teacher assistance. It requires students to reexamine Do this now 5.51 or the problem just before this one. They must recognize that the centroid of a semidisk occurs along the perpendicular bisector of the straight edge at a distance $\frac{4}{3\pi}$ of the radius from the center. The radius of the disk in (f) is 1.5 in. Its center, and the midpoint of the straight edges of the semidisks in (g) and (h), is 2.5 in. from the axis. Thus, the volume of the full torus is

$$\pi \cdot 1.5^2 \cdot 2\pi \cdot 2.5 = 11.25\pi^2 \approx 111.0 \text{ cu. in.}$$

Students who did the last part of problem 4 already computed this.

The centroid of semidisk (g) is $\frac{4}{3\pi} \cdot 1.5 = \frac{2}{\pi}$ *closer* to the axis; that is,

its distance from the axis is $2.5 - \frac{2}{\pi} \approx 1.86$ in. Thus, the volume of the solid is approximately

$$\left(\frac{\pi}{2}\right)1.5^2 \cdot 2\pi \cdot 1.86 \approx 41.3 \text{ cu. in.}$$

This is only 37.2% $\left(= \frac{41.3}{111}\right)$ of the volume of the entire torus.

The approximate volume of the other piece, (h), can be obtained by simple subtraction: $111.0 - 41.3 = 69.7$ cu. in., 62.8% of the volume of the torus.

For students who work this one out directly from the centroid of semidisk (h): The centroid of this semidisk is $\frac{2}{\pi}$ farther from the axis, so its distance from the axis is approximately 3.14 inches. Therefore, the volume of the solid is

$$\left(\frac{\pi}{2}\right)1.5^2 \cdot 2\pi \cdot 3.14 \approx 69.7 \text{ cu. in.}$$

Chapter 5

5.6 Space by the Numbers

Learning Outcomes

After studying this section you will be able to:

Describe locations in 3-space by using coordinates;

Use the coordinates of two points in 3-space to find the distance between them.

5.52

Earlier In MATH *Connections* you learned about Cartesian coordinate systems for the plane. What are they? Answer these questions about them.

1. Why are they called *Cartesian*?

2. What is another name for this kind of system?

3. What is a *point* in such a system?

4. What is a *straight line* in such a system?

5. What is a coordinate?

6. What is an axis? How many are there?

7. How is a Cartesian coordinate system like the Window of a graphing calculator? How is it different?

In this chapter, we have been talking about how to generalize the ideas of 2-space to our three dimensional world. One of the most powerful ways of doing this is by using coordinates.

- We begin by extending the floor beneath your feet to form an unbounded plane. This floor plane is a two dimensional space, so, by using a Cartesian coordinate system, we can label each of its points with an ordered pair of real numbers.

- Any point in the universe that is not on this floor plane must be directly above it or directly below some point of it. To label any point, *P*, in the universe, then, we need just two things,

  – the ordered pair for the point on the plane that *P* is directly over or under;

  – third number to indicate exactly how far *P* is directly above or below the plane.

If we think like Cavalieri, then 3-space is just a big stack of copies of the floor plane—*lots* of copies—above and below it. An ordered pair of numbers pinpoints the same location on *each* plane; a third number is needed to tell you which copy to look at.

5.6 Space by the Numbers

This section reviews the Cartesian coordinate system that was introduced in Chapter 3 of **MATH** *Connections* Year 1 and extends it to 3-space. A main idea of the section is the distance formula for 3-space, which will be used in the Section 5.8 to develop the equation for a sphere. Each major 3-space concept presented in this section and the next is derived from the analogous 2-space concept. This process of generalization to the next higher dimension lays the groundwork for a future discussion of four and more dimensions.

5.52

This should be done right away as a class discussion to refresh students' memories about the rectangular coordinate system for a plane. You might have to help them along in places. The goal is to make the students active participants in reviewing these ideas.

Cartesian coordinate systems for the plane are ways of labeling each point in the plane with a two number "address." The numbers are locations on two perpendicular reference lines, called *coordinate axes*.

1. They are named after René Descartes, a 17th century French mathematician and philosopher, who gave the first comprehensive, useful description of such a system (in 1637).

2. They are also called *rectangular coordinate systems*.

3. A "point" is an ordered pair of real numbers.

4. A straight line is the set of all ordered pairs of real numbers that satisfy a particular first degree equation. Students should realize that the line is *not* the equation itself, even though we sometimes speak loosely about it that way. Many different equations can describe the same line.

5. The word *coordinate* is used in two senses. Its more usual meaning is one of the numbers used to specify a point. In the plane, the *x-coordinate* and *y-coordinate* of a point are the two numbers used to specify its location. Less frequently, the word *coordinate* is used to refer to a coordinate axis.

6. An *axis* is one of the reference lines used to determine which numbers label which points. In plane (two dimensional) coordinate geometry there are two axes. As we shall see in this section, the number of coordinate axes is the same as the dimension of the space.

7. This question is fairly open-ended; the similarities and the differences can be described in lots of different ways. Here are a few examples.

Chapter 5

Assessments   Blackline Masters   Extensions   Supplements
For Additional Support Materials see page T-335

A third coordinate axis perpendicular to the plane at the origin of the plane's coordinate system can be used to measure the height above or below the plane. In this way, every location in our three dimensional universe can be labeled by an ordered triple of numbers. Display 5.56 illustrates this for two specific points— P, which is 4 units above the point $(2, 5, 0)$ of the xy-plane, and Q, which is 3 units below the point $(2, 1, 0)$ of the xy-plane.

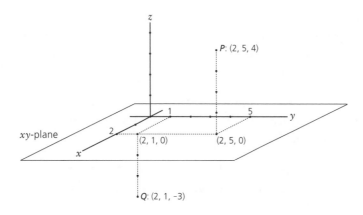

Two points in 3-space

Display 5.56

As in 2-space, this arrangement of perpendicular axes is called a **Cartesian coordinate system**. The third coordinate axis is usually called the z-**axis**. Recall that a plane that contains a pair of coordinate axes is called a coordinate plane and is named by the axis letters. In 3-space there are three coordinate planes—the xy-plane, the xz-plane, and the yz-plane. The one pictured in Display 5.56 is the xy-plane.

You are a radar spotter with the Coast Guard, watching coastal traffic.

5.53

1. Your instruments show two ships converging on the point 37° N (latitude) and 75° W (longitude). Should you be worried? Why or why not?

2. Your instruments show two airplanes converging on the point 37° N (latitude) and 75° W (longitude). One is flying level at 10,000 feet altitude; the other is at 25,000 feet. Should you be worried? Why or why not?

Published by IT'S ABOUT TIME, Inc. © 2000 MATHconx, LLC

439

Similarities
- There are two perpendicular coordinate axes.

- Each location on the screen is determined by an ordered pair of numbers.

- Straight lines correspond to first degree equations.

Differences
- The screen is finite, so the lengths of the coordinate axes are finite; they must be set by choice.

- The scales on the two axes may be set differently. This may be done, but is not usual, in traditional Cartesian geometry. It distorts shapes.

- There are only finitely many different screen locations (pixels). Thus, the correspondence between ordered pairs of numbers and dots displayed is not one-to-one. The TRACE function partially compensates for this in analyzing graphs.

Students might come up with other similarities and/or differences in the course of the discussion.

5.53

This is a concrete example to help students visualize how coordinates work. Having your students answer these orally in class should not take long, and you may be able to spot, early on, any confusion with the underlying idea of specifying location by coordinates.

The first two questions make the basic point. The third points out an important difference between this example and the 3-space described by the rectangular coordinate system. Part 4 is an optional curiosity question.

1. Probably. The two ships are on course for a collision.

2. Probably not. Unless one of the planes changes altitude abruptly, they will miss each other by almost 3 miles.

Chapter 5

3. Is this situation an example of a Cartesian coordinate system? Explain.

4. Where is the nearest Coast Guard station you should notify?

The ordered triple (x, y, z) specifies locations on the x-, y-, and z-axes. As Display 5.57 shows, the location of the point $(3, 4, 2)$ can be found by moving 3 units along the x-axis from the origin to $(3, 0, 0)$, then 4 units parallel to the y-axis in the positive direction to $(3, 4, 0)$, then 2 units parallel to the z-axis in the positive direction to $(3, 4, 2)$.

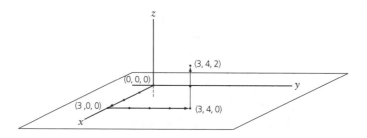

The point (3, 4, 2)

Display 5.57

5.54

Face the front of your classroom. Think of the back left floor corner of the room as the origin of a rectangular coordinate system like the one shown in Display 5.57. Think of the floor as the xy-plane and the line where the back wall meets the floor as the positive x-axis.

1. Relative to the coordinates, in which direction are you facing?

2. Which axis is the line where the left wall meets the floor?

3. In this context, what is the z-axis?

4. Assume that the unit of measure is 1 foot. Estimate to the nearest foot the coordinates of the point at the right front corner of your desktop. Describe what you would have to measure to check your estimate. If your teacher says so, do the measuring.

440

3. Not exactly. It is similar to one, in that each location on the floor plane (the surface of the Earth) is specified by an ordered pair of numbers and the height above (or below) that surface is measured at right angles to the surface. In fact, locally (within a radius of a few miles of anywhere) it is approximately a Cartesian system. However, the curvature of the Earth makes the floor plane a finite sphere (roughly), rather than the infinite flat surface envisioned by the Cartesian system.

4. For this question you will need to obtain a map of the United States with latitude and longitude marked on it. Otherwise, students should not be expected to know the answer. The Coast Guard station is somewhere near Norfolk, Virginia.

5.54

The assumption underlying this visualization exercise is that your classroom is rectangular. If it is not, you'll have to ask students to imagine being in a rectangular room. This probably is best handled as a brief all class discussion.

1. In the positive y direction.

2. The y-axis.

3. The line where the back wall meets the left side wall.

4. This is the most important question of the exercise. It is a personalized version of Display 5.57. If you are doing this as a group discussion, you might want to check two or three students' estimates by measuring. The exact measurements are not important, as long as the students recognize that they must measure each distance in a direction perpendicular to a coordinate plane. They need not describe it that way.

Chapter 5

This kind of coordinate system is called *rectangular* because the axes meet at right angles. There are coordinate systems for which this is not true, but they are not as easy to work with. Although such systems are sometimes useful in higher mathematics and science, they will not concern us here. We shall insist that all coordinate axes meet at right angles. Our reason for this insistence is The Pythagorean Theorem.

1. **What does the Pythagorean Theorem say? What does it have to do with right angles?**

5.55

2. **Think about how many of the geometric ideas you have studied depend on knowing the distance between two points. Describe at least five of them.**

Perpendicular axes allow us to use the Pythagorean Theorem to calculate the distance between two points directly from their coordinates. Here is a typical example from 2-space.

The straight line distance d between $(2, 1)$ and $(6, 4)$ is the length of the hypotenuse of a right triangle whose right angle vertex is $(6, 1)$. (See Display 5.58a.) The lengths of the other two sides of this triangle are 4 (the difference between the first coordinates) and 3 (the difference between the second coordinates). The Pythagorean Theorem tells us that

$$d^2 = 4^2 + 3^2$$

so

$$d = \sqrt{16 + 9} = \sqrt{25} = 5$$

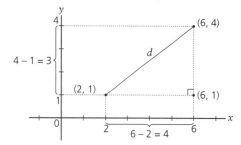

Finding the distance between (2,1) and (6, 4)

Display 5.58(a)

Published by IT'S ABOUT TIME, Inc. © 2000 MATHconx, LLC

441

5.55

This discussion question need not take much time. It is intended to help your students to focus on these two ideas.

1. The Pythagorean Theorem applies to *right* triangles only. The second part of this question is intended to check on the understanding of students whose response to the first part is just "$a^2 + b^2 = c^2$".

2. Distance is a fundamental spatial concept. The length of a line segment is the distance between its endpoints, so all the congruence, similarity, area, and volume results we have seen depend on distance. The radius of a circle or a sphere is a distance between two points. And so on.

Note that in keeping with the theme of generalizing from 2-space to 3-space, the following discussion begins by showing how the Pythagorean Theorem—a planar statement—provides the formula for distance in a plane. This is immediately generalized to the analogous formula for distance in 3-space, again by way of the Pythagorean Theorem.

Chapter 5

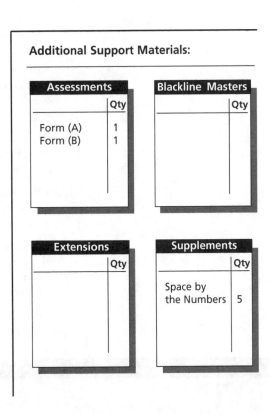

Additional Support Materials:

Assessments	Qty
Form (A)	1
Form (B)	1

Blackline Masters	Qty

Extensions	Qty

Supplements	Qty
Space by the Numbers	5

A Fact to Know: The distance between two points (x_1, y_1) and (x_2, y_2) in 2-space is

$$\sqrt{(x_2 - x_1)^2 + (y_2 - y_1)^2}$$

We shall call this the **2-space distance formula**.

5.56

1. Write the 2-space distance formula in words, *without using any algebraic symbols.* Use more than one sentence, if you want.

2. Find the distance between each of these pairs of points.

   (a) (2, 1) and (8, 9)   (b) (2, 3) and (4, 5)
   (c) (–2, 6) and (7, –5)`   (d) (9, –1) and (9, 4)

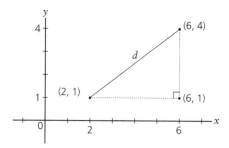

Finding the distance between (2,1) and (6, 4)

Display 5.58(b)

5.57

1. If the plane of Display 5.58(b) is the *xy*-plane of a three dimensional coordinate system, what are the coordinates of the three vertices of the triangle? Why?

2. If the plane of Display 5.58(b) is the *yz*-plane of a three dimensional system, what are the coordinates of the three vertices of the triangle? Why?

3. If the plane of Display 5.58(b) is the *xz*-plane of a three dimensional system, what are the coordinates of the three vertices of the triangle? Why?

4. If the plane of Display 5.58(b) is the plane parallel to the *xy*-plane of a three dimensional system and 3 units above it, what are the coordinates of the three vertices of the triangle? Why?

Published by IT'S ABOUT TIME, Inc. © 2000 MATHconx, LLC

5.56

The first question is a check to see if students are reading the symbols properly. The others are routine reinforcement of the formula.

1. One possible answer: To find the distance between two points using their coordinates, first find the differences between their first coordinates and their second coordinates. Then square those two differences, add them, and take the square root of the sum.

2. (a) $\sqrt{(8-2)^2 + (9-1)^2} = \sqrt{36+64} = \sqrt{100} = 10$
   (b) $\sqrt{(4-2)^2 + (5-3)^2} = \sqrt{4+4} = \sqrt{8} \approx 2.83$
   (c) $\sqrt{(7-(-2))^2 + (-5-6)^2} = \sqrt{81+121} = \sqrt{202} \approx 14.21$
   (d) $\sqrt{(9-9)^2 + (-1-4)^2} = \sqrt{0+25} = \sqrt{25} = 5$

5.57

The explanations here can be brief and simple. There are four ideas to get across.

- Embedding a plane in 3-space requires a third coordinate for each of its points.

- One coordinate of any point on a coordinate plane must be 0.

- All points on the a plane parallel to a coordinate plane have one coordinate in common.

- The distance between two given points in a plane is not affected by the position of that plane.

1. (2, 1, 0), (6, 4, 0), (6, 1, 0)

2. (0, 2, 1), (0, 6, 4), (0, 6, 1)

3. (2, 0, 1), (6, 0, 4), (6, 0, 1)

4. (2, 1, 3), (6, 4, 3), (6, 1, 3)

5. (2, 1, -17), (6, 4, -17), (6, 1, -17)

6. No; the lengths are always the same. Positioning a plane in space does not distort it in any way.

Chapter 5

5. If the plane of Display 5.58(b) is the plane parallel to the *xy*-plane of a three dimensional system and 17 units below it, what are the coordinates of the three vertices of the triangle? Why?

6. Do the lengths of the sides of the triangle change in any of these cases? Why or why not?

The formula for finding the distance between any two points in 3-space is very much like the distance formula for 2-space. As a typical example, we'll compute the distance between (1, 2, 3) and (5, 7, 9). You can think of these points as the diagonally opposite corners of a rectangular box with sides parallel to the coordinate planes, as shown in Display 5.59.

Note that in Display 5.59, we have rotated the coordinate system 90° counterclockwise around the *z*-axis. We did this so that you can see all three axes without interference from the box. Think of the page as the *xz*-plane, with the positive *y* direction *behind* the page. The near face of the box is 2 units behind the plane of the page.

What are the coordinates of the two corner points on the lower right edge of the rectangular box in Display 5.59?

5.58

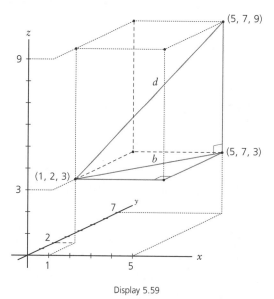

Display 5.59

5.58

This is another quick visualization check. The rear lower right corner is at (5, 7, 3); the front lower right corner is at (5, 2, 3). We use both of these points in the next paragraph. Help students to see why these points are labeled this way. If necessary, ask them about the coordinates of the other corners of the box. Note that the x-axis and y-axis have been rotated 90°; this was done to make the display easier for students to read.

NOTES

The length d of the diagonal between $(1, 2, 3)$ and $(5, 7, 9)$ can be found by applying the Pythagorean Theorem twice. First we find the length b of the diagonal of the bottom of the box—the distance between $(1, 2, 3)$ and the lower right rear corner, $(5, 7, 3)$. Both points are on a plane parallel to the xy-plane, so we can apply the 2-space distance formula to their x- and y-coordinates to find the distance between them:

$$b = \sqrt{(5 - 1)^2 + (7 - 2)^2}$$

Now, the distance we want, d, is the hypotenuse of the right triangle with vertices $(1, 2, 3)$, $(5, 7, 3)$ and $(5, 7, 9)$.

How do you know it's a right triangle? One leg of this triangle has length b. The length of the other leg is just the difference in the z-coordinates of $(5, 7, 3)$ and $(5, 7, 9)$. By the Pythagorean Theorem (again)

$$d = \sqrt{b^2 + (9 - 3)^2}; \text{ since } b^2 = (5 - 1)^2 + (7 - 2)^2$$
$$d = \sqrt{[(5 - 1)^2 + (7 - 2)]^2 + (9 + 3)^2}$$
$$d = \sqrt{4^2 + 5^2 + 6^2}$$
$$d = \sqrt{77}$$

Notice that the second of these equations expresses d in terms of the x-, y-, and z-coordinate differences between the two endpoints. This is a typical example of the following formula.

A Fact to Know: The distance between two points (x_1, y_1, z_1) and (x_2, y_2, z_2) in 3-space is

$$\sqrt{(x_2 - x_1)^2 + (y_2 - y_1)^2 + (z_2 - z_1)^2}$$

We shall call this the **3-space distance formula.**

5.59

1. Write the 3-space distance formula in words, *without using any algebraic symbols.* Use more than one sentence, if you want.

2. Find the distance between each of these pairs of points.

   (a) $(2, 1, 1)$ and $(3, 8, 5)$
   (b) $(4, 6, -1)$ and $(-7, 7, 1)$
   (c) $(7, 5, 2)$ and $(7, 8, 4)$
   (d) $(0, 0, 0)$ and $(\sqrt{2}, \sqrt{3}, \sqrt{4})$

5.59

These questions are analogous to those of Do this now 5.56, emphasizing the similarity between the 2-space and 3-space distance formulas.

1. One possible answer: To find the distance between two points using their coordinates, first find the differences between their first coordinates, their second coordinates, and their third coordinates. Then square those three differences, add them, and take the square root of the sum.

2. (a) $\sqrt{(3-2)^2 + (8-1)^2 + (5-1)^2} = \sqrt{1 + 49 + 16} = \sqrt{66} \approx 8.1$

    (b) $\sqrt{(-7-4)^2 + (7-6)^2 + (1-(-1))^2} = \sqrt{121 + 1 + 4} = \sqrt{126} \approx 11.2$

    (c) $\sqrt{(7-7)^2 + (8-5)^2 + (4-2)^2} = \sqrt{0 + 9 + 4} = \sqrt{13} \approx 3.6$

    (d) $\sqrt{(\sqrt{2}-0)^2 + (\sqrt{3}-0)^2 + (\sqrt{4}-0)^2} = \sqrt{2 + 3 + 4} = \sqrt{9} = 3$

NOTES

Chapter 5

Problem Set: 5.6

1. Some computers play board games. If you program a computer to play Tic-Tac-Toe, you have to tell it how to recognize each of the 9 different boxes of the game diagram. This can be done by using coordinate pairs of numbers. One way to do it is shown in Display 5.60. In each pair, the first number identifies the column (counting from left to right) and the second number identifies the row (counting from bottom to top).

(1, 3)	(2, 3)	(3, 3)
(1, 2)	(2, 2)	(3, 2)
(1, 1)	(2, 1)	(3, 1)

Display 5.60

(a) To signal a win, the computer needs to "see" three boxes in a straight line. How can it recognize such a combination from the coordinates?

(b) Some game companies make a 3-D Tic-Tac-Toe game, formed by three layers of the two-dimensional game. Extend the coordinate pattern of Display 5.60 so that the computer can recognize each of the different three dimensional boxes in the 3-D game. How many boxes are there?

(c) Winning the 3-D game also depends on getting three boxes in a straight line. How can the computer recognize such a combination from the coordinates? Try to answer this with as few different cases as possible.

2. A standard checkerboard is an 8 by 8 array of squares in an alternating red and black pattern.

(a) Extend the coordinate pattern of Display 5.60 so that the computer can recognize each of the 64 different squares on the board.

(b) The game of checkers is played on only the squares of one color. How can the computer tell which squares are red and which squares are black?

Problem Set: 5.6

1. The coordinate pattern used here mimics the pattern of the first quadrant of the Cartesian plane, rather than the more usual row column pattern of matrices.

   (a) There are three kinds of straight lines to account for.
   - If three pairs all have the same first coordinate and all different second coordinates, they form a vertical straight line.
   - If three pairs all have the same second coordinate and all different first coordinates, they form a horizontal straight line.
   - The lower left to upper right diagonal consists of the three pairs whose first and second coordinates are equal (the $y = x$ line). The other diagonal consists of the pairs $(1, 3)$, $(2, 2)$ and $(3, 1)$.

   (b) There are 27 boxes (three layers of nine). Add a third coordinate for the height of the layer, from bottom to top. Then, for instance, each box in the bottom layer will have the same first two coordinates as in Display 5.60 and 1 as its third coordinate, and so on.

   (c) This is a much more difficult problem than its two dimensional analogue. Students can gain insight about the logical symmetry of coordinate axes by trying to break this into cases. One valuable insight is that three points (coordinate triples) with the same value in any one coordinate place lie in the same 2-D "slice," so the test of part (b) can be applied to them. The only other possibilities are the four main diagonals of the $3 \times 3 \times 3$ box. Each of these four sets of three points must contain $(2, 2, 2)$ (the center box). The other two points in each case are

   $(1, 1, 1)$, $(3, 3, 3)$ $(1, 3, 1)$, $(3, 1, 3)$ $(1, 1, 3)$, $(3, 3, 1)$
   $(3, 1, 1)$, $(1, 3, 3)$

2. (a) Starting with $(1, 1)$ in the front left corner, each square is labeled (x, y), where x and y are whole numbers between 1 and 8, inclusive. The x-value designates the column (the *file* in chess); the y-value designates the row (the *rank* in chess).

   (b) An easy way is by adding the coordinates of the square. An odd sum represents one color; an even sum represents the other. In particular, if the front left corner square, $(1, 1)$, is black (as it must be in chess), then the black squares are precisely the squares with an even coordinate sum.

(c) What does a three dimensional checkerboard look like? Extend the coordinate pattern of part (a) so that the computer can recognize each of the different three dimensional boxes in the 3-D game. How many boxes are there?

(d) How can the computer use coordinates to distinguish between the red squares and the black squares of the 3-D checkerboard?

3. In a 2-space coordinate system, (−1, 5) and (7, 2) are the diagonally opposite corners of a rectangle with all its sides parallel to the axes?

   (a) What are the coordinates of the other corners of this rectangle? How many other corners are there? (Drawing a sketch might help.)

   (b) How long are the sides of this rectangle?

   (c) How long is its diagonal (the line segment between the two given points)?

   (d) How long is its other diagonal? Justify your answer.

4. In a 3-space coordinate system, (6, −3, 5) and (9, 7, −2) are the diagonally opposite corners of a rectangular box with all its sides parallel to the coordinate planes,

   (a) What are the coordinates of the other corners of this box? How many other corners are there?

   (b) What are the dimensions (length, width, height) of this box?

   (c) How long is its main diagonal (the line segment between the two given points)?

   (d) How many other diagonals of the same length does this box have? What are the coordinates of their endpoints?

5. This problem is about the sequence of points

   $$(0, 1, 2), (0, 2, 3), (0, 3, 4), (0, 4, 5), ...$$

   (a) All these points lie in a coordinate plane. Which one? How do you know?

   (b) In this sequence, what is the next point? What is the 10th point? What is the 100th point? What is the nth point?

   (c) Calculate the distance between (0, 0, 1) and each of the first five points of this sequence.

Published by IT'S ABOUT TIME, Inc. © 2000 MATHconx, LLC

(c) It is a stack of eight layers of 8×8 of alternately red and black boxes. The alternating pattern extends vertically, as well as horizontally. Suggest that students think of arranging 64 alternating red and black cubes on a regular checkerboard. Now take a second, identical layer of these cubes and place them on top of the first, but rotated 90°. Then reds will be on blacks, and vice versa. Continue stacking the remaining six layers this way. There are $8^3 = 512$ boxes in all.

(d) This is more difficult to visualize at first, but the solution is surprisingly easy. Summing the coordinates works again. Think of the bottom layer as before. The front left box, (1, 1, 1), is black and the sum of its coordinates is odd, so all the boxes on the bottom layer that have an odd coordinate sum must be black. Now, when you move up a layer, you just add 1 to the coordinate sum for each position within that layer, reversing the parity. That is, all the black positions become red, and all the red positions become black, as required! Each step up to the next layer adds 1 to the third coordinate, thereby reversing the parity of each position.

3. (a) (–1, 2) and (7, 5)
   (b) The x distance is $|7 - (-1)| = 8$, the y distance is $|5 - 2| = 3$, so the rectangle is 8 by 3 units.
   (c) $\sqrt{8^2 + 3^2} = \sqrt{73} \approx 8.544$
   (d) It is the same length as the diagonal of part (c); the two diagonals of a rectangle are always equally long. There are many ways to justify this; we suggest that you accept anything that (at least) makes good intuitive sense. For instance.
   (i) A rectangle is symmetric with respect to 180° rotation, so the diagonals must match.
   (ii) The diagonals are the hypotenuses of two right triangles that have one leg in common and the other legs equally long. Hence, the triangles are congruent, so the hypotenuses must be equally long. There are other legitimate arguments, as well.

4. (a) There are six other corners.
   (9, –3, 5), (6, 7, 5), (6, –3, –2), (6, 7, –2), (9, –3, –2), (9, 7, 5)
   (b) The x distance is $|9 - 6| = 3$; the y distance is $|7 - (-3)| = 10$; the z distance is $|-2 - 5| = 7$. Thus, the box measures 3 by 10 by 7 units.
   (c) $\sqrt{3^2 + 10^2 + 7^2} = \sqrt{158} \approx 12.57$
   (d) There are three other diagonals of the same length, between these pairs of points
   (9, –3, 5) and (6, 7, –2)   (6, 7, 5) and (9, –3, –2)
   (6, –3, –2) and (9, 7, 5)

Chapter 5

(d) Do you see a pattern in these distances? If so, use it to predict the distance between (1, 0, 1) and the 10th point in the sequence. Then calculate that distance.

(e) In terms of n, what is the distance between (0, 0, 1) and the nth term of this sequence? Use your answer to find the distance between (0, 0, 1) and the 100th point in the sequence. Also use it to find the distance between (0, 0, 1) and the 537th point in the sequence.

6. This problem is about the sequence of points,

$$(1, 2, 2), (2, 3, 4), (3, 4, 6), (4, 5, 8), (5, 6, 10), \ldots.$$

(a) In this sequence, what is the next point? What is the 10th point? What is the 100th point? What is the nth point?

(b) Calculate the distance between (0, 1, 0) and each of the first five points of this sequence.

(c) Do you see a pattern in these distances? If so, use it to predict the distance between (0, 1, 0) and the 10th point in the sequence. Then calculate that distance.

(d) In terms of n, what is the distance between (0, 1, 0) and the nth term of this sequence? Use your answer to find the distance between (0, 1, 0) and the 100th point in the sequence. Also use it to find the distance between (0, 1, 0) and the 279th point in the sequence.

Published by IT'S ABOUT TIME, Inc. © 2000 MATHconx, LLC

5. This problem combines the ideas of this section with patterns and sequences studied in **MATH** *Connections* Year 1 and some algebra.
   (a) Because the x-coordinate of all these points is 0, they all lie in the yz-plane.
   (b) $(0, 5, 6)$; $(0, 10, 11)$; $(0, 100, 101)$; $(0, n, n + 1)$
   (c) $\sqrt{(0 - 0)^2 + (1 - 0)^2 + (2 - 1)^2} = \sqrt{2}$ Similarly,
   $\sqrt{0^2 + 2^2 + (3 - 1)^2} = \sqrt{8}$;
   $\sqrt{0^2 + 3^2 + (4 - 1)^2} = \sqrt{18}$;
   $\sqrt{0^2 + 4^2 + (5 - 1)^2} = \sqrt{32}$;
   $\sqrt{0^2 + 5^2 + (6 - 1)^2} = \sqrt{50}$;
   (d) If students don't see a pattern yet, don't push them to find it in this part. Part (e) makes the pattern explicit. The distance is $\sqrt{200}$.
   (e) The algebra makes the pattern clear. The distance is

   $$\sqrt{0^2 + n^2 + ((n + 1) - 1)^2} = \sqrt{2n^2} = n\sqrt{2}$$

   Thus, the distance between $(0, 0, 1)$ and the 100th point is $100\sqrt{2}$. The distance between $(0, 0, 1)$ and the 537th point is $537\sqrt{2}$.

6. This problem is like problem 5, but the sequence of points does not lie in a single coordinate plane. The patterns are a little less obvious in this case.
   (a) $(6, 7, 12)$; $(10, 11, 20)$; $(100, 101, 200)$; $(n, n + 1, 2n)$
   (b) $\sqrt{(1 - 0)^2 + (2 - 1)^2 + (2 - 0)^2} = \sqrt{6}$ Similarly,
   $\sqrt{2^2 + (3 - 1)^2 + 4^2} = \sqrt{24}$;
   $\sqrt{3^2 + (4 - 1)^2 + 6^2} = \sqrt{54}$;
   $\sqrt{4^2 + (5 - 1)^2 + 8^2} = \sqrt{96}$;
   $\sqrt{5^2 + (6 - 1)^2 + 10^2} = \sqrt{150}$
   (c) If students don't see a pattern, don't push them to find it in this part. Part (d) makes the pattern explicit. The distance is $\sqrt{600}$.
   (d) The algebra makes the pattern clear. The distance is

   $$\sqrt{n^2 + ((n + 1) - 1)^2 + (2n)^2} = \sqrt{6n^2} = n\sqrt{6}$$

   Thus, the distance between $(0, 1, 0)$ and the 100th point is $100\sqrt{6}$. The distance between $(0, 1, 0)$ and the 279th point is $279\sqrt{6}$.

Chapter 5

5.7 Stacks of Lines

Learning Outcomes

After studying this section you will be able to:

Write coordinate descriptions of lines and planes that are parallel to coordinate planes;

Write coordinate descriptions of lines and planes that are parallel to coordinate axes.

When René Descartes proposed his system of coordinate geometry in 1637, the power of algebra for solving problems was just beginning to be understood. The symbols $+$, $-$, $=$, $<$ and the use of letters to stand for unknown quantities were new then. These symbols were not used consistently until the late 1500s, almost 100 years after Columbus discovered America! Descartes' coordinate geometry linked the patterns of algebra with the pictures of geometry, so that each aspect of mathematics could use the tools of the other. The power of this combination opened the way for calculus less than 40 years later. Calculus, in turn, unlocked many mysteries of physics and astronomy during the next 200 years.

Today, the realistic pictures and exciting real world action scenes displayed by televisions, computers, electronic games, and virtual reality devices all depend on coordinate geometry. All the data for those displays are stored and processed *as numbers*. These machines deal with each location of each point in each picture *as a string of coordinates*. The shapes and movements are created by *algebraic* manipulation of these coordinates, done millions of times faster than humans could ever do it.

Most of the algebra needed for computer graphics is far beyond what we can show you here. However, we *can* show you the fundamentals of this connection between geometry and algebra. You saw the beginning of this story earlier in **MATH** *Connections*, when you studied descriptions of lines in a coordinate plane. We shall build on those ideas here. In the rest of this chapter you will see how to describe some basic 3-space shapes by algebraic formulas. You will also see how to represent certain kinds of algebraic expressions by geometric shapes. Chapter 6 will show you how this sort of thing is used in real world problem situations.

5.60

These review exercises about coordinates in 2-space will help you refresh your memory.

1. Describe the coordinates of all the points on the *x*-axis.

2. What inequality statement describes the segment between 3 and 8 on the *x*-axis?

5.7 Stacks of Lines

This short section is a steppingstone for the rest of the chapter. Its main idea is the description of a plane as a stack of lines, which recalls the underlying intuition of Cavalieri's Principle. This imagery allows students to visualize planes that are parallel to a coordinate axis in 3-space. The stacking concept is the basis for the next section.

5.60

This is a review of some elementary ideas from Section 3.2 of **MATH** *Connections* Year 1. You may have to help your students recall them. It's better for them to refresh their memories actively, with your help, than for the text simply to restate the necessary ideas and language. In this way, your students will realize that these are things they *should* remember from last year.

1. All points (x, y) such that $y = 0$.

2. $3 \leq x \leq 8$. That is, this segment is the set of all points (x, y) such that $y = 0$ and $3 \leq x \leq 8$. Equivalently, it is the set of all points $(x, 0)$ such that $3 \leq x \leq 8$.

Chapter 5

Additional Support Materials:

Assessments	Qty
Form (A)	1
Form (B)	1

Blackline Masters	Qty

Extensions	Qty

Supplements	Qty
Stacks of Lines	1

3. Use an inequality statement to describe the segment between 3 and 8 on the y-axis. Write your answer in set-builder notation.

4. On a coordinate plane mark the inequality

$$\{(x, 3) \mid -4 \le x \le 6\}$$

Set-builder notation saves a lot of unnecessary writing. Instead of writing

"the set of all (x, y) such that x and y satisfy (some condition)"

just write

$$\{(x, y) \mid \text{(some condition)}\}$$

For instance, the x-axis of 3-space can be written as

$$\{(x, 0, 0) \mid x \text{ is a real number}\}$$

This means "the set of all triples of real numbers with second and third coordinates 0." In this case, the condition is just a reminder of something already implied: unless they are restricted by the condition after the "$\mid$", the letters x, y, and z can stand for any real number.

The segment of the x-axis between 2 and 9, inclusive, is written

$$\{(x, 0, 0) \mid 2 \le x \le 9\}$$

1. Write set-builder notation for the y-axis of 3-space.

5.61

2. Write set-builder notation for the segment of the y-axis between 2 and 9, inclusive.

3. Write set-builder notation for the segment of the z-axis between –7 and 4, inclusive.

Lines and segments parallel to the coordinate axes are easy to visualize and to describe in set-builder notation. For example, the line l_1 in Display 5.61 is parallel to the z-axis and goes through the point $(2, 4, 0)$ on the xy-plane. It is the set of points

$$\{(2, 4, z) \mid z \text{ is a real number}\}$$

The segment on it between $z = -1.5$ and $z = 3$ is

$$\{(2, 4, z) \mid -1.5 \le z \le 3\}$$

Published by IT'S ABOUT TIME, Inc. © 2000 MATHconx, LLC

449

3. $\{(0, y) \mid 3 \leq y \leq 8\}$

4. This is a line segment 10 units long, 3 units above the x-axis and parallel to it, beginning above $x = -4$ and ending above $x = 6$. See Display 5.21T.

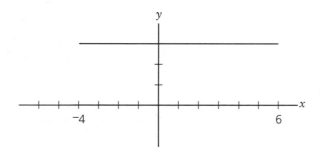

Display 5.21T

5.61

This is routine practice.

1. $\{(0, y, 0) \mid y \text{ is a real number}\}$

2. $\{(0, y, 0) \mid 2 \leq y \leq 9\}$

3. $\{(0, 0, z) \mid -7 \leq z \leq 4\}$

NOTES

Chapter 5

The way to describe all lines and segments parallel to axes are shown in this example and the ones that follow.

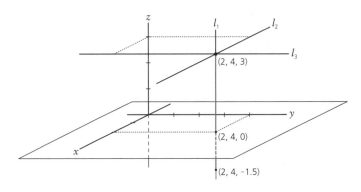

Display 5.61

5.62

These questions refer to Display 5.61.

1. The line l_3 is parallel to a coordinate axis; to which one? Write a set-builder description of l_3.

2. Write a set-builder description of the segment of l_3 that is 10 units long and is centered at $(2, 4, 3)$.

3. The line l_2 is also parallel to an axis; to which one? Write a set-builder description of l_2.

4. Write a set-builder description of the segment of l_2 that is 10 units long and is centered at $(2, 4, 3)$.

5. The lines l_2 and l_3 are in a plane that is parallel to the xy-plane. How would you write a set-builder description of this plane? Explain your answer.

In 2-space, the xy-plane (the only plane) is

$$\{(x, y) \mid x \text{ and } y \text{ are real numbers}\}$$

In 3-space, the xy-plane is just a copy of this set at the $z = 0$ level.

$$\{(x, y, 0) \mid x \text{ and } y \text{ are real numbers}\}$$

Any plane parallel to this one can be described just by specifying its z-level. The z coordinate tells you how far above or below the coordinate plane it is. The plane just half a unit above the xy-plane, for instance, is

$$\{(x, y, \tfrac{1}{2}) \mid x \text{ and } y \text{ are real numbers}\}$$

450

5.62

Students who can do questions 1–4 easily have a good visual grasp of lines parallel to the axes. Question 5 asks them to stretch a little and anticipate the next step in the development. It might make a good (brief) discussion question, if it doesn't seem obvious to the class.

1. l_3 is parallel to the y-axis. It is $\{(2, y, 3) \mid y$ is a real number$\}$

2. $\{(2, y, 3) \mid -1 \leq y \leq 9\}$

3. l_2 is parallel to the x-axis. It is $\{(x, 4, 3) \mid x$ is a real number$\}$

4. $\{(x, 4, 3) \mid -3 \leq x \leq 7\}$

5. $\{(x, y, 3) \mid x$ and y are real numbers$\}$ All the points must be at z height 3, and any such point is on this plane.

NOTES

Chapter 5

The plane 3 units below the xy-plane is

$$\{(x, y, -3) \mid x \text{ and } y \text{ are real numbers}\}$$

The other two coordinate planes and the planes parallel to them are described similarly.

5.63

Visualize the 3-space coordinate system as it appears in Display 5.61: The x-axis is front-back, the y-axis is right-left, and the z-axis is up-down.

1. Where is the plane $\{(x, y, \sqrt{2}) \mid x \text{ and } y \text{ are real numbers}\}$?

2. Write set-builder descriptions for the yz-plane and the xz-plane (the other two coordinate planes).

3. Write a set-builder description for the plane that is parallel to the xz-plane and 5 units to its right.

4. Write a set-builder description for the plane that is parallel to the yz-plane and 2 units behind it.

5. Where are each of these planes?

   (a) $\{(x, -1, z) \mid x \text{ and } z \text{ are real numbers}\}$
   (b) $\{(7.25, y, z) \mid y \text{ and } z \text{ are real numbers}\}$
   (c) $\{(x, y, -3.8) \mid x \text{ and } y \text{ are real numbers}\}$

5.64

These questions ask you to recall some things that you studied earlier in MATH *Connections*.

1. What kind of equation represents a straight line in the coordinate plane? Give an example. Use your example to describe the line in set-builder notation.

2. What is the *slope* of a straight line? What is its *y-intercept*? How can you find the slope and the y-intercept of a straight line from its equation? What are the slope and the y-intercept of your example from question 1?

Now let's build on these ideas. Remember that each coordinate plane is just a copy of the 2-space coordinate plane. In 2-space, the equation $y = 2x$ describes the line with slope 2 through the origin, $(0, 0)$. In set-builder notation, this line is

$$\{(x, y) \mid y = 2x\}$$

Published by IT'S ABOUT TIME, Inc. © 2000 MATHconx, LLC

451

5.63

These questions reinforce the previous paragraph and show what the phrase described similarly means here.

1. It is parallel to the xy-plane and $\sqrt{2}$ units above it.

2. The xz-plane: $\{(x, 0, z) \mid x \text{ and } z \text{ are real numbers}\}$

   The yz-plane: $\{(0, y, z) \mid y \text{ and } z \text{ are real numbers}\}$

3. $\{(x, 5, z) \mid x \text{ and } z \text{ are real numbers}\}$

4. $\{(-2, y, z) \mid y \text{ and } z \text{ are real numbers}\}$

5. (a) Parallel to the xz-plane and 1 unit to the left of it
   (b) Parallel to the yz-plane and 7.25 units in front of it
   (c) Parallel to the xy-plane and 3.8 units below it

5.64

These discussion questions should be done right at this point, if they are to be done at all. Their purpose is to refresh students' memories about equations for straight lines in 2-space from Chapter 3 of **MATH** *Connections* Year 1.

1. A first degree equation in x and y (i.e., in two variables). For example, $y = 2x + 3$. Using this example as typical, there are two different, equally useful ways to represent the line in set-builder notation

   $$\{(x, y) \mid y = 2x + 3\} \quad \text{or} \quad \{(x, 2x + 3) \mid x \text{ is a real number}\}$$

2. The slope of a straight line is the ratio of its "vertical" change to its "horizontal" change. It's the amount the dependent variable (y) changes for each unit change in the independent variable (x). The y-intercept is the point (or the y-value) at which the line intersects the y-axis. If the equation is in the form $y = ax + b$, the slope is a and the y-intercept is determined by b. The y-intercept is the point with coordinates $(0, b)$. For our example, the slope is 2 and the y-intercept is the point $(0, 3)$.

Chapter 5

This same line in the xy-plane of 3-space can be described in just about the same way.

$$\{(x, y, 0) \mid y = 2x\}$$

Its slope is 2 and its y-intercept is $(0, 0, 0)$.

5.65

1. Where (in 3-space) is the line $\{(x, y, 3) \mid y = 2x\}$? What is its slope? What is its y-intercept?

2. Where is the line $\{(x, y, -4) \mid y = 2x\}$? What is its slope? What is its y-intercept?

3. Write the set-builder description of a line parallel to $\{(x, y, 0) \mid y = 2x\}$ and 5.7 units above it (in the z-direction).

4. What geometric figure is $\{(x, y, z) \mid y = 2x\}$? Describe it and its location as precisely as you can.

5. Display 5.62 illustrates the answers to questions 1-4. Explain how each question matches some part of the figure.

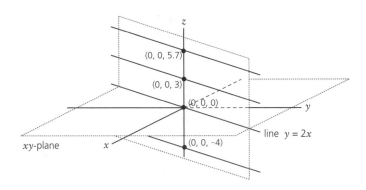

Display 5.62

As you can see from Display 5.62, the stack of all lines parallel to $\{(x, y, 0) \mid y = 2x\}$, above and below it in the z-direction, form a plane. This plane meets the xy-plane at right angles, along the line $y = 2x$. If we shift this line over so that its y-intercept is 3, for example, then its equation becomes $y = 2x + 3$. As a set of points in 3-space, it is

$$\{(x, y, 0) \mid y = 2x + 3\}$$

Published by IT'S ABOUT TIME, Inc. © 2000 MATHconx, LLC

5.65

These questions show how a plane can be viewed as a stack of parallel lines, each one parallel to a given line in a coordinate plane. This is a two dimensional version of Cavalieri's Principle. The idea can be confusing at first, but once students get the visual image clear, the coordinate adjustments should fall right into place.

1. It is parallel to the line $\{(x, y, 0) \mid y = 2x\}$, three units directly above it (in the z direction). It is in the $z = 3$ plane, so its slope and y-intercept must be understood in relation to the x-axis and y-axis in the $z = 3$ plane. Its slope is 2. Its y-intercept, the point at which it intersects the y-axis in the $z = 3$ plane, is $(0, 0, 3)$.

2. It is parallel to the line $\{(x, y, 0) \mid y = 2x\}$, four units directly below it (in the z direction). It is in the $z = -4$ plane. Its slope is 2. Its y-intercept, the point at which it intersects the y-axis in the $z = -4$ plane, is $(0, 0, -4)$.

3. $\{(x, y, 5.7) \mid y = 2x\}$

4. It is a plane that is perpendicular to the xy-plane. It intersects the xy-plane along the line $\{(x, y, 0) \mid y = 2x\}$. Since there is no restriction on z in the set-builder description, z can take on every real number value. Thus, the set contains a copy of the line $y = 2x$ at *every* height z. The plane is the union of all these lines, as Cavalieri might have imagined it.

5. Explanations may vary, but the basic ideas should be pretty much the same from student to student. Watch for any problems in misinterpreting the figure. Some students have difficulty translating such two dimensional drawings into three dimensional imagery.

Chapter 5

The same equation describes copies of this line on planes parallel to the xy-plane. That is, for each value of z,

$$\{(x, y, z) \mid y = 2x + 3\}$$

is a parallel copy of the original line, directly above or below it. Each copy has slope 2 and y-intercept 3 on its z-level plane. Display 5.63 shows this original line and three copies of it— at $z = -4$, $z = 3$, and $z = 5.7$. The four points marked by large dots are the y-intercepts of these lines on their z-level planes. For instance,

$$\{(x, y, 5.7) \mid y = 2x + 3\}$$

is the topmost line of Display 5.63. On the $z = 5.7$ plane, the line's slope is 2 and its y-intercept is $(0, 3, 5.7)$. All these lines taken together, one for each value of z, form a plane. The z-axis is not on this plane, but is parallel to it.

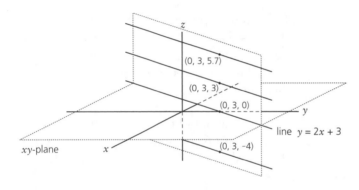

Display 5.63

All the lines and planes in 3-space that you have studied so far are special in a way that makes them fairly easy to describe. A more general treatment of lines and planes in space is better done by a different approach, which we leave for a later time in your mathematical studies.

NOTES

Problem Set: 5.7

1. This problem refers to Display 5.61.

   (a) Compute the distance between $(2, 4, 3)$ and the origin, $(0, 0, 0)$.

   (b) The line l_1 through $(2, 4, 3)$ is parallel to the z-axis. Imagine the point moving along l_1, through all the other points $(2, 4, z)$ on this line. Express the distance between each of these points and the origin as a function of z; call it $d_1(z)$. For which z does this function have the smallest value? That is, which point on l_1 is closest to the origin? How do you know? How close is it?

   (c) The line l_2 through $(2, 4, 3)$ is parallel to the x-axis. Express the distance between each point on this line and the origin as a function of x; call it $d_2(x)$. Which point on l_2 is closest to the origin? How do you know? How close is it?

   (d) The line l_3 through $(2, 4, 3)$ is parallel to the y-axis. Express the distance between each point on this line and the origin as a function of y — call it $d_3(y)$. Which point on l_3 is closest to the origin? How do you know? How close is it?

   (e) Of all the points on the lines l_1, l_2, and l_3, which is closest to the origin? What is its distance from the origin?

   For problems 2 and 3, visualize the 3-space coordinate system as it appears in Displays 5.61, 5.62, and 5.63: The x-axis is front-back, the y-axis is right-left, and the z-axis is up-down.

2. Write a set-builder description for each of the following planes.

   (a) the plane parallel to the xy-plane and 3 units below it
   (b) the plane parallel to the yz-plane and 7 units behind it
   (c) the plane parallel to the xz-plane and 8.62 units to its right

3. Write a set-builder description for each of the following lines.

   (a) the line parallel to the x-axis through the point $(7, 5, 8)$
   (b) the line parallel to the y-axis through the point $(6, -3, \sqrt{5})$
   (c) the line parallel to the z-axis through the point $(5.8, -6.01, -2.33)$

Published by IT'S ABOUT TIME, Inc. © 2000 MATHconx, LLC

MATH *Connections*: A Secondary Mathematics Core Curriculum

Problem Set: 5.7

1. This exercise combines several ideas in a single setting. It provides practice with the distance formula, reviews the idea of a function, and asks students to use their common sense and their visualization skills to interpret algebraic expressions.

   (a) $\sqrt{2^2 + 4^2 + 3^2} = \sqrt{29}$

   (b) $d_1(z) = \sqrt{2^2 + 4^2 + z^2} = \sqrt{20 + z^2}$. Since z^2 is always nonnegative, the smallest value of this function is at $z = 0$; that is, the point $(2, 4, 0)$ is closest. Its distance from the origin is $\sqrt{20}$, approximately 4.47 units.

   (c) $d_2(x) = \sqrt{x^2 + 4^2 + 3^2} = \sqrt{x^2 + 25}$. Since x^2 is always nonnegative, the smallest value of this function is at $x = 0$; that is, the point $(0, 4, 3)$ is closest. Its distance from the origin is $\sqrt{25} = 5$ units.

   (d) $d_3(y) = \sqrt{2^2 + y^2 + 3^2} = \sqrt{13 + y^2}$. Since y^2 is always nonnegative, the smallest value of this function is at $y = 0$; that is, the point $(2, 0, 3)$ is closest. Its distance from the origin is $\sqrt{13}$, approximately 3.6 units.

   (e) By the answers to the previous three parts, $(2, 0, 3)$ is the closest of all the points; it is $\sqrt{13}$ units (about 3.6 units) away from the origin.

2. (a) $\{(x, y, -3) \mid x$ and y are real numbers$\}$

   (b) $\{(-7, y, z) \mid y$ and z are real numbers$\}$

   (c) $\{(x, 8.62, z) \mid x$ and z are real numbers$\}$

3. (a) $\{(x, 5, 8) \mid x$ is a real number$\}$

   (b) $\{(6, y, \sqrt{5}) \mid y$ is a real number$\}$

   (c) $\{(5.8, -6.01, z) \mid z$ is a real number$\}$

Chapter 5

4. Write a set-builder description for each of the following.

   (a) the plane parallel to the z-axis that intersects the
   xy-plane along the line $y = 5x - 2$

   (b) the plane parallel to the y-axis that intersects the
   xz-plane along the line $z = \frac{2}{3}x + 0.65$

   (c) the line on the $z = 3$ plane that is parallel to the line
   $y = -x$ on the xy-plane

   (d) the line on the $x = -4$ plane that is parallel to the line
   through $(1, 2, 3)$ and $(1, 5, 8)$

   (e) the line on the $y = 5$ plane that is parallel to the line
   through $(7, 2, 1)$ and $(8, 2, -1)$

5. If the coordinate axes for a plane (or for 3-space) are not
   perpendicular, they form an **oblique coordinate system**. How
   does such a system work? In particular, how do you think
   such axes can be used to label points with coordinates?
   How much of what we have done with rectangular
   coordinates can be reconstructed for oblique coordinates?
   What goes bad? Can it be fixed? If so, how? Investigate
   these ideas and write a brief paper on what you find.

4. Parts (a) and (b) are straightforward examples of techniques presented in this section. Parts (d) and (e) are fairly challenging. They build on the idea in part (c) and require students to recall how to find the equation of a line from two points, which was done in Chapter 3 of **MATH** Connections Year 1.

   (a) $\{(x, y, z) \mid y = 5x - 2\}$

   (b) $\{(x, y, z) \mid z = \frac{2}{3}x + 0.65\}$

   (c) $\{(x, y, 3) \mid y = -x\}$ (or $\{(x, -x, 3) \mid x$ is a real number$\}$)

   (d) The two given points, and hence the line, are in the $x = 1$ plane. Thinking of that as a copy of 2-space, the first step here is to get the equation for the line through $(2, 3)$ and $(5, 8)$ in 2-space, using y and z as the variables. By the method of Chapter 3 of **MATH** Connections Year 1, this equation is $z = \frac{5}{3}y - \frac{1}{3}$. The line in the $x = -4$ plane parallel to

   this one is $\{(-4, y, z) \mid z = \frac{5}{3}y - \frac{1}{3}\}$

   or $\{(-4, y, \frac{5}{3}y - \frac{1}{3}) \mid y$ is a real number$\}$

   (e) This is like the previous part. The two given points, and hence the line, are in the $y = 2$ plane. Thinking of that as a copy of 2-space, the first step here is to get the equation for the line through $(7, 1)$ and $(8, -1)$ in 2-space, using x and z as the variables. This equation is $z = -2x + 15$. The line in the $y = 5$ plane parallel to this one is $\{(x, 5, z) \mid z = -2x + 15\}$, or $\{(x, 5, -2x + 15) \mid x$ is a real number$\}$.

5. This is a challenging project. It is suitable for your very best students and also for small groups.

   The first part is straightforward, but might not be obvious. In 2-space, to label a point in terms of positions (distances from 0) on the coordinate axes, draw through the point the two lines parallel to the axes. The places where each of the lines intersect the other axis gives you a coordinate for the point. The 3-space generalization is the obvious thing. Beyond that essential first step, this investigation is quite open-ended. One critical idea that goes bad is the distance formula because the triangles in the derivation no longer are *right* triangles. This can be fixed by what amounts to a particular case of the Law of Cosines (though students don't have to recognize it as such), which will allow them to find the third side of any oblique triangle with angles determined by the angle at which the axes intersect. Such a system is theoretically sound as an alternative description of the plane or of 3-space, but it is not nearly as intuitively comfortable or as simple in its computations.

   For students capable of dealing with uncertainty, this project provides a challenge that will pay off in an enhanced appreciation of the power of rectangular coordinates, regardless of what else they are able to do with it.

Chapter 5

5.8 The Algebra of Three Dimensional Shapes

Learning Outcomes

After studying this section you will be able to:

Use coordinates to describe some solid and some hollow three dimensional shapes;

Recognize and use equations that describe circles anywhere in 2-space;

Recognize and use equations that describe spheres anywhere in 3-space.

At the end of Section 5.7, you saw that a plane can be formed by stacking up lines. For instance, you can think of the plane, P, perpendicular to the xy-plane along the line $y = x$ as an infinite stack of copies of this line, one for each possible value of z. In set-builder notation,

$$P = \{(x, y, z) \mid y = x\}$$

If you don't use any negative values for z, you get a half-plane

$$\{(x, y, z) \mid y = x \text{ and } z \geq 0\}$$

(See Display 5.64)

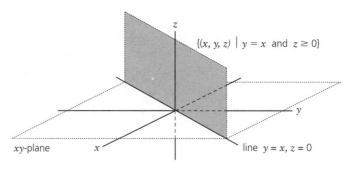

Display 5.64

Restricting z in other ways gives you different pieces of the plane. For instance, if you stack only the copies of this line for z between -1 and 1, you get a strip of this plane 2 units wide, extending 1 unit above and 1 unit below the xy-plane.

$$\{(x, y, z) \mid y = x \text{ and } -1 \leq z \leq 1\}$$

(See Display 5.65)

456

5.8 The Algebra of Three Dimensional Shapes

This section is devoted to describing two and three dimensional shapes in terms of equations and inequalities. The set theoretic union of set-builder descriptions of shapes is described informally after 5.68. Its usage is consistent with the earlier definition of union of sets, in Chapter 7 of **MATH** *Connections* Year 1. If you choose, you might also remind students of set theoretic intersection. The set theoretic intersection of two figures, considered as sets of points, agrees with the usual geometric meaning of *intersection*.

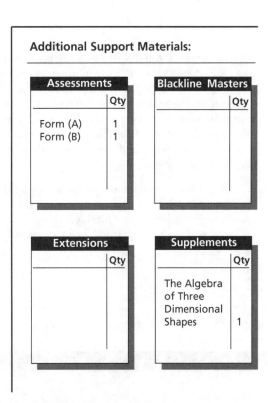

Additional Support Materials:

Assessments	Qty
Form (A)	1
Form (B)	1

Blackline Masters	Qty

Extensions	Qty

Supplements	Qty
The Algebra of Three Dimensional Shapes	1

Chapter 5

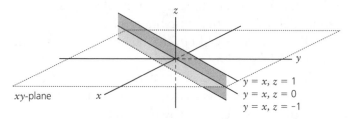

$y = x, z = 1$
$y = x, z = 0$
$y = x, z = -1$

xy-plane   *x*

Display 5.65

We can extend this idea to form other figures.

> **Think of each of these 2-space shapes in the *xy*-plane. In each case, describe the 3-space shape you get if you stack copies of the given 2-space shape for $0 \leq z \leq 3$.**
>
> 1. a line segment
>
> 2. a rectangle
>
> 3. a rectangular region
>
> 4. a right triangle
>
> 5. a right triangular region
>
> 6. a circle
>
> 7. a disk
>
> 8. an annulus

5.66

How would you describe to a computer each of the shapes above? That is, what coordinate information would you have to supply so that the computer would "know" the size and shape of the figure you have in mind? In these cases, the *z*-coordinate information for all of them is already given: $0 \leq z \leq 3$. The rest of the information is just the coordinate description of the particular two dimensional shape in the *xy*-plane. Let's work through these shapes one at a time.

(1)   Begin with a line segment 5 inches long. The easiest way to describe this in 3-space coordinates is to choose inches as the unit length and mark off this segment on an axis—the *x*-axis—starting at 0. Then the segment is

$$\{(x, 0, 0) \,|\, 0 \leq x \leq 5\}$$

Published by IT'S ABOUT TIME, Inc. © 2000 MATHconx, LLC

457

5.66

This is a simple exercise to give visual meaning to the algebraic variation of a coordinate between two boundary values. In each case, the resulting 3-space figure is 3 units high.

1. a rectangular region, one side of which is 3 units long
2. a hollow rectangular box (without top or bottom)
3. a solid rectangular box; i.e., a right rectangular prism
4. a right prism with a right triangular base
5. a right prism with an isosceles triangular base
6. a right cylinder (hollow)
7. a solid cylinder
8. a cylinder with a thick wall (like a 3 inch piece of pipe, if the unit is inches)

NOTES

Chapter 5

The stack is a rectangular region in the xz-plane

$$\{(x, 0, z) \,|\, 0 \le x \le 5 \text{ and } 0 \le z \le 3)\}$$

If you want the region to be parallel to the xz-plane, but not on it, just move it over by picking a value for y. This one is 4 inches away from the xz-plane

$$\{(x, 4, z) \,|\, 0 \le x \le 5 \text{ and } 0 \le z \le 3)\}$$

(See Display 5.66)

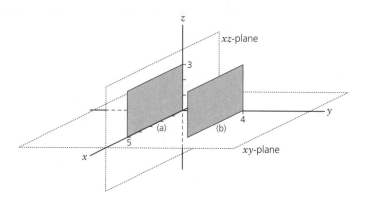

Display 5.66

a
5.67

1. If you want this 5 by 3 inch rectangular region to be in the yz-plane, how would you write its coordinate description?

2. Write the coordinate description for a 5 by 3 inch rectangular region that is parallel to the yz-plane and 10 inches away from it.

(2) Now let's describe a 5 by 7 inch rectangle. A rectangle is just four line segments connected at right angles, so we can build on the work we just did.

b
5.68

1. Use set-builder notation to describe a 5 inch segment on the x-axis and a 7 inch segment on the y-axis, each with an endpoint at the origin.

2. Describe the two other segments that are needed to complete the rectangle. Move over the two segments you described in part 1.

Published by IT'S ABOUT TIME, Inc. © 2000 MATHconx, LLC

458

5.67

This mimics what was just done, substituting y for x.

1. If you want the region to be in the yz-plane, just start along the y-axis.

$$\{(0, y, z) \mid 0 \leq y \leq 5 \text{ and } 0 \leq z \leq 3\}$$

2. $\{(10, y, z) \mid 0 \leq y \leq 5 \text{ and } 0 \leq z \leq 3\}$ or $\{(-10, y, z) \; 0 \leq y \leq 5$ and $0 \leq z \leq 3\}$, or , assuming the segment is placed along the positive y-axis starting at the origin.

5.68

This rectangle is easiest to describe if it is positioned in the first quadrant, so that no negative numbers are needed. These questions extend the previous example in a straightforward way. Question 3 might cause a little difficulty at first. If necessary, help students to see that they can stack each side of the rectangle separately and then put them together.

1. $\{(x, 0, 0) \mid 0 \leq x \leq 5\}$ and $\{(0, y, 0) \mid 0 \leq y \leq 7\}$

2. $\{(x, 7, 0) \mid 0 \leq x \leq 5\}$ and $\{(5, y, 0) \mid 0 \leq y \leq 7\}$

3. $\{(x, 0, 0) \mid 0 \leq x \leq 5 \text{ and } 0 \leq z \leq 3\} \cup \{(0, y, 0) \mid 0 \leq y \leq 7 \text{ and } 0 \leq z \leq 3\} \cup \{(x, 7, 0) \mid 0 \leq x \leq 5 \text{ and } 0 \leq z\} \leq 3 \cup \{(5, y, 0) \mid 0 \leq y \leq 7 \text{ and } 0 \leq z \leq 3\}$. There's no need to require a precise definition of union here. The $\cup$ symbol is just an abbreviation for "take all these points together." That works quite well enough and is compatible with the formal definition.

Chapter 5

NOTES

3. Now write a set-builder description of the stack, which is a 3 inch high rectangular box without top or bottom. To put the four pieces together, use ∪—union of sets. Recall that earlier in MATH *Connections* we combined sets of elements with the set operation union.

(3) Describing a rectangular region by coordinates is simpler than describing a rectangle. For example, a point (x, y) is within the 5 by 7 inch rectangular region of part (2) if, and only if, x is between 0 and 5, inclusive, and y is between 0 and 7, inclusive. The set-builder form of this description is

$$\{(x, y, 0) \mid 0 \leq x \leq 5 \text{ and } 0 \leq y \leq 7)\}$$

The 3 inch high stack of these rectangular regions is

$$\{(x, y, z) \mid 0 \leq x \leq 5, 0 \leq y \leq 7, 0 \leq z \leq 3)\}$$

A comma between conditions in set-builder notation is a simpler way to write *and*. The result is a rectangular block, outlined in Display 5.67. Which block is it? How can you tell?

If you want this block placed somewhere else, just translate it. Figure out where you want the $(0, 0, 0)$ corner to end up, and then add the coordinates of that point to every point of the figure. The easiest way to do this is to adjust the inequality boundary values by these amounts. For instance, if the block just described is moved 4 inches back and 8.5 inches to the right, the point $(0, 0, 0)$ becomes $(-4, 8.5, 0)$. (See Display 5.67.) In set-builder notation, this block is

$$\{(x, y, z) \mid -4 \leq x \leq 1, 8.5 \leq y \leq 15.5, 0 \leq z \leq 3)\}$$

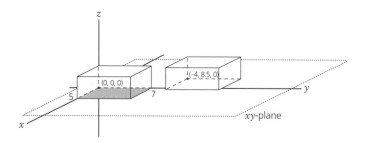

Display 5.67

Published by IT'S ABOUT TIME, Inc. © 2000 MATHconx, LLC

459

NOTES

Chapter 5

5.69

These questions refer to Display 5.67 (shown again below). The original block described in (3) is on the left; the translated block is on the right.

1. List the coordinates of each of the other seven corners of the translated block.

2. Write the coordinates of two points that are inside the translated block and of two points that are outside both blocks.

3. For each of these points, tell whether it is inside, on the surface of, on an edge of, or outside the original block. Justify your choice.

(a) (3, 5, 2)   (b) (5, 2, 3)   (c) (2, 3, 5)
(d) (5, 3, 2)   (e) (1, 1, 1)   (f) (3, 3, 3)
(g) (0, 7, 1.4) (h) (−1, 4, 3)

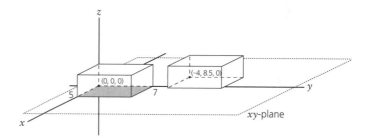

Display 5.67 (duplicate)

5.70

Adapt the ideas of (1)–(3) above to answer these questions about right triangles.

1. What shape is $\{(x, -x, 0) \mid 0 \le x \le 5\}$? What is its size? Where is it?

2. What shape is $\{(x, -x + 5, 0) \mid 0 \le x \le 5\}$? What is its size? Where is it?

3. Describe in set-builder form a right isosceles triangle with 5 inch legs. Extend your answer to a set-builder description of a 3 inch high stack of these triangles.

4. Describe in set-builder form the region enclosed by a right isosceles triangle with 5 inch legs. Extend your answer to a set-builder description of a 3 inch high stack of these regions.

Published by IT'S ABOUT TIME, Inc. © 2000 MATHconx, LLC

460

5.69

These questions are straightforward and simple. If your students are having no trouble making the connection between pictures and coordinate descriptions, you can skip some of them. However, they might help students, if they are having trouble, to build some confidence in their work.

1. (1, 8.5, 0), (1, 15.5, 0), (-4, 15.5, 0), (-4, 8.5, 3), (1, 8.5, 3), (1, 15.5, 3), (-4, 15.5, 3)

2. Answers will vary.

3. (a) It's inside; all three coordinates are properly between the boundary values.
   (b) It's on the upper front edge, 2 inches to the right.
   (c) It's outside (above) the block; the z-coordinate is greater than 3.
   (d) It's on the front, 3 inches to the right and 2 inches up.
   (e) It's inside the block.
   (f) It's on the top surface (because $z = 3$).
   (g) It's on the back right edge, 1.4 inches up.
   (h) It's outside (behind) the block.

5.70

These questions should also be done now. They are listed as discussion questions because they are less routine than the inside-outside on questions about the rectangular boxes. The combined wisdom of several students might get them by the puzzling spots. If not, help them along.

1. It's a line segment $5\sqrt{2}$ units long. One end is at the origin; the other is at (5, -5, 0).

2. It's a line segment $5\sqrt{2}$ units long. One end is at the (0, 5, 0); the other is at (5, 0, 0). This is the previous line segment translated up 5 units—that is, by adding 5 to the y-coordinate of each point.

3. The simplest way to envision such a triangle is by using the origin as the right angle vertex and 5 units of each positive axis in the xy-plane as the legs. The previous question describes the hypotenuse of this triangle. The triangle is the union of three line segments.
   $$\{(x, 0, 0) \mid 0 \leq x \leq 5\} \cup \{(0, y, 0) \mid 0 \leq y \leq 5\} \cup$$
   $$\{(x, -x + 5, 0) \mid 0 \leq x \leq 5\}$$
   The 3 inch high stack is
   $$\{(x, 0, z) \mid 0 \leq x \leq 5, 0 \leq z \leq 3\} \cup \{(0, y, z) \mid 0 \leq y \leq 5, 0 \leq z \leq 3\} \cup$$
   $$\{(x, -x + 5, z) \mid 0 \leq x \leq 5, 0 \leq z \leq 3\}$$

4. As in the case of the rectangle, the region has a more compact description than its boundary. There is an important new idea here, however, which might warrant some discussion and perhaps a sketch.

Chapter 5

5. Part 4 describes a right prism with a triangular base. Specify the coordinates of

   (a) a point inside the prism (not on any face or edge);
   (b) a point on the xy-plane that is outside the prism;
   (c) a point on the top face of the prism, but not on an edge;
   (d) a point on the hypotenuse edge of the base triangle.

That takes care of 4 and 5 on the list of eight 2-space shapes that we have been working through. Stacks of shapes 6, 7, and 8 depend on describing circles and disks in 2-space. These coordinate descriptions come from the 2-space distance formula $\sqrt{(x_2 - x_1)^2 + (y_2 - y_1)^2}$. Let's apply the distance formula to a circle. Recall that a circle is the set of all points in a plane that are a fixed distance (radius) from a particular point (the center). If we know the radius and the coordinates of the center, then we ought to be able to write a coordinate description of the circle.

We begin by describing a circle of radius 6 cm in terms a computer can handle. The simplest way to do this is to treat the origin of the (two dimensional) coordinate system, $(0, 0)$, as the center of the circle and measure the coordinates in centimeters. Then the circle is the set of all points (x, y) that are 6 cm from $(0, 0)$. (See Display 5.68(a)) By the distance formula, the distance between $(0, 0)$ and (x, y) is

$$\sqrt{(x - 0)^2 + (y - 0)^2} = \sqrt{x^2 + y^2}$$

The circle is the set of all points for which this distance equals 6

$$\{(x, y) \,|\, \sqrt{x^2 + y^2} = 6\}$$

We can simplify the condition by squaring both sides of the equation,

$$\{(x, y) \,|\, x^2 + y^2 = 36\} \qquad \text{(5.71)}$$

This example illustrates how to describe any circle centered at the origin.

A Fact to Know: The equation $x^2 + y^2 = r^2$ describes a circle with center $(0, 0)$ and radius r.

Published by IT'S ABOUT TIME, Inc. © 2000 MATHconx, LLC

461

The hypotenuse lies along the line $y = -x + 5$. A point $(x, y, 0)$ is *above* this line on the xy-plane if $y > -x + 5$; the point is below this line if $y < -x + 5$.

The region in the xy-plane is
$$\{(x, y, 0) \mid 0 \le x \le 5, 0 \le y \le -x + 5\}$$

The 3 inch high right prism (the stack) is
$$\{(x, y, 0) \mid 0 \le x \le 5, 0 \le y \le -x + 5, 0 \le z \le 3\}$$

5. These answers will vary. In the event of confusion, check the coordinates against the defining conditions.

5.71

Some students might wonder if squaring might hurt something. This is a legitimate algebraic worry. However, in this situation we are dealing only with distances, so all the numbers are nonnegative. Thus, there is a one-to-one correspondence between the distances and their squares. This ensures that nothing goes wrong when we square both sides of the equation.

Chapter 5

NOTES

Describing a circle of radius 6 cm centered somewhere else in the coordinate plane is almost as easy. If you want the circle to be centered at (5, 4), for instance, you follow the same thought process as before. This circle is the set of all points (x, y) that are 6 cm from (5, 4). (See Display 5.68(b).)

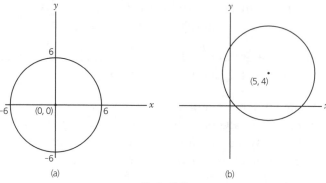

(a)   (b)

Display 5.68

By the distance formula, the distance between (5, 4) and (x, y) is

$$\sqrt{(x - 5)^2 + (y - 4)^2}$$

The circle is the set of all points for which this distance equals 6,

$$\{(x, y \mid \sqrt{(x - 5)^2 + (y - 4)^2} = 6\}$$

We can simplify the condition by squaring both sides of the equation,

$$\{(x, y \mid (x - 5)^2 + (y - 4)^2 = 36\}$$

This illustrates a general way of describing a circle of any radius centered anywhere.

A Fact to Know: A circle of radius r centered at (a, b) is the set of all points (x, y) that satisfy the equation

$$(x - a)^2 + (y - b)^2 = r^2$$

Display 5.69 illustrates this fact

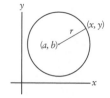

Display 5.69

Published by IT'S ABOUT TIME, Inc. © 2000 MATHconx, LLC

NOTES

5.72

For these questions, assume that you have a 2-space coordinate system measured in centimeters. It might help to draw a sketch.

1. Write an equation that describes a circle with radius 10 cm centered at the origin. Call this Circle 1.

2. Write an equation that describes a circle with radius 10 cm centered at (8, −6). Call this Circle 2.

3. For each of the following points, say whether it is inside, outside, or on Circle 1. Also say whether it is inside, outside, or on Circle 2. Justify your answers.

   (a) (0, 0)   (b) (9, 5)   (c) (9, −5)
   (d) (6.6, −2.1) (e) (−6, −8)   (f) (8, −16)

4. Find a number x such that (x, 4) is on Circle 1. Explain how you found it.

5. Find a number x such that (x, 4) is on Circle 2. Explain how you found it.

6. Find a number y such that (4, y) is on Circle 2. Explain how you found it.

7. Write set-builder descriptions of the disks enclosed by Circles 1 and 2. Explain them briefly.

   Now that you have the keys to the 2-space shapes 6, 7, and 8, you should be able to unlock them on your own. We leave them for the Problem Set (see problem 3) and turn, instead, to the 3-space analogue of what we have just done.

   The three dimensional analogue of a circle is a sphere. It is the set of all points in 3-space that are a fixed distance (radius) from a particular point (the center). It is easy to write an equation to describe any sphere centered at the origin.

Published by IT'S ABOUT TIME, Inc. © 2000 MATHconx, LLC

5.8 The Algebra of Three Dimensional Shapes

463

5.72

Most of these are routine reinforcement questions for the coordinate description of a circle. Question 6 requires a little algebraic common sense. *The last question is important, even though it's a simple extension of the ideas already presented. Please make sure that your students understand how to write the algebraic expression for a disk.*

1. Circle 1. $x^2 + y^2 = 100$

2. Circle 2. $(x - 8)^2 + (y + 6)^2 = 100$

3. (a) inside Circle 1; on Circle 2
   (b) outside both circles
   (c) outside Circle 1; inside Circle 2
   (d) inside both circles
   (e) on Circle 1; outside Circle 2
   (f) outside Circle 1; on Circle 2

4. Solve $x^2 + 4^2 = 100$ for x: $x = \pm\sqrt{84}$

5. Solve $(x - 8)^2 + (4 + 6)^2 = 100$ for x: $x = 8$

6. Solve $(4 - 8)^2 + (y + 6)^2 = 100$ for y: $y = -6 \pm\sqrt{84} \approx 3.165, -15.165$

7. Since a disk is all points whose distance from the center point is less than or equal to the radius, just replace $=$ in the equations for these circles with $\leq$ to get the disks
   Disk 1: $\{(x, y) \mid x^2 + y^2 \leq 100\}$
   Disk 2: $\{(x, y) \mid (x - 8)^2 + (y + 6)^2 \leq 100\}$

NOTES

Chapter 5

5.8 The Algebra of Three Dimensional Shapes

If we let r stand for the radius, then such a sphere is the set of all points (x, y, z) exactly r units away from $(0, 0, 0)$, as in Display 5.70(a). The 3-space distance formula does the rest.

A Fact to Know: A sphere of radius r centered at $(0, 0, 0)$ is the set of all points (x, y, z) that satisfy the equation

$$x^2 + y^2 + z^2 = r^2$$

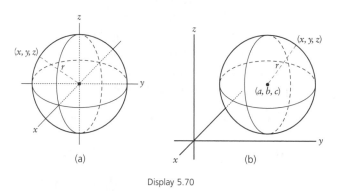

Display 5.70

If the center of the circle is at some point (a, b, c), as in Display 5.70(b), the same kind of reasoning produces a similar result.

A Fact to Know: A sphere of radius r centered at (a, b, c) is the set of all points (x, y, z) that satisfy the equation

$$(x - a)^2 + (y - b)^2 + (z - c)^2 = r^2$$

5.73

For these questions, assume that you have a 3-space coordinate system measured in centimeters.

1. Write a set-builder description of a sphere with radius 11 cm centered at the origin. Call this Sphere 1.

2. Write a set-builder description of a sphere with radius 11 cm centered at $(9, -6, 2)$. Call this Sphere 2.

3. For each of the following points, tell whether it is inside, outside, or on Sphere 1. Also tell whether it is inside, outside, or on Sphere 2. Justify your answers.

   (a) $(0, 0, 0)$ (d) $(\sqrt{50}, \sqrt{40}, \sqrt{31})$
   (b) $(10, -5, 3)$ (e) $(4, -3, -1)$
   (c) $(10, 5, 3)$ (f) $(9, -17, 2)$

5.73

This set of questions is deliberately analogous to the questions of 5.72 in order to emphasize the direct connection between circles in 2-space and spheres in 3-space. However, the pattern of the inside-outside on answers is different so that students who see the analogy can't use the old answers without working out these questions.

1. Sphere 1. $\{(x, y, z) \mid x^2 + y^2 + z^2 = 121\}$

2. Sphere 2. $\{(x, y, z) \mid (x - 9)^2 + (y + 6)^2 + (z - 2)^2 = 121\}$

3. (a) inside Sphere 1; on Sphere 2
    (b) outside Sphere 1; inside Sphere 2
    (c) outside both spheres
    (d) on Sphere 1; outside Sphere 2
    (e) inside both spheres
    (f) outside Sphere 1; on Sphere 2

NOTES

Chapter 5

4. Solve $x^2 + 4^2 + 3^2 = 121$ for x: $x = \pm\sqrt{96}$

5. Solve $(0 - 9)^2 + (y + 6)^2 + (4 - 2)^2 = 121$ for y. $(y + 6)^2 = 36$, so $y + 6 = \pm 6$, implying $y = 0$ or $y = -12$.

6. This question is a bit different than the two before it, in that there is a lot of free choice here. We want x and z such that $x^2 + 7^2 + z^2 = 121$, that is, $x^2 + z^2 = 72$. A whole circle of points fits this condition! Choose x to be anything between $-\sqrt{72}$ and $\sqrt{72}$, and then solve for z. For instance, $(5, 7, \sqrt{47})$ is on Sphere 1; so are $(6, 7, 6)$, $(6, 7, -6)$, $(0, 7, \sqrt{72})$, $(1, 7, \sqrt{71})$, etc.

7. Since a solid ball is all points whose distance from the center point is less than or equal to the radius, just replace $=$ in the equations for these spheres with $\leq$ to get the solid balls:
   Ball 1. $\{(x, y, z) \mid x^2 + y^2 + z^2 \leq 121\}$
   Ball 2. $\{(x, y, z) \mid (x - 9)^2 + (y + 6)^2 + (z - 2)^2 \leq 121\}$

5.74

This really isn't much of a writing exercise. There's only a step or two in each case, but it's important that students see the link between the distance formula and the equations.

Sphere at origin. By the distance formula, the distance between $(0, 0, 0)$ and (x, y, z) is

$$\sqrt{(x - 0)^2 + (y - 0)^2 + (z - 0)^2} = \sqrt{x^2 + y^2 + z^2}$$

The sphere is the set of all points for which this distance equals r

$$\{(x, y, z) \mid \sqrt{x^2 + y^2 + z^2} = r\}$$

Simplify the condition by squaring both sides of the equation

$$\{(x, y, z) \mid x^2 + y^2 + z^2 = r^2\}$$

Sphere at (a, b, c). By the distance formula, the distance between (a, b, c) and (x, y, z) is

$$\sqrt{(x - a)^2 + (y - b)^2 + (z - c)^2}$$

The sphere is the set of all points for which this distance equals r

$$\{(x, y, z) \mid \sqrt{(x - a)^2 + (y - b)^2 + (z - c)^2} = r\}$$

Simplify the condition by squaring both sides of the equation

$$\{(x, y, z) \mid (x - a)^2 + (y - b)^2 + (z - c)^2 = r^2\}$$

NOTES

Chapter 5

4. Find a number x such that $(x, 4, 3)$ is on Sphere 1. Explain how you found it.

5. Find a number y such that $(0, y, 4)$ is on Sphere 2. Explain how you found it.

6. Find numbers x and z such that $(x, 7, z)$ is on Sphere 1. Explain how you found them.

7. Write set-builder descriptions of the solid balls enclosed by Spheres 1 and 2. Explain them briefly.

Write out the details of how the distance formula leads to the equations for a sphere (1) centered at the origin and (2) centered at any point (a, b, c). (Mimic what was done for the circle.)

5.74

Problem Set: 5.8

1. Using a rectangular coordinate system with inches as the unit of measure, describe each of these objects in set-builder notation.

   (a) a solid 10 by 8 by 4 inch rectangular block with one corner at $(0, 0, 0)$ (There is more than one way to do this. Pick the one that seems simplest to you.)
   (b) the surface of the rectangular block of part (a)
   (c) the block of part (a) translated so that its $(0, 0, 0)$ corner becomes $(5, -6, 9)$, but not rotated or reflected in any way

2. Each of the following algebraic expressions represents a circle, a disk, a sphere, or a solid ball. For each one, say which of these four shapes it represents. Then specify the center, the radius, and one point of the figure.

   (a) $x^2 + y^2 = 9$
   (b) $x^2 + y^2 + z^2 = 27$
   (c) $x^2 + y^2 - 12 = 0$
   (d) $x^2 + y^2 + z^2 \leq 100$
   (e) $(x - 1)^2 + (y - 5)^2 \leq 18$
   (f) $(x - 5)^2 + (y - 3)^2 + (z + 4)^2 = 7$
   (g) $(x + 7)^2 + y^2 + z^2 \leq 1$
   (h) $(x + 4)^2 + (y - 4)^2 \leq 4$

3. These questions refer to a 3-space coordinate system measured in centimeters. They are examples of questions 6, 7, and 8 in the first Discuss this now of this section. Write each description in set-builder notation.

Published by IT'S ABOUT TIME, Inc. © 2000 MATHconx, LLC

465

Problem Set: 5.8

1. (a) Here is one simple way. $\{(x, y, z) \mid 0 \le x \le 10, 0 \le y \le 8, 0 \le z \le 4\}$
   (b) This one is not particularly difficult, but it is annoyingly long. Each of the six faces must be described separately. Note that the specific answer in this part depends on how the student answered part (a).
   $\{(0, y, z) \mid 0 \le y \le 8, 0 \le z \le 4\} \cup \{(x, 0, z) \mid 0 \le x \le 10, 0 \le z \le 4\}$
   $\cup \{(x, y, 0) \mid 0 \le x \le 10, 0 \le y \le 8\}$
   $\cup \{(10, y, z) \mid 0 \le y \le 8, 0 \le z \le 4\}$
   $\cup \{(x, 8, z) \mid 0 \le x \le 10, 0 \le z \le 4\}$
   $\cup \{(x, y, 4) \mid 0 \le x \le 10, 0 \le y \le 8\}$
   (c) Add these coordinate values to each boundary of the inequalities, as follows.
   $$\{(x, y, z) \mid 5 \le x \le 15, -6 \le y \le 2, 9 \le z \le 13\}$$

2. These are straightforward examples of the relevant formulas, but some of the numbers will trap unwary students. Of course, the test for a student's "one point of the figure" in each case is whether or not it satisfies the given equation or inequality.

   (a) A circle with center $(0, 0)$ and radius 3
   (b) A sphere with center $(0, 0)$ and radius $\sqrt{27}(= 3\sqrt{3})$
   (c) A circle with center $(0, 0)$ and radius $\sqrt{12} \ (= 2\sqrt{3})$
   (d) A solid ball with center $(0, 0, 0)$ and radius 10
   (e) A disk with center $(1, 5)$ and radius $\sqrt{18} \ (= 3\sqrt{2})$
   (f) A sphere with center $(5, 3, -4)$ and radius $\sqrt{7}$
   (g) A solid ball with center $(-7, 0, 0)$ and radius 1
   (h) A disk with center $(-4, 4)$ and radius 2

3. Each of these is pretty straightforward, but they lead, step by step, from simple situations to less obvious ones. We suggest that you cover them all, rather than just picking out a few parts to assign. Make sure that students understand right from the start that even in the xy-plane all points have *three* coordinates. The last part might require some discussion.

   (a) circle: $\{(x, y, 0) \mid x^2 + y^2 = 49\}$ disk: $\{(x, y, 0) \mid x^2 + y^2 \le 49\}$
   (b) Note that the stack of circles is a cylinder 3 cm high (just the surface), open at top and bottom. The stack of disks is the solid for which this cylinder is the vertical "skin."
   cylinder: $\{(x, y, 0) \mid x^2 + y^2 = 49, 0 \le z \le 3\}$
   solid: $\{(x, y, 0) \mid x^2 + y^2 \le 49, 0 \le z \le 3\}$
   The tops of these figures are a circle and a disk, respectively. Only the z-coordinate changes
   circle: $\{(x, y, 3) \mid x^2 + y^2 = 49\}$ disk: $\{(x, y, 3) \mid x^2 + y^2 \le 49\}$
   (c) circle: $\{(x, y, 0) \mid (x - 4)^2 + (y - 9)^2 = 49\}$
   disk: $\{(x, y, 0) \mid (x - 4)^2 + (y - 9)^2 \le 49\}$

(a) Describe a circle in the xy-plane that has radius 7 cm and is centered at the origin, (0, 0, 0). Then describe the disk it encloses.

(b) Describe the 3-space shapes you get when you make 3 cm high stacks of the figures in part (a). Then describe just the top of each of these figures. (Also say in words what these tops look like.)

(c) Describe a circle in the xy-plane that has radius 7 cm and is centered at (4, 9, 0). Then describe the disk it encloses.

(d) Describe the 3-space shapes you get when you make 3 cm high stacks of the figures in part (c). Then describe just the top of each of these figures. Describe what these tops look like.

(e) Describe an annulus in the xy-plane, centered at the origin, with inner radius 2 cm and outer radius 5 cm. Then describe the annulus you get when this one is translated so that its center point is (12, 20, 0).

(f) Describe the 3-space shapes you get when you make 3 cm high stacks of the figures in part (e). Then describe just the top of each of these figures. Also say in words what these tops look like.

(g) Can you describe the circle and the disk you get when the circle and the disk of part (a) are translated so that their center point is (25, 35, 45)? If you can, do it. If you cannot, explain what more information you think you need to be able to do it.

(h) Can you describe the shapes you get when the shapes of part (b) are translated so that its base center point is (25, 35, 45)? If you can, do it. If you cannot, explain what more information you think you need to be able to do it.

4. In each part, the 2 three dimensional figures are very much alike, but they differ in one particular way. Identify the shape of the figures in each case. Then describe the difference between the two figures in terms of size, orientation and/or location.

(a) $\{(x, y, z) \mid 0 \le x \le 2, 0 \le y \le 2, 0 \le z \le 7)\}$
$\{(x, y, z) \mid 0 \le x \le 3, 0 \le y \le 3, 0 \le z \le 7)\}$

(b) $\{(x, y, z) \mid 0 \le x \le 2, 0 \le y \le 2, 0 \le z \le 7)\}$
$\{(x, y, z) \mid 5 \le x \le 7, 4 \le y \le 6, 0 \le z \le 7)\}$

Published by IT'S ABOUT TIME, Inc. © 2000 MATHconx, LLC

(d) cylinder: $\{(x, y, 0) \mid (x - 4)^2 + (y - 9)^2 = 49, 0 \le z \le 3\}$
solid: $\{(x, y, 0) \mid (x - 4)^2 + (y - 9)^2 \le 49, 0 \le z \le 3\}$
Tops: circle: $\{(x, y, 3) \mid (x - 4)^2 + (y - 9)^2 = 49\}$
disk: $\{(x, y, 3) \mid (x - 4)^2 + (y - 9)^2 \le 49\}$

(e) at origin: $\{(x, y, 0) \mid 4 \le x^2 + y^2 \le 25\}$
translated: $\{(x, y, 0) \mid 4 \le (x - 12)^2 + (y - 20)^2 \le 25\}$

(f) They both are the same size and shape—a solid cylinder of radius 5 cm with a 2 cm radius hole down through the center.
at origin: $\{(x, y, 0) \mid 4 \le x^2 + y^2 \le 25, 0 \le z \le 3\}$
translated: $\{(x, y, 0) \mid 4 \le (x - 12)^2 + (y - 20)^2 \le 25, 0 \le z \le 3\}$
In each case, the top is an annulus of the same size and shape as the original, but 3 cm higher (in the z direction):
at origin: $\{(x, y, 3) \mid 4 \le x^2 + y^2 \le 25\}$
translated: $\{(x, y, 3) \mid 4 \le (x - 12)^2 + (y - 20)^2 \le 25\}$

(g) A possible source of confusion in this part and the next is the precise meaning of *translate* in this context. If properly understood (as exemplified in the section), it is easy. But the correct visual image is critical, and expressing it in terms of coordinates may puzzle some students at first. Just as in the plane (in Chapters 2 and 3), a translation is a movement of the figure that does not rotate it in any way. It is particularly simple to describe in terms of coordinates: a fixed quantity, one for each coordinate direction, is added to each coordinate of every point of the figure. In this case, if the center $(0, 0, 0)$ is translated to $(25, 35, 45)$, then 25 must be added to each x-coordinate, 35 to each y-coordinate, and 45 to each z-coordinate. This can be done in two different, but equally easy ways in set-builder notation.
One way,
circle: $\{(x + 25, y + 35, 45) \mid x^2 + y^2 = 49\}$
disk: $\{(x + 25, y + 35, 45) \mid x^2 + y^2 \le 49\}$
Another way,
circle: $\{(x, y, 45) \mid (x - 25)^2 + (y - 35)^2 = 49\}$
disk: $\{(x, y, 45) \mid (x - 25)^2 + (y - 35)^2 \le 49\}$

(h) As in the previous part, there are two obvious ways to do this.
One way,
cylinder: $\{(x + 25, y + 35, 45) \mid (x - 4)^2 + (y - 9)^2 = 49, 0 \le z \le 3\}$
solid: $\{(x + 25, y + 35, 45) \mid (x - 4)^2 + (y - 9)^2 \le 49, 0 \le z \le 3\}$
Another way,
cylinder: $\{(x, y, 0) \mid (x - 29)^2 + (y - 44)^2 = 49, 45 \le z \le 48\}$
solid: $\{(x, y, 0) \mid (x - 29)^2 + (y - 44)^2 \le 49, 45 \le z \le 48\}$

4. This is an exercise in reading and interpreting algebraic descriptions of 3-space figures.

(a) Solid rectangular boxes with square ends. The first measures $2 \times 2 \times 7$; the second measures $3 \times 3 \times 7$.

(c) $\{(x, y, z) \mid 0 \le x \le 2, 0 \le y \le 2, 0 \le z \le 7)\}$
$\{(x, y, z) \mid 0 \le x \le 7, 0 \le y \le 2, 0 \le z \le 2)\}$

(d) $\{(x, y, z) \mid x^2 + y^2 = 4, 0 \le z \le 5\}$
$\{(x, y, z) \mid x^2 + y^2 = 4, -2 \le z \le 3\}$

(e) $\{(x, y, z) \mid x^2 + y^2 = 4, 0 \le z \le 5\}$
$\{(x, y, z) \mid x^2 + y^2 \le 4, 0 \le z \le 5\}$

(f) $\{(x, y, z) \mid x^2 + y^2 \le 4, 0 \le z \le 5\}$
$\{(x, y, z) \mid x^2 + z^2 \le 4, 0 \le y \le 5\}$

(g) $\{(x, y, z) \mid x^2 + y^2 \le 4, 0 \le z \le 9\}$
$\{(x, y, z) \mid x^2 + y^2 \le 9, 0 \le z \le 4\}$

(h) $\{(x, y, 3) \mid x^2 + y^2 \le 4\}$
$\{(x, y, 7) \mid x^2 + y^2 \le 4\}$

5. This problem refers to Sphere 1 and Sphere 2, the two spheres that you were asked to describe in questions 1–7 at the end of this section.

(a) Write a program for your calculator that will compute the distance between two points in 3-space. Call it DISTANCE.

(b) Use the DISTANCE program to compute the distances between these pairs of points.

(i) (24, 93, –76.2) and (–31, 250, 42.7)
(ii) $(0.93, -2.11, \sqrt{0.31})$ and $(1.04, \sqrt{0.58}, -1.15)$

(c) Write a program for your calculator that will take the three coordinates of a point, in x-y-z order, and tell you whether that point is inside, outside, or on Sphere 1. Call this program SPHERE1.

(d) Write a program for your calculator that will take the three coordinates of a point, in x-y-z order, and tell you whether that point is inside, outside, or on Sphere 2. Call this program SPHERE2.

(e) Use the programs SPHERE1 and SPHERE2 to check your answers to question 3 at the end of this section.

(f) Use the three programs you wrote to help you find the vertices of a triangle that has no side shorter than 9 cm and is *completely contained inside both* spheres. Explain your solution.

Published by IT'S ABOUT TIME, Inc. © 2000 MATHconx, LLC

467

(b) Solid $2 \times 2 \times 7$ rectangular boxes with square ends. They are standing up with one square end on the xy-plane, but in different places. The back left bottom corner of the first is at $(0, 0, 0)$; the back left bottom corner of the second is at $(5, 4, 0)$.

(c) Solid $2 \times 2 \times 7$ rectangular boxes with square ends and one corner at the origin. The first is standing on end on the xy-plane; the second is lying down on the xy-plane.

(d) Hollow cylinders of radius 2 and length 5. One is standing on end on the xy-plane with its base centered at the origin; the other is 2 units lower (in the z direction.)

(e) Two cylinders of radius 2 and length 5, both in the same location and position. The first is hollow; the second is solid.

(f) Two solid cylinders of radius 2 and length 5. The first is standing on end on the xy-plane (with its base centered at the origin); the second is lying down, with one end on the xz-plane (centered at the origin).

(g) Two solid cylinders standing at the origin of the xy-plane. The first has radius 2 and length 9; the second has radius 3 and length 4.

(h) Two disks of radius 2, centered at a point on the z-axis and parallel to the xy-plane. The first is 3 units above the xy-plane (in the z direction); the second is 7 units above the xy-plane (in the z direction).

5. Besides being good exercises in programming calculators, these questions reinforce the algebraic roles of variables and literal constants. These programs can be written in many ways, some more elegant and user friendly than others. The following programs, which were written for TI-82 (TI-83) calculators, are typical, but they are not the only possibilities.

Note that although programming a TI-82 (TI-83) calculator is much like programming a computer in BASIC, there are a few simple differences that might be confusing at first. For instance, the Input command can be used only for one storage variable at a time. Other comments appear in later parts, as needed.

(a)
```
PROGRAM:DISTANCE
:Disp "1ST PT"
:Input  A
:Input  B
:Input  C
:Disp "2ND PT"
:Input  D
:Input  E
:Input  F
:√((D−A)^2+(E−B)^2+(F−C)^2)→L
:Disp "DISTANCE:",L
```

Chapter 5

(b) (i) 204.4778961 (ii) 3.342325845

(c) This program uses If and Then from the CTL menu of PGRM. Several features distinguish their TI calculator usage from that in usual computer programming.

- Each of these commands must be on a separate line.
- The condition appears right after If on the same line.
- However, Then appears on a line by itself; the consequent command is on the next line.
- The end of each conditional sequence is signaled by End, on a separate line.

PROGRAM:SPHERE1
:Input X
:Input Y
:Input Z
:X^2+Y^2+Z^2→S
:If S = 121
:Then
:Disp "ON"
:End :If S > 121
:Then
:Disp "OUTSIDE"
:End :If S < 121
:Then
:Disp "INSIDE"
:End

(d) SPHERE2 is exactly the same as SPHERE1 except for the formula line (right after the three coordinate inputs). This line becomes
$(X-9)^2+(Y+6)^2+(Z-2)^2→S$
Note that students do *not* have to re-key the entire program. They can use 2nd RCL to copy all of SPHERE1 into a new program called SPHERE2, then go back and change the one line. This is done by creating the new program from PRGM, by pressing PRGM right arrow NEW; name the program SPHERE2. Press ENTER RCL PRGM left arrow. When the RCL prompt appears, select SPHERE1 from the EXEC menu of PRGM. ENTER and change the line.

(e) This is routine practice.

(f) This can be done either with careful planning or with trial-and-error experimentation (and some luck). As a first step, students need to be able to visualize the intersection of two spheres well enough to see that, if two points are in the intersection, then so is the entire line segment between them. Thus, this problem reduces to finding in the intersection 3 points that are at least 9 cm away from each other. Infinitely many different correct answers are possible. Here is one of them (with all coordinates integers, just for a touch of elegance!).

$$(1, 1, 1), (8, -5, 1), (9, 3, -3)$$

NOTES

Chapter 5

5.9 Time and Other Dimensions

Learning Outcomes

After studying this section you will be able to:

Represent time, temperature, and other non-spatial measurement scales geometrically as coordinate axes;

Interpret two and three dimensional pictures that represent non-spatial measurements;

Use equations and inequalities to describe some shapes in two and three dimensional time-space;

Explain how the world we live in can be viewed as a four dimensional space.

Up to now, we have focused on ways to represent geometric shapes by numbers and algebra. But that's only half of the story of coordinate systems. The other half is about picturing numerical relationships by geometric shapes. Here's the simple, basic, very important idea.

A coordinate axis is a number line, so *any* numerical scale can be pictured by a coordinate axis.

Lots of things are represented by numerical scales—by numbers relative to some unit of measure. Here are a few common examples, along with various units of measure for each.

temperature — measured in degrees Celsius, degrees Fahrenheit, etc.

money — measured in dollars, pesos, yen, marks, francs, lire, etc.

weight — measured in grams, ounces, kilograms, pounds, etc.

time — measured in seconds, minutes, hours, days, years, etc.

5.75

Can you think of other things that are measured by numerical scales? Name at least three more such things, along with a common unit of measure for each one.

Many graphs of relationships between two of these things use coordinate geometry. Look at the graph of the Dow Jones Industrial Average (DJIA) in Display 5.71, for example. It pictures about 18 months in the "life" of the DJIA by using two coordinate axes. The horizontal axis represents time (in months) and the vertical axis represents the DJIA (in points), a money-based index number for the stock market. This coordinate plane captures the time-money relationship as a single region in 2-space. Each point of the region specifies a time and a point value. For instance, the highest point of the jagged line, near the far right, is (roughly) the point

([early August, 1997], 8250)

The shaded region shows us, all at once, how the DJIA changed over 18+ months of time.

Published by IT'S ABOUT TIME, Inc. © 2000 MATHconx, LLC

5.9 Time and Other Dimensions

This section introduces the idea of representing nonspatial numerical measurement scales as coordinate axes. This general idea is then focused on time as a coordinate axis, leading to some simple examples of illustrating as a single 3-space shape a two dimensional figure that changes over a period of time. These ideas provide the gateway for future study of four and more dimensions. Nothing beyond three dimensions is actually covered, but the section ends with a historical note about the idea of four dimensional space in physical science.

5.75

This need not be done if you are pressed for time, but it does serve to sharpen students' understanding of the idea of a numerical scale.
There are many such things; here are a few more.

light (brightness)—in lumens

electric power—in Watts, kiloWatts

barometric pressure—in inches of mercury

sound (volume)—in decibels

Students may come up with various other suggestions.

Chapter 5

Additional Support Materials:

Assessments	Qty
Form (A)	1
Form (B)	1
Chapter Test(A)	1
Chapter Test(B)	1

Blackline Masters	Qty

Extensions	Qty
5.9 A Glimpse of...	1
Following 5.9 Equations for...	1
Geometric...	1
Why Headlights...	1

Supplements	Qty

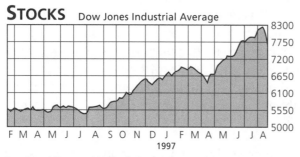

STOCKS Dow Jones Industrial Average

Display 5.71

What are the approximate coordinates of the lowest point of the jagged line?

5.76

The only annoying difference between Display 5.71 and the figures we have been studying in Cartesian 2-space is that the exact position of the axes is unclear from this figure. In particular, the location of the origin is uncertain. We can infer from the vertical scale that the 0 level of the DJIA is about 1.8 inches below the bottom of the picture, but where is the beginning of time? In order for the first coordinate of a point to represent the horizontal distance from 0, some particular date must be chosen as the 0 of the time axis.

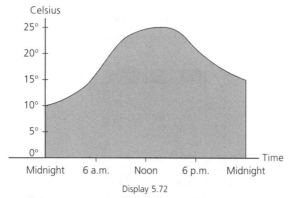

Display 5.72

Display 5.72 shows a clearer example. It is a simplified drawing of a day's outdoor temperature. Drawings like this often are made by instruments in weather stations or scientific laboratories. A piece of graph paper is wrapped around a slowly rotating drum and a pen point moves up or down as the temperature rises or falls. At the end of the day, the line traced by the pen shows you the entire day's temperature all at once.

469

5.76

This is a quick check to see if students understand Display 5.71 and the coordinate description of it. The approximate coordinates are

([late July, 1996], 5400)

If your students are interested in this example, you might extend it by bringing in a copy of the current *Wall Street Journal* and asking them to interpret some of its other graphs. The DJIA graph and others normally appear on page 1 of the Money & Investing section.

NOTES

Display 5.72 tells you that, during this particular day, the temperature rose from a midnight low of 10° to a high of 25° at 2 p.m., then dropped back to 15° by the following midnight. Is this more likely a winter day or a summer day in the Northwest?

5.77 Find a scientific instrument (in your school's science lab or somewhere else in your community) that measures the change of some quantity over time by tracing a line on a rotating drum. Write a brief, clear description of what it is and what it does.

This next example uses a 2-space coordinate picture to relate one spatial dimension to time. In this situation, the vertical axis represents height (in inches) and the horizontal axis represents time (in minutes). Suppose that you have a very thin 3 inch candle that burns at the constant rate of 1 inch every 2 minutes. Set up the candle (but don't light it) and start your stopwatch. At the end of exactly 3 minutes the unlit candle will have "lived through" a 3 by 3 square in the height-time plane. After 3 minutes light the candle. As it burns, its height decreases by 1 inch every 2 minutes, so that it will be completely gone 6 minutes after you light it. The candle, set up for 9 minutes and burning from the end of the third minute until the end of the ninth minute, is represented by the two dimensional region shown in Display 5.73.

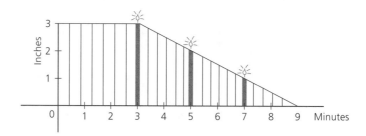

Display 5.73

5.78 The last 9 minutes in the life of the candle of Display 5.73 are shown as an odd shaped quadrilateral. What are the coordinates of its four vertices?

Published by IT'S ABOUT TIME, Inc. © 2000 MATHconx, LLC

5.77

If your science department doesn't have an instrument that traces quantity on a rotating drum and there is no place in your area where one can be observed, you could change this to a library research question. Seismographs and recording thermometers are perhaps the most common of such instruments, but there are others, including medical instruments for EKG readouts, etc. Some auto repair shops have electronic diagnosis machines or emissions testing devices that print readouts like this.

5.78

The purpose of this brief, simple exercise is to get students to think of this time space object as a single geometric figure. The vertices are $(0, 0)$, $(0, 3)$, $(3, 3)$, and $(9, 0)$.

NOTES

Chapter 5

Now let's move this idea up a dimension, to 3-space. We have already seen that a cube can be thought of as a stack of (infinitely many) square regions. (See Display 5.74.) Now, if the vertical direction is taken to represent time, then the cubic solid represents the existence of the square region during a period of time. Each horizontal cross section is the square region *at a particular instant of time.*

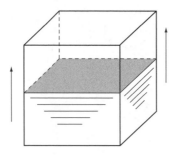

Display 5.74

What if the shape of the planar region were to change at some moment? How would that affect the three dimensional time-space figure? Here's an example.

A square piece of paper, 6 centimeters on each side, is lying on a desk during a 10 minute observation period. After 6 minutes, someone comes along and cuts the paper in half (top to bottom), and throws one half away. The life of the paper on the desk during the 10 minute observation period can be represented by a single solid region in 3-space. Two coordinates measure length (in centimeters) and one measures time (in minutes). The first 6 minutes of the paper's 10 minute existence on the desk are represented by a cube and the last 4 minutes by a rectangular box, as is shown in Display 5.75.

471

NOTES

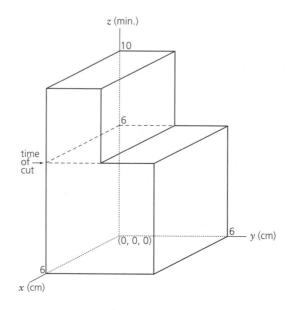

Display 5.75

5.79

These questions refer to the three dimensional solid in Display 5.75.

1. How many vertices does this solid figure have? List the coordinate names for each one.

2. Describe the cross section at time 4 minutes. What shape is it? How many vertices does it have? What are the coordinates of these vertices? What does this cross section represent in time-space?

3. Describe the cross section at time 7 minutes. What shape is it? How many vertices does it have? What are the coordinates of these vertices? What does this cross section represent in time-space?

4. Describe the cross section for $y = 2$ cm. What shape is it? How many vertices does it have? What are the coordinates of these vertices? What does this cross section represent in time-space?

5. Describe the cross section for $x = 4$ cm. What shape is it? How many vertices does it have? What are the coordinates of these vertices? What does this cross section represent in time-space?

Published by IT'S ABOUT TIME, Inc. © 2000 MATHconx, LLC

5.79

These questions require varying degrees of insight. Question 1 is completely straightforward. Questions 2 and 3 require a basic understanding of the time axis representation. Questions 4 and 5 reach beyond the illustrations of previous examples, requiring a more perceptive grasp of that representation. You can regard questions 4 and 5 as optional, if you like, but your students' insight might surprise you.

1. There are 12 in all, in three different time layers.
   Beginning: (0, 0, 0), (0, 6, 0), (6, 0, 0), (6, 6, 0)
   At time of cut: (6, 6, 6), (0, 6, 6), (6, 3, 6), (0, 3, 6)
   End: (0, 0, 10), (6, 0, 10), (6, 3, 10), (0, 3, 10)

2. It is a square 6 cm on a side. It has four vertices: (0, 0, 4), (0, 6, 4), (6, 0, 4), (6, 6, 4). It represents the sheet of paper on the desk at exactly 4 minutes from the start of the observation period.

3. It is a 3 by 6 (cm) rectangle. It has four vertices: (0, 0, 7), (6, 0, 7), (6, 3, 7), (0, 3, 7). It represents the sheet of paper on the desk at exactly 7 minutes from the start of the observation period.

4. It is a 6 by 10 (cm by min.) rectangle. It has four vertices: (0, 2, 0), (6, 2, 0), (0, 2, 10), (6, 2, 10). It represents the existence of a single top to bottom line of the paper (at $y = 2$) that is on the desk for the entire 10 minute observation period.

5. It is a 6 by 10 (cm by min.) rectangle with a 3 by 4 rectangle cut out of one corner. It has six vertices: (4, 0, 0), (4, 3, 6) (4, 6, 6), (4, 6, 0), (4, 3, 10), (4, 0, 10). It represents the 10 minute existence of a single left to right line of the paper (at $x = 4$) that is cut in half exactly 6 minutes into the 10 minute observation period.

Chapter 5

A disk of radius 5 cm stays in the same position, unchanged, for a 10 minute observation period.

5.80

1. To give a coordinate description of this situation in time-space, how many dimensions do you need?

2. What point do you think is the most convenient choice for the origin?

3. If you use the z-axis for time, what unit of measure would you choose for it? What unit of measure would you choose for the other axes?

4. Using the z-axis for time, write a set-builder description for this 10 minute period in the life of the disk. What shape is this? Where is it relative to the coordinate planes?

5. How would the set-builder description of the previous part change if you used the x-axis for time? How would the visual image change?

6. How would your two set-builder descriptions change if you were dealing with a circle, instead of with a disk?

Coordinate descriptions of time-space figures get more interesting when the size or shape of the figure changes as time passes. Such situations often are complicated. Here is a simple example to show you the general idea.

When a pebble is dropped into still water, a circular ripple spreads outward from the splash point. The circle expands at an approximately constant rate; that is, its radius is a constant multiple of the time since the splash. If the expansion rate is 2 feet per second, for instance, then the radius of the outermost ripple is 2 feet at the end of 1 second, 4 feet at the end of 2 seconds, and so on. If we took a series of pictures of that part of the pond, one per second, we'd see something like Display 5.76. Taking a series of pictures that shows how something changes over a period of time is called time-lapse photography.

Published by IT'S ABOUT TIME, Inc. © 2000 MATHconx, LLC

473

5.9 Time and Other Dimensions

5.80

These are marked as discussion questions because they provide an opportunity for students to talk out any misunderstandings about how to use time as an axis. You should do these in class right away. The last example of the section will be much easier to understand if the ideas in these questions are clear. Encourage students to talk out any ambiguities and misunderstandings on their own, as much as your patience and the time available will allow. You might have to help them make their visual descriptions clear.

1. You need three dimensions, two for the disk and one for time.

2. The center of the disk at time 0 is the most convenient choice of origin.

3. Measure z in minutes; measure x and y in cm. Other units could be used, but the description of the situation makes these the obvious choices.

4. With the origin and units chosen as in the previous parts, this 10 minute period in the life of the disk is $\{(x, y, z) \mid x^2 + y^2 \leq 25, 0 \leq z \leq 10\}$. This is a solid right cylinder. If you assume that the z-axis is vertical, this cylinder is standing upright on the xy-plane, on a disk base centered at $(0, 0, 0)$.

5. The set-builder description becomes $\{(x, y, z) \mid y^2 + z^2 \leq 25, 0 \leq x \leq 10\}$. It's still a cylinder, but—assuming that the orientation of the axes is the same as in the previous part—this cylinder is "lying down." Its flat "base" is on the yz-plane, again centered at $(0, 0, 0)$.

6. The distance inequalities become equalities; everything else stays the same. That is, the two descriptions become
    $\{(x, y, z) \mid x^2 + y^2 = 25, 0 \leq z \leq 10\}$ and
    $\{(x, y, z) \mid y^2 + z^2 = 25, 0 \leq x \leq 10\}$

Chapter 5

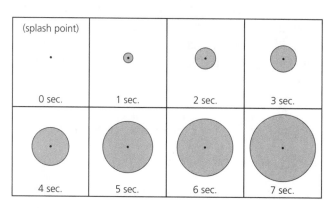

Display 5.76

Of course, the disk of rippled water actually is expanding continuously, not just in one-second jumps. In fact, the radius, r, of the disk is a simple function of the time, t.

$$r = 2t$$

where r is in feet and t is in seconds. We can use this relationship to describe the 3-space figure that represents this expanding disk from the instant the pebble hits the water to the end of the 7th second, as pictured in Display 5.76.

For convenience, we'll choose the origin of our time-space system to be the splash point at the instant the pebble hits. This location is the center of the expanding disk. Relabel the z-axis as t (for time). Then each point of our figure has coordinates (x, y, t), where $0 \leq t \leq 7$. Combining this with the fact that $r = 2t$, we get

$$\{(x, y, t) \mid x^2 + y^2 \leq (2t)^2, 0 \leq t \leq 2, 0 \leq z \leq 7)\}$$

Can you picture this 3-space object? Think about it for a minute or two. If you imagine the xy-plane as the surface of the still water at time 0, then this figure is a solid cone with its point at the origin, as shown in Display 5.77.

These questions refer to the time-space cone of Display 5.77.

5.81

1. What is the cross section of this figure for $t = 3$? Give a geometric (pictorial) description and an algebraic (set-builder) description.

5.81

Part 4 is likely to cause some trouble. The others are straightforward.

1. It is a disk of radius 6 feet: $\{(x, y, 3) \mid x^2 + y^2 = 6^2\}$

NOTES
_____

Chapter 5

2. What is the cross section of this figure for $t = 4$? Give a geometric description and an algebraic description.

3. Give the coordinates of five specific points on the top edge of this cone.

4. Give a geometric description and an algebraic description of the cross section for $y = 0$.

5. What time-space figure represents just the outer edge of the expanding ripple during these 7 seconds? Give a geometric description and an algebraic description.

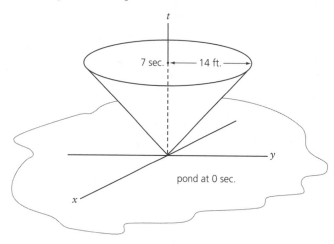

Display 5.77

These 3-space examples of time as a coordinate axis and of time-space figures barely scratch the surface of one of the deepest insights of 20th century science – the realization that *we live in a four dimensional world*. That is, a proper understanding of physical laws depends on treating time as a dimension just like length, width, and depth.

Newton

Isaac Newton's Laws of Motion and Gravity, formulated in the late 1600s, treated the universe as existing in stationary, uniform three dimensional space. For Newton, time was an absolute concept, moving onward at a steady pace that is the same everywhere in the universe. Space was a fixed frame of reference within which gravity determined the motions of all objects, from atoms to planets. This view of the universe dominated physical science for 200 years.

Published by IT'S ABOUT TIME, Inc. © 2000 MATHconx, LLC

475

2. It is a disk of radius 8 feet: $\{(x, y, 3) \mid x^2 + y^2 = 8^2\}$

3. There are four easy ones: $(14, 0, 7)$, $(0, 14, 7)$, $(-14, 0, 7)$, $(0, -14, 7)$ A fifth one is simply a matter of making the equation work, e.g., $(1, \sqrt{195}, 7)$

4. It is an upside down isosceles triangular region of height 7 (sec.) and base 28 (ft.)
$$\{(x, 0, t) \mid x^2 \leq (2t)^2, 0 \leq t \leq 7\}$$
which reduces to
$$\{(x, 0, t) \mid -2t \leq x \leq 2t, 0 \leq t \leq 7\}$$

5. It is the outside skin of this cone: $\{(x, y, t) \mid x^2 + y^2 = (2t)^2, 0 \leq t \leq 7\}$

NOTES

_____

_____

_____

_____

_____

_____

_____

_____

_____

_____

_____

_____

_____

Chapter 5

In 1881, the work of two American physicists, Michelson and Morley, cast doubt on Newton's laws. The questions they raised led eventually to Einstein's theory of relativity. Einstein rejected the notions of absolute space and time, claiming that length, width, depth, *and* duration could only be measured *in relation to some observer,* and that these dimensions varied from observer to observer. To locate an object or an event in our universe accurately, we must be able to say where *and when* it is, relative to some frame of reference. Thus, four coordinates are needed to describe the when-where location of each object in time-space.

Representing non-spatial quantities on coordinate axes also leads beyond four dimensional time-space. Geometric principles, translated into algebra by means of coordinates, combine the power of both algebra and geometry to explore spaces of five dimensions, or six, or ten, or more. Later in your mathematical studies you will see how coordinates can open up these higher dimensional worlds that our limited human imaginations cannot visualize.

5.82   Go to the library or resource center and look up information about the Michelson-Morley experiment of 1881. Write a one page description of what they did and what their results meant. Think of your reader as someone who hasn't studied much science.

REFLECT

In this chapter you have seen how to extend 2-space geometry to the 3-space world in which we live. You constructed three dimensional shapes out of two dimensional ones by folding, stacking, and spinning. Then you learned how to describe and analyze three dimensional figures algebraically by adding a third axis to a coordinate system. Finally, you saw how coordinate geometry can be extended beyond three visual dimensions to describe four dimensional time-space and connections between other things that are measured by real numbers. The next chapter uses this coordinate approach to 3-space to show you how geometry helps to describe and solve algebra problems involving several linear equations.

Published by IT'S ABOUT TIME, Inc. © 2000 MATHconx, LLC

5.82

Briefly, A. A. Michelson and E. W. Morley developed an apparatus that could measure the speed of a light beam to within a fraction of a mile per second. Newton's laws of motion as applied to light rays flashed from a point on our moving planet, Earth, assert that the speed of those light rays (186,284 miles per second) must vary some 20 miles per second one way or the other, depending on whether the light is flashed with or against the direction of the Earth's motion. The Michelson-Morley experiment showed that this variation does not occur!

NOTES

Chapter 5

Problem Set: 5.9

1. (a) Draw a two dimensional time-space region to represent the following.

   A 5 inch candle that burns at a constant rate of 1 inch every 3 minutes is lighted, burns for 6 minutes, and is blown out. Five minutes later, it is relighted and burns until nothing is left.

   (b) Your answer to part (a) should be a polygonal region. How many vertices does it have? Write the coordinates of all the vertices.

2. A circle of radius 6 cm is drawn with its center at the origin of a rectangular coordinate system. It exists, unchanged in size or shape, for 10 minutes; then it is erased. The life of this circle can be represented by a time-space figure.

   (a) The space of smallest dimension that contains this time-space figure is

   $$1\text{-space} \quad 2\text{-space} \quad 3\text{-space} \quad 4\text{-space}$$

   (b) Describe this figure algebraically, using set-builder notation.

   (c) Its shape is
   sphere cylinder circle line segment parallel segments

   (d) The shape of its cross section at $x = 0$ cm is
   sphere cylinder circle line segment parallel segments
   Use your set-builder description to justify your choice.

   (e) The shape of its cross section at 7 minutes is
   sphere cylinder circle line segment parallel segments
   Use your set-builder description to justify your choice.

   (f) The shape of its cross section at $x = 6$ cm is
   sphere cylinder circle line segment parallel segments
   Use your set-builder description to justify your choice.

3. A cookbook contains the diagram shown in Display 5.78 and this statement

   These recipes were tested at sea level. If you are cooking at high altitudes, you may have to adjust them for the change in atmospheric pressure .... Generally speaking, the boiling point of a liquid is lowered by 1° C for every 960 feet above sea level.

477

Problem Set: 5.9

1. (a) See Display 5.22T.
   (b) It has five vertices: (0, 0), (0, 5), (6, 3), (11, 3), (20, 0)

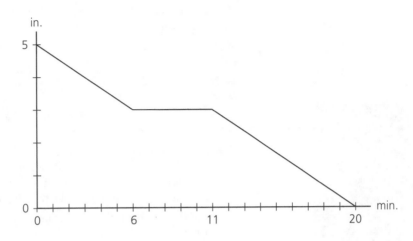

Display 5.22T

2. (a) 3-space
   (b) $\{(x, y, t) \mid x^2 + y^2 = 36, 0 \leq t \leq 10\}$
   (c) It is a cylinder.
   (d) Parallel segments. It is the set of points
   $\{(0, y, t) \mid y^2 = 36, 0 \leq t \leq 10\}$. But $y^2 = 36$ means that
   $y = 6$ or $y = -6$, so this set of points is just
   $\{(0, 6, t) \mid 0 \leq t \leq 10\} \cup \{(0, -6, t) \mid 0 \leq t \leq 10\}$
   (e) A circle: $\{(x, y, 7) \mid x^2 + y^2 = 36\}$
   (f) A single line segment: $\{(6, y, t) \mid 6^2 + y^2 = 36, 0 \leq t \leq 10\}$
   But $6^2 + y^2 = 36$ means that y must be 0, so this is the set
   $\{(6, 0, t) \mid 0 \leq t \leq 10\}$

3. The statement is a condensed paraphrase of information supplied
   in a real cookbook. The diagram is not in the real book.

Chapter 5

5.9 Time and Other Dimensions

Refer to Display 5.78 as you answer these questions.

(a) What are the coordinates of the vertices of the triangular region?

(b) Which axis measures distance? What does the other axis measure?

(c) Is the point (95°, 3500) inside or outside the region? What does it represent?

(d) Is the point (98°, 4000) inside or outside the region? What does it represent?

(e) You are camping on a mountain 5300 feet above sea level. When the water in your cooking pot boils, how hot will it be? What point of the diagram represents this? Is it inside, outside, or on an edge of the triangular region?

(f) What do the points on the hypotenuse of the triangle represent?

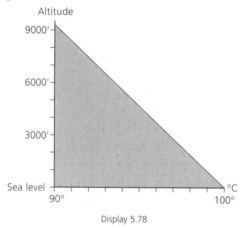

Display 5.78

4. (a) Explain how a hemisphere can be thought of as a circle that expands during a period of time.

(b) Explain how a cone can be thought of as a circle that expands during a period of time.

(c) In terms of the way in which the circle is expanding, what is the essential difference between part (a) and part (b)?

(d) In your school library or local library, find a book called *Flatland*, by Edwin A. Abbott. Look up the part where A. Square meets the Sphere. Then write a paragraph explaining how that description in the book is related to the other parts of this problem. Also make a note of the year in which *Flatland* was first published.

Published by IT'S ABOUT TIME, Inc. © 2000 MATHconx, LLC

(a) (90, 0), (90, 9600), (100, 0)

(b) The vertical axis measures distance. The horizontal axis measures temperature.

(c) Inside. It represents 95° C at a place 3500 feet above sea level.

(d) Outside. It represents 98° C at a place 4000 feet above sea level.

(e) It boils at 94.7° C. The point is (94.7, 5300); it is on the hypotenuse of the triangle.

(f) The points on the hypotenuse represent the boiling point of a liquid at every altitude between sea level and 9600 feet above sea level.

4. (a) Imagine the hemisphere resting on its "pole" on a horizontal plane at time 0 and let time be measured vertically upward from the plane. Each instant of time after 0 determines a horizontal cross section of the hemisphere; that cross section is a circle. As time progresses upward, the circular cross sections increase in diameter, until the equator is reached. After that, the circle disappears.

(b) This description is almost identical to part (a), except that you start by imagining a cone balanced on its point on the time-0 plane.

(c) The essential difference is the rate of expansion of the circle. In part (a), the circle expands rapidly at first, then slows its expansion as its size approaches maximum. In part (b), the rate of expansion is constant.

(d) This library assignment shows students that the visual images we are discussing here have been in use for a very long time. Abbott's description is quite good, but not explicit enough to distinguish between a sphere and a double cone passing through Flatland.

Chapter 5

NOTES

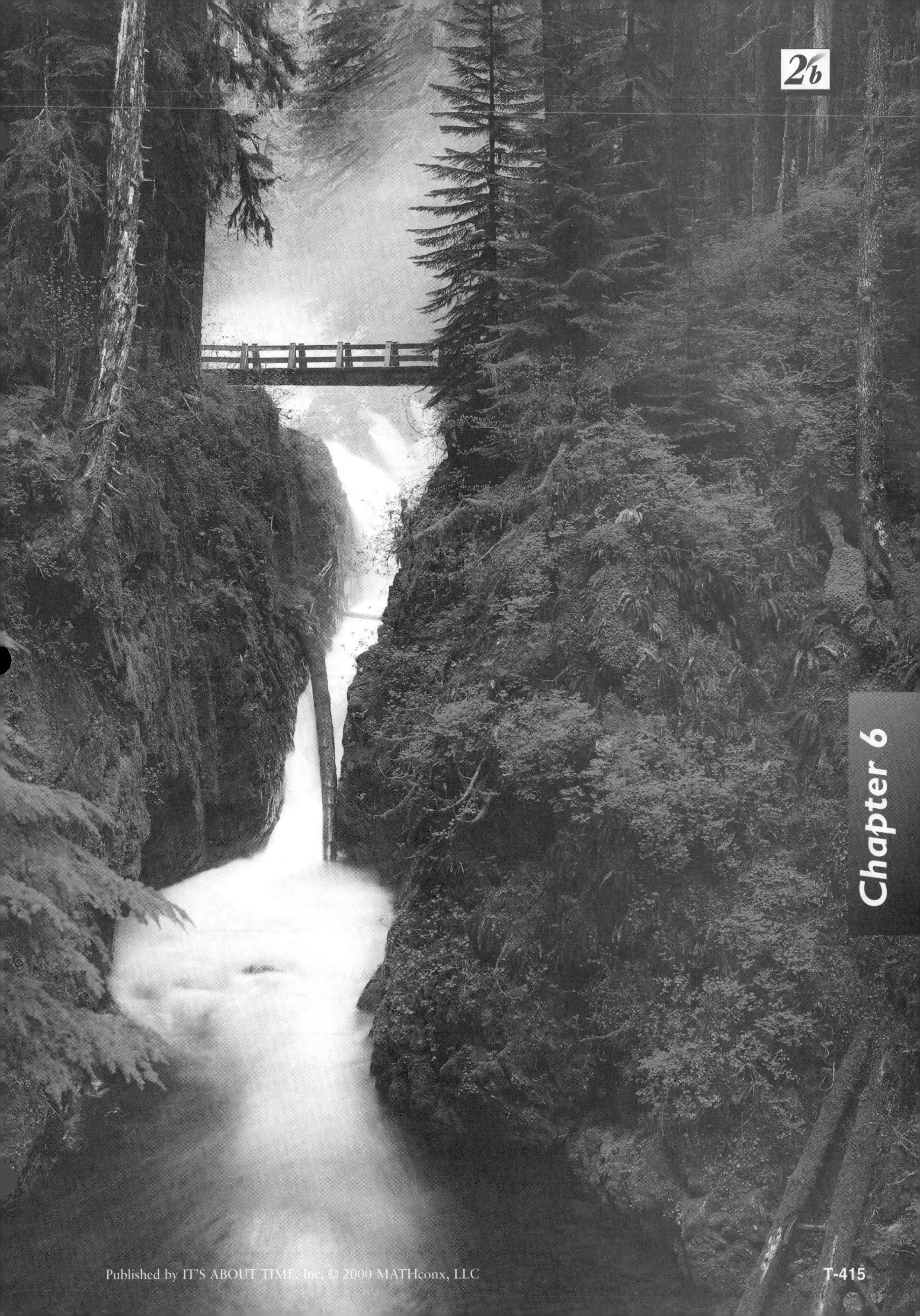
2b
Chapter 6

Chapter 6 Planning Guide

Chapter 6 Linear Algebra and Matrices

Starting with planes in 3-space, Chapter 6 treats solutions of systems of linear equations in a way designed to promote understanding of the process. The geometrical representation is linked with algebraic manipulation by means of matrix operations, including Gaussian elimination and matrix multiplication.

Assessments Form (A)	Assessments Form B (B)	Blackline Masters
Quiz 6.1-6.2(A) Quiz 6.3(A) Quiz 6.4-6.5(A) Chapter Test (A)	Quiz 6.1-6.2(B) Quiz 6.3(B) Quiz 6.4-6.5(B) Chapter Test (B)	Student p. 505

Extensions	Supplements for Chapter Sections	Test Banks
Following 6.5 The Inverse of a Matrix Systems of Linear Inequalities	6.1 Systems of Equations and Graphs — 1 Supplement 6.2 Consistent and Inconsistent Systems — 4 Supplements 6.3 Solving Systems of Equations — 2 Supplements 6.4 Solving Linear Systems Using Matrices — 4 Supplements 6.5 Matrix Operations — 4 Supplements	To be released

Pacing Range 4-5 weeks including Assessments

Teachers will need to adjust this guide to suit the needs of their own students. Not all classes will complete each chapter at the same pace. Flexibility — which accommodates different teaching styles, school schedules and school standards — is built into the curriculum.

Teacher Commentary is indexed to the student text by the numbers in the margins (under the icons or in circles). The first digit indicates the chapter — the numbers after the decimal indicate the sequential numbering of the comments within that unit. Example:

6.9

(6.37)

Student Pages in Teacher Edition

6.9

(6.37)

Teacher Commentary Page

Observations

Bryan Morgan, Teacher
Oxford Hills Comprehensive High School, ME

"In general this chapter is a nice finish to geometry and a good bridge to the algebra 2 concepts. Also, it utilizes the material about 3-space that has just been covered in the previous chapter to visualize solutions to systems of equations with two and three variables. Matrices and operations on them are developed as a method for organizing, categorizing, and combining data.

"I really like one section in the chapter where the students work on the same problem three times, but are directed through it with new skills each time. They end up solving the same problem in three different ways. Each way is progressively more sophisticated and less cumbersome than the one before, and the students therefore are able to master even more complicated work."

Chapter 6

Lisa Jones
Crashes Planes to Help Save Lives

Lisa Jones is the lead engineer for the Impact Dynamics Research Facility at NASA's Langley Research Center in Virginia. She conducts full-scale aircraft crash tests so that better designs can be developed for future aircraft.

"Basically," explains Lisa, "this means getting the aircraft instrumented and rigged for impact conditions. I also do the calculations for the velocities we want tested." Using a pendulum method, the aircraft are suspended by cables and swung into the impact surface at various altitudes and velocities. Data is collected and analyzed to study the aircraft's crashworthiness. Crashworthiness is the degree of an aircraft's ability to protect occupants from serious or fatal injuries.

Lisa always loved math and science. At first, she wanted to be a commercial pilot, but was talked out of it. "It was probably gender related," she reflects. "So I decided that if I couldn't fly aircraft, I'd build them."

She graduated from Georgia Institute of Technology with one degree in Aerospace Engineering and another in Mathematics.

"Before each test," Lisa continues, "we run a computer analytical study modeling the event. We compare that with the data from the actual test. Our computer models allow us to predict what will happen at given points on the aircraft. Matrices are developed. Then we download data from the onboard 'occupants'. (These are dummies fitted with various instrumentation.) From this information, we can determine the types of 'injuries' that occurred."

"The computer solves the matrix equations with the variables we input," Lisa states. "Years ago, before computers, NASA had rooms full of mathematicians who manually solved the matrix equations that the engineers needed."

"I love my work," Lisa exclaims. "We're helping to develop the safest aircraft possible using current technologies. You might say it brings out the mother in me. I want my children and all people to be safe when they fly."

480

Published by IT'S ABOUT TIME, Inc. © 2000 MATHconx, LLC

Chapter 6 Linear Algebra and Matrices

This chapter contains many traditional topics such as the solution of systems of equations, but it treats them in a way designed to promote understanding of the process in concert with algebraic manipulation. In addition, students are exposed to matrices and this is linked to the work they have done early in the chapter with systems of equations. Wherever possible, students are encouraged to link the algebra with the geometrical interpretation of that algebra—this helps to make what is often just symbol manipulation into something that can be visualized and talked about pictorially.

Chapter 6

Linear Algebra and Matrices

CHAPTER 6

6.1 Systems of Equations and Graphs

Earlier in **MATH** *Connections* you learned about the graphs of linear equations; and, earlier in this book you studied a number of geometrical concepts. In this section we are going to look at how these ideas can be linked together and how they can be used to solve realistic problems. Some of the ideas will be similar to concepts and techniques you have seen before, and some will be brand new. Let's look at a problem that helps to review some of the ideas that you have seen before.

Krelling Industries is a small manufacturer of wooden furniture. They have two plants, one in California and one in Maine. The plants can make several kinds of furniture, but at the moment the only item they make is a wooden desk. Each plant has certain fixed costs that have to be paid regardless of how many desks are produced each month, for example, heat, electricity and taxes. The plant in California has fixed monthly costs of $4,000 and the plant in Maine has fixed monthly costs of $10,000. The Maine plant is newer and can manufacture the desks more economically, so they produce a profit of $150 on each desk manufactured. The California plant is older

Published by IT'S ABOUT TIME, Inc. © 2000 MATHconx, LLC

481

6.1 Systems of Equations and Graphs

The first section of the chapter poses a real life problem which can be solved by using previously learned ideas and the graphing calculator. An apparently simple extension to the original problem then leads into a discussion of three dimensional geometry which reviews some of the ideas that were discussed in previous chapters. However, the development here is more detailed and makes use of a very important idea—that of changing a problem that is unfamiliar to one that is known. In this case, the graphs of shapes in three dimensional space is achieved by fixing one of the variables and then proceeding to establish a table of values. It is extremely difficult for students to draw such graphs; however, the diagrams that are illustrated here are designed to show them exactly what is going on each step of the way. The discussion then allows for a resolution of the difficulty with the problem situation with which the previous section ends.

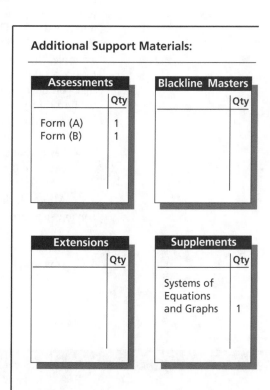

Additional Support Materials:

Assessments	Qty
Form (A)	1
Form (B)	1

Blackline Masters	Qty

Extensions	Qty

Supplements	Qty
Systems of Equations and Graphs	1

Chapter 6

and only makes a profit of $130 on each desk. Which plant is more profitable?

When doing a problem such as this it is often helpful to establish when the costs are the same and then work from there.

a
6.1

1. Use the information in the problem to write an equation for the monthly profit for the Maine plant. Use *y* for the monthly profit, in dollars, and *x* for the number of desks produced.

2. Use the information in the problem to write an equation for the monthly profit for the California plant. Use *y* for the monthly profit, in dollars, and *x* for the number of desks produced.

3. Graph both of these equations on your graphing calculator and establish where they intersect. How many desks per month does this profit correspond to?

4. For what levels of production is the Maine plant more profitable?

5. For what levels of production is the California plant more profitable?

The graphing calculator allows us to quickly find the point where the two lines representing these equations intersect. Algebraically, this gives us the value of *x* and *y* that will satisfy both equations. That is, the values of *x* and *y* that we have found should make both equations true statements. For example, the intersection of these two lines was at the point (300, 35,000). If we substitute the *x*-value into the equation that you developed for the Maine plant, $y = 150x - 10,000$, we find that $y = (150) * (300) - 10,000 = 35,000$ and this is exactly what *y* is supposed to be.

b
6.2

Verify that these *x* and *y* values also satisfy the equation that you developed for the California plant.

Although the graphing calculator is very handy for solving such problems, it is not without its weaknesses.

6.3

Give two advantages and two disadvantages of using your graphing calculator to solve this problem.

6.1

1. $y = 150x - 10,000$

2. $y = 130x - 4000$

3. (300, 35,000), 300 desks per month production

4. For productions of more than 300 desks per month

5. For productions of less than 300 desks per month (67–299 inclusive)

6.1

$y = 130*300 - 4000 = 35,000$

6.3

Advantages: (i) allows for a visual representation of the problem (ii) quick (iii) does not require the student to know algebraic solution of equations. Disadvantages: (i) can take time to establish an accurate intersection point with many zooms required (ii) need to have a calculator!

NOTES

Chapter 6

Another way to establish where the lines intersect is to try to find the values of x and y that the two equations have in common algebraically. For example, we know that the equations representing the monthly profit for the Maine and California plants are

$$(1) \quad y = 150x - 10,000$$
$$(2) \quad y = 130x - 4000$$

Recall that in the Electric Company problem we did earlier, we had two expressions equal to the same variable. We used subsitution to solve the problem.

We want to find out for what value of x and y both equations are satisfied. Because the left sides of each of these equations are equal to y, the right sides must be equal. So, we can set the right sides of each equation equal to each other. This gives

$$150x - 10,000 = 130x - 4000$$

which has only one variable, x.

1. Solve this equation for x.

2. Substitute the value of x that you found into either equation (1) or equation (2) to find y.

6.4

3. Compare your values for x and y with the values that you found by using the calculator. Are they the same? Explain.

We now know that both companies will produce the same profit for a production of 300 desks.

Give two advantages and two disadvantages of solving this system of equations using algebra.

6.5

We call a pair of equations that has two variables a *system of two equations in two unknowns.*

Thinking Tip

Sometimes when we are trying to solve a problem it is helpful to reduce it to a problem we already know how to solve.

1. $x = 300$

2. $y = 35,000$

3. They should be the same. This represents the point on the two lines where the x and the y values are equal—in other words the intersection point.

Advantages: (i) don't need a calculator (ii) exact answer possible
Disadvantages: (i) need to know formal methods of equation solving
(ii) sometimes involves complex arithmetic that make errors likely.

NOTES

Chapter 6

a
6.6

1. Use the algebraic method of substitution to solve each of the following systems of two equations in two unknowns.

(a) $y = x + 1$
$y = 3x - 5$

(b) $y = 2x - 1$
$y = x + 2$

(c) $y = x$
$y = 3x + 7$

2. Use your graphing calculator to find the point of intersection of each pair of lines in problem 1 to verify your answers.

The directors of Krelling Industries are pleased with sales and decide to expand their product line. They begin to make wooden tables that match the desks that they already produce. The California plant can produce tables with a profit of $80 each, while the newer Maine plant produces them with a profit of $100 each. For what production of tables and chairs are the profits of both plants equal?

b
6.7

1. Use the information above to develop an equation for the monthly profit for the California plant. Use x for the number of desks produced each month, y for the number of tables produced each month, and z for the monthly profit.

2. Use the information above to develop an equation for the monthly profit for the Maine plant. Use x for the number of desks produced each month, y for the number of tables produced each month, and z for the monthly profit.

3. In what way are these equations different from the previous ones that you developed for the monthly profit of the two plants?

Earlier in **MATH** *Connections* you studied functions, and the equations you just developed also represent functions. They are different from ones you have seen before because in this case z (monthly profit) is a function of two variables, x (monthly desk production) and y (monthly table production). This makes the problem of finding when the monthly profits will be equal a little different, because our graphing calculators can only graph functions of one variable. Fortunately, we have

484

6.6

1. (a) $x = 3$, $y = 4$
   (b) $x = 3$, $y = 5$
   (c) $x = -3.5$, $y = -3.5$

6.7

1. $z = 130x + 80y - 4000$

2. $z = 150x + 100y - 10,000$

3. These equations involve three variables not two, alternatively, the profit is a function of two variables rather than just one.

NOTES

Chapter 6

our algebraic technique of substitution to fall back on, so let's use it here.

To find out when the monthly profits are equal we put the right sides of the equations equal to each other to produce

$$150x + 100y - 10,000 = 130x + 80y - 4000$$

If we now add 10,000 to both sides we will get

$$150x + 100y = 130x + 80y + 6000$$

Subtracting $130x$ and $80y$ from both sides produces the equation

$$20x + 20y = 6000 \text{ or } x + y = 300$$

But what kind of a solution is that? In the previous example we were able to find a value for the variable and then look at how much profit was produced by that number of desks. Here we don't have specific values for the variables.

1. Calculate the profit for both plants if they produce 100 desks and 200 tables.

2. Calculate the profit for both plants if they produce 50 desks and 250 tables.

3. Explain why the profits are the same in both cases.

4. Give two other monthly productions for tables and desks that would produce the same profit.

6.8

Because we have three variables rather than just two, we must consider the graph of this equation in three dimensional space. Let's review some ideas about graphing in three dimensional space and see how we can use them to help graph these profit equations. The profit equations were

(1) $z = 150x + 100y - 10,000$
(2) $z = 130x + 80y - 4000$

Let's consider the graph of the first equation. This can be visualized by looking at the corner of the room. The x-axis is the line between the floor and the wall on the left. The y-axis is the line between the floor and the wall on the right. The z-axis is the line between the two walls. Normally when we draw graphs we get a table of values and then plot points. In this case, it is easier to first change the equation into something like the ones with

Published by IT'S ABOUT TIME, Inc. © 2000 MATHconx, LLC

6.8

1. $25,000; $25,000

2. $22,500; $22,500

3. $z = 150*100 + 100*200 - 10,000 = 25,000$
   $z = 130*100 + 80*200 - 4000 = 25,000$

4. 1 and 299, 300 and zero. Any sum of two natural numbers that equals 300.

NOTES

Chapter 6

which we are more familiar. Let's fix the value of one of the variables, say x, at 200. Now equation (1) becomes,

$$z = 150x + 100y - 10,000$$
$$z = 150 * 200 + 100y - 10,000$$
$$z = 100y + 20,000$$

We can now proceed to fill in a table of values and find some ordered triples that satisfy this equation. We know that in each equation the value of x will be 200 because that is what we have chosen it to be.

1. **Make a copy of this table in your notebook and complete it, using the equation $z = 100y + 20,000$.**

6.9

x	200	200	200	200	200
y	50	100	150	200	250
z					

Now we can graph these points. The result is illustrated in Display 6.1. In this graph each unit on the x-axis represents 100, and each unit on the y-axis represents 50. Consequently, to plot the point (200, 150, 35,000) we move two units in the x-direction, 3 units in the y-direction and then up to the appropriate value of z. The arrows are used to illustrate how to plot this point but they are not required, and they are not part of the graph. Similarly, the other points can be plotted. We can see that these points appear to lie on a straight line. Imagine that we now repeated the process except this time let $x = 300$. Then do it again with $x = 400$. The result is illustrated in Display 6.2. You can see what is happening —as we pick different fixed values for x, we get a series of straight lines, all of which are in the same plane.

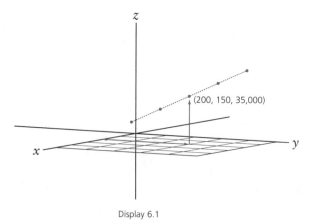

Display 6.1

6.9

x	200	200	200	200	200
y	50	100	150	200	250
z	25,000	30,000	35,000	40,000	45,000

NOTES

Chapter 6

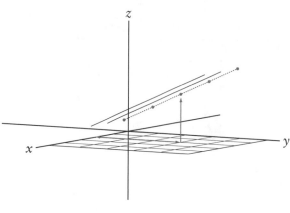

Display 6.2

What would have happened if we had picked a fixed value for y in the first equation, say 150? In that case equation (1) would have become,

$$z = 150x + 100y - 10,000$$
$$z = 150x + 100 * 150 - 10,000$$
$$z = 150x + 5000$$

1. **Make a copy of this table in your notebook and complete it using the equation $z = 150x + 5000$.**

6.10

x	100	150	200	250	300
y	150	150	150	150	150
z					

Once again we can plot the ordered triples from the table and this time the points can be joined by a dotted line as illustrated in Display 6.3. Just as before we could choose different values for y, for example, 100, and set up a new table of values and plot the resulting points. In Display 6.3 the other two lines are those established by choosing y to be 100 and 200.

6.10

x	100	150	200	250	300
y	150	150	150	150	150
z	20,000	27,500	35,000	42,500	50,000

NOTES

Chapter 6

6.1 Systems of Equations and Graphs

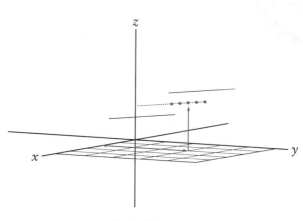

Display 6.3

Note that the point (200, 150, 35,000) is still part of the graph. If we put these two separate pictures together, as in Display 6.4, we begin to see the graph of the equation

$$z = 150x + 100y - 10,000.$$

In fact, if we graph many different points, we eventually form a good picture of the plane that this equation represents. (See Display 6.5.) Remember that this diagram really only shows a portion of the actual plane.

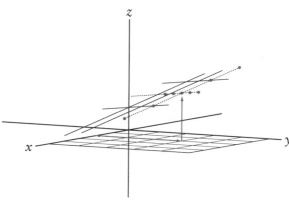

Display 6.4

488

NOTES

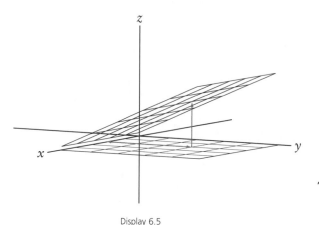

Display 6.5

Diagrams like this are called *wire frame pictures*. Computers use this method to generate pictures of 3-D objects because it quickly gives a good idea of what the object looks like. If we want more detail we just need to add more "wires" to the picture. In this way computers are able to use mathematics to describe very complex shapes. And this is the basis of all computer aided design (CAD) systems. Even a more complicated 3-D shape such as the one in Display 6.6 can be described by a simple mathematical equation. When you drive in a car, the comfort and styling you enjoy is partly due to the power that such CAD systems put in the hands of talented engineers!

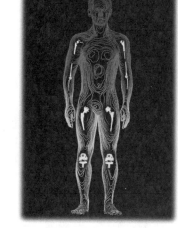

Display 6.6

This exercise has allowed us to see a very important mathematical idea. In two dimensions the graph of a linear equation such as $y = 3x + 5$ is a line. In three dimensions the graph of a linear equation in three variables such as $z = x + y - 5$ is a plane!

Now, when we tried to solve the two equations

(1) $z = 150x + 100y - 10,000$
(2) $z = 130x + 80y - 4000$

NOTES

what we were doing was finding the intersection of the two planes that these equations represent. When two planes are not parallel or do not coincide, they intersect in a line. And, the line of intersection consists of all the points on the original planes that satisfy the condition that $x + y = 300$. For example, the point (100, 300, 35,000) is on the plane $z = 150x + 100y - 10,000$ because it satisfies the equation of this plane. It is not on the line of intersection because it does not satisfy the condition that $x + y = 300$. The point (100, 200, 25,000) *is* on the plane and it *does* satisfy the condition that $x + y = 300$. Therefore, this point of the plane is also on the line of intersection. This is similar to the situation in two dimensions where lines consist of all points in the xy-plane satisfying some condition such as $y = x + 5$. The graphs of both planes and their line of intersection are illustrated in Display 6.7.

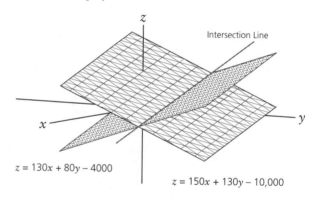

Display 6.7

A Fact to Know: In three dimensional space, a plane will have an equation of the form $ax + by + cz + d = 0$ where a, b, *and* c, cannot all equal zero.

6.11

....

6.12

In your previous work you considered some special planes, for example, the xy-plane. The equation of the xy-plane was $z = 0$. In this case, where $ay + by + cz + d = 0$, the values of a, b and d would be zero, while the value of c would be 1.

1. Find the value(s) for a, b, c and d that would give the equation of each of the following planes.

   (a) the yz-plane

6.11 The equation of a plane can be confusing because it is important to realize that even equations that have only two variables still represent planes when graphed in three dimensions. This is the reason that we state that not *all* of a, b, and c can be zero; however, some of them certainly can. This makes the earlier discussion about the line of intersection of the two planes very important.

6.12

(a) $x = 0$, $\quad a = 1$, $\quad b = 0$, $\quad c = 0$, $\quad d = 0$

NOTES

Chapter 6

(b) the xz-plane

(c) a plane parallel to the xy-plane, 4 units above it

(d) a plane parallel to the xz-plane, 2 units to the left of it

So where does all this leave the directors of Krelling Industries? The fact is that there is no way to determine a single specific value that makes the profits of the two plants equal. It is not possible to solve two equations in three unknowns and get *a specific value*, although it is possible to establish *many values* that satisfy both equations. This raises an important question: If we can't find a specific solution for two equations with three unknowns, can we ever solve a system of equations that have three unknowns and get a specific value for each? The answer to this question will be considered in the next two sections.

Problem Set: 6.1

1. Find the point of intersection of each of the following pairs of lines using your graphing calculator.

    (a) $y = 7x - 3$
        $y = 4x + 1$
    (b) $y = x + 4$
        $y = -2x - 5$
    (c) $y = x$
        $y = 3$

2. Verify each of the intersection points found in problem 1. by solving each of the systems of equations using substitution.

3. Two different companies drill water wells for people with country homes. The first charges $500 plus $10 per foot to drill the well. The second charges $1,000 plus $8 per foot to drill the well.

    (a) For what depth(s) of well would the cost for drilling be the same for both companies?
    (b) For what depth(s) of well would the cost for drilling by the first company cost less?
    (c) For what depth(s) of well would the cost for drilling by the second company cost less?

(b) $y = 0$, $b = 1$, $a = 0$, $c = 0$, $d = 0$
(c) $z = 4$, $a = 0$, $b = 0$, $c = 1$, $d = -4$
(d) $y = -2$, $a = 0$, $b = 1$, $c = 0$, $d = 2$

Problem Set: 6.1

1. (a) $x = \dfrac{4}{3}, y = \dfrac{19}{3}$
   (b) $x = -3, y = 1$
   (c) $x = 3, y = 3$

2. Calculator results should confirm these values.

3. (a) For a depth of 250 feet the costs are the same.
   $C = 500 + 10f$; $C = 1000 + 8f$
   (b) For wells shallower than 250 feet, the first company is less costly.
   (c) For wells deeper than 250 feet, the second company is less costly.

NOTES

4. The two lines represented by the system of equations

$$y = 2x + b$$
$$y = -x + a$$

intersect at the point (–1, 3). Find the value(s) of a and b that make this true.

5. In news reports, the profits of a company for the present year (P) are often given as a comparison with the profits of the previous year (L). If a company made $30 million more this year than last and profits were up by 20% over last year, what is this year's profit?

6. A catering company charges a fixed fee and then a charge per plate for each person who eats at the banquet. If it costs $2,000 for 100 people and $3,500 for 200 people, what is the fixed fee and the price per plate?

7. Find the condition for points to be on the line of intersection of each of the following pairs of planes. Leave your answer in the form $f(x, y) = 0$.

   (a) $z = x + y - 3$
   $\quad z = 2x - 4y + 9$
   (b) $z = -x + 5$
   $\quad z = 7x + 5y - 10$
   (c) $z = 5x + y$
   $\quad z = x - y + 9$

8. (a) Find the condition for points to be on the line of intersection of the planes represented by the following pair of equations.

   $$z = 2x - 5y + 3$$
   $$z = 2x - 5y + 1$$

   (b) Explain what you think this result might mean geometrically.

9. Recall that a plane in 3-D will have an equation of the form $ax + by + cz + d = 0$, where a, b, c are not all zero. For what values of a, b, c, d, will this equation describe each of the following planes?

   (a) the xy-plane
   (b) a plane parallel to the xz-plane, 4 units to the right of it
   (c) a plane parallel to the xy-plane, 2 units below it

Published by IT'S ABOUT TIME, Inc. © 2000 MATHconx, LLC

4. $b = 5, a = 2$

5. The equations are
$$P = L + 30$$
$$P = 1.2L$$
Solving yields a profit for the year of $150 million.

6. The price per plate is $15, while the fixed charge is $500.

7. (a) $x - 5y + 12 = 0$
   (b) $8x + 5y - 15 = 0$
   (c) $4x + 2y - 9 = 0$

8. (a) The resulting equation is $3 = 1$. (b) This is impossible and means that the two planes do not have any intersection points. In other words they must be parallel.

9. (a) $z = 0$, $a = 0$, $b = 0$, $c = 1$, $d = 0$
   (b) $y = 4$, $a = 0$, $b = 1$, $c = 0$, $d = -4$
   (c) $z = -2$, $a = 0$, $b = 0$, $c = 1$, $d = 2$

NOTES

Chapter 6

6.2 Consistent and Inconsistent Systems

Before we consider systems of equations with three variables let's reconsider for a moment what it means to solve a system of two equations in two unknowns. When we solve a system of two linear equations in two unknowns we are finding if the two straight lines intersect, and if so, where they intersect. This is easy to do with a graphing calculator and by substitution, if the equations are given to us in the $y = $ [something] form.

1. Use your graphing calculator to find the point(s) of intersection of each of the following pairs of lines.

    6.13

    (a) $y = 3x + 2$
       $y = -x + 6$

    (b) $y = 2x + 2$
       $y = 2x + 6$

    (c) How many solutions are there for each system of equations?

2. If you graph the system of equations listed below, how many points of intersection are there?

    $y = 3x + 2$
    $2y = 6x + 4$

In the systems of equations of the previous problems, there were either no solutions, one solution or infinitely many solutions. In part 1(a) there was exactly one point where the lines intersected, so this system has a unique solution. A system that has one or more solutions is called a **consistent** system. In part 1(b) the lines are parallel, so no intersection is possible and this system has no solutions. A system that has no solution is called an **inconsistent** system. In part 2 the two lines are really the same line, and so they intersect everywhere, therefore there are many, many solutions for this system. The system is consistent, but not very interesting! Sometimes however, we can graph lines that are the same without knowing it.

Learning Outcomes

After studying this section you will be able to:

Demonstrate understanding the geometric interpretation of consistent and inconsistent systems;

Identify systems of two equations in two unknowns as consistent or inconsistent, algebraically;

Solve two equations in two unknowns by the method of substitution.

Thinking Tip

When attacking a new problem it is often useful to look at a simpler problem first and then see if the ideas can be applied to the more complicated one.

493

6.2 Consistent and Inconsistent Systems

The ability to solve systems of equations, although important, is not particularly useful if the algebraic methods provide us with results that we cannot understand or interpret. This section helps students to see the geometrical significance of algebraic results and to learn what is meant by consistent, inconsistent and dependent systems. Although the geometric interpretation cannot be extended easily into dimensions higher than 3, at least they can see that in two and three dimensions there are geometrical interpretations for various unusual algebraic results. And, that these same concepts apply to systems of equations with many more than 3 variables.

6.13

1. (a) $x = 1, y = 5$
   (b) None
   (c) 1 in (a) and none in (b)

2. The lines are coincident, so they intersect everywhere.

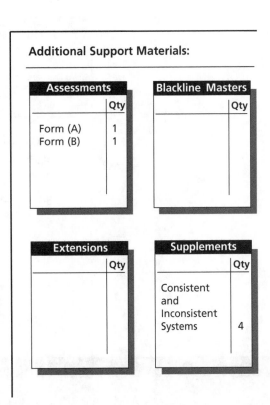

Additional Support Materials:

Assessments	Qty
Form (A)	1
Form (B)	1

Blackline Masters	Qty

Extensions	Qty

Supplements	Qty
Consistent and Inconsistent Systems	4

Chapter 6

a
6.14

Questions 1–2 refer to the equations in the previous
Do this now.

1. (a) Solve this system of equations by substitution
$$y = 3x + 2$$
$$y = -x + 6$$

   (b) How many solutions are there for this system of equations?

   (c) Is this system consistent? Explain.

   (d) Explain why the algebraic solution of this system confirms what you established by looking at the graphs of the two equations.

2. (a) Solve this system of equations by substitution.
$$y = 2x + 2$$
$$y = 2x + 6$$

   (b) How many solutions are there for this system of equations?

   (c) Is this system consistent? Explain.

   (d) Explain why the algebraic solution of this system confirms what you established by looking at the graphs of the two equations.

The results of previous exercises are summarized in this chart.

Graphs	Number of Solutions	Classification
Two lines intersect in one point	1 solution	Consistent
Two lines are parallel	No solutions	Inconsistent
Two lines are actually the same line	Infinitely many	Consistent

b
6.15

Solve each of the following systems of equations algebraically and classify each system as consistent or inconsistent.

(a) $y = 2x + 2$
    $y = x + 6$

494

Published by IT'S ABOUT TIME, Inc. © 2000 MATHconx, LLC

6.14

1. (a) (1,5)
   (b) 1
   (c) Yes, it has one unique solution.
   (d) One solution corresponds to one point of intersection.

2. (a) No solution
   (b) None
   (c) No, the solution represents an impossible situation. Parallel lines do not intersect.
   (d) The lines are parallel. They have the same slope.

6.15

(a) $x = 4$, $y = 10$, consistent, exactly one solution

NOTES

Chapter 6

(b) $y = -2x - 5$
$y = 3x + 5$

(c) $y = 3x + 2$
$y = 3x + 6$

(d) $y = x - 9$
$y = -4x + 6$

Now let's turn to the problem of systems of equations with three unknowns. Once again we will consider the geometrical possibilities first. The graphing calculator is not useful here because of its inability to graph planes in 3-D. Luckily we have a great example of 3-D space all around—your classroom!

Use one corner of the room as a model (starting point) and sketch the following 3-D coordinate system. Denote the wall to the right by the letter R, the wall to the left by the letter L, the floor by the letter F and the ceiling by the letter C. Imagine that each of these extends in all directions, then each is a plane surface. We can describe the intersection of these surfaces. For example, the intersection of L and R is a line which happens to form the corner of the room. Likewise, the intersection of C and R and L is a point which happens to be the top corner of the room.

6.16

1. Describe in your own words the intersection of each of the following.

(a) L and F
(b) F and R
(c) F and C
(d) F and C and R
(e) F and C and L
(f) F and R and L
(g) C and R and L

2. If we were to assume that each of these surfaces (C, R, L and F) were represented by an equation, which system(s) would be consistent?

3. Which system(s) would be inconsistent?

4. Which system(s) would have a unique solution?

5. Which system(s) would have many solutions?

(b) $x = -2$, $y = -1$, consistent, exactly one solution
(c) No solution, inconsistent, the lines are parallel
(d) $x = 3$, $y = -6$, consistent, exactly one solution

6.16

1. (a) Line
   (b) Line
   (c) No intersection because the planes are parallel
   (d) Two parallel lines of intersection (no points in common to F, C, R)
   (e) Two parallel lines of intersection (no points in common to F, C, R)
   (f) A single point (the corner)
   (g) A single point

2. (a), (b), (f) and (g) would represent consistent systems.

3. (c), (d) and (e) would be inconsistent.

4. (f) and (g) have one solution.

5. (a) and (b) would have infinitely many solutions.

NOTES

Chapter 6

The ideas about consistent and inconsistent systems that we learned from the discussion of two equations in two unknowns still hold. If two planes do not intersect then the system is inconsistent. If three planes intersect in a line the system is consistent and has many solutions. If three planes intersect in a point the system is consistent and has a unique solution. If three planes intersect in two different lines or three different lines then the system is inconsistent because there is no point that is common to all three. This information is summarized in this chart.

Graphs	Number of Solutions	Classification
Two planes are parallel	None	Inconsistent
Two or more planes intersect in a line	Infinitely many	Consistent
Three planes intersect in a point	One	Consistent
Three planes intersect in two different lines	None	Inconsistent

Unfortunately, not every possible situation can be modeled by using the corner of the room.

6.17

In Display 6.8 there are five diagrams, each one representing a system of three equations in three unknowns. These are the only possibilities for the intersection of three planes.

(a) **Classify each system as consistent or inconsistent. Give a reason in each case.**

(b) **Which system(s) would have a unique solution? Explain.**

(c) **Which system(s) would have many solutions? Explain.**

(d) **Which system(s) would have no solutions? Explain.**

(i)

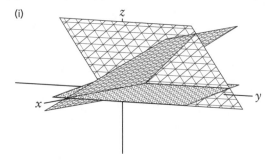

Published by IT'S ABOUT TIME, Inc. © 2000 MATHconx, LLC

MATH *Connections*: A Secondary Mathematics Core Curriculum

6.17

(a) The diagrams in (i) and (v) represent consistent systems because all three planes have either a point or a line in common. The others are inconsistent because there is nothing common to all three of the planes.

(b) Only part (i) would have a unique solution because it is the only situation where the planes meet in a common point.

(c) The diagram in part (v) represents a system with many solutions because the intersection of the planes is a line.

(d) The diagrams in parts (ii), (iii) and (iv) represent system with no solutions.

NOTES

Chapter 6

(ii)

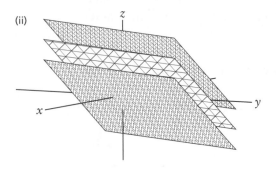

(iii)

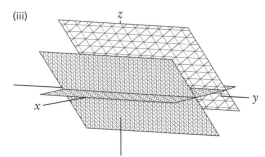

(iv)

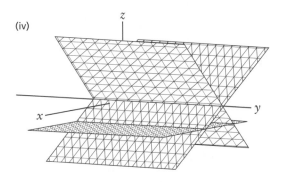

(v)

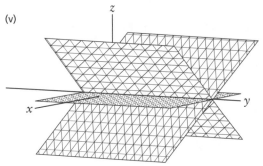

Display 6.8

Published by IT'S ABOUT TIME, Inc. © 2000 MATHconx, LLC

NOTES

Algebraically, the same kinds of things occur that we noticed when solving two equations in two unknowns. For example, we have already shown that when two planes intersect in a line there are infinitely many solutions. Also, if a system is inconsistent we get results that are nonsense. Consistent systems have one or many solutions. Unfortunately, at this time we have no way to solve three equations in three unknowns.

However, in the next section we will look at another method of solving two linear equations in two unknowns that can also be used for solving three linear equations in three unknowns and even more complex systems. We will also return to our old friend Krelling Industries to consider yet another real life problem, this time one where the solution of three equations in three unknowns is important.

Problem Set: 6.2

1. (a) Is this system consistent or inconsistent? Explain.

   $y = 2x + 5$
   $y = -x + 4$

   (b) Give a geometric interpretation of this situation.
   (c) Graph both lines on your calculator to verify that your geometric interpretation was correct.

2. (a) Is this system consistent or inconsistent? Explain.

   $y = x + 4$
   $y = x - 3$

   (b) Give a geometric interpretation of this situation.
   (c) Graph both lines on your calculator to verify that your geometric interpretation was correct.

3. Given the following system of equations

   (1) $z = x + y - 5$
   (2) $z = 3x - 3y + 3$
   (3) $z = 2x - y + 1$

Published by IT'S ABOUT TIME, Inc. © 2000 MATHconx, LLC

Problem Set: 6.2

1. (a) Consistent because we are able to solve for x and y.
      The solution is $\left(\dfrac{-1}{3}, \dfrac{13}{3}\right)$.
   (b) This means the lines intersect at one point.
   (c) Students should see that the lines intersect at one point.

2. (a) Solving produces the statement that $4 = -3$ or the slopes are both 1. This means the system is inconsistent.
   (b) This means that the lines have no intersection point, that is, they are parallel.

3. (a) $2x - 4y = -8$
   (b) $x - 2y = -2$
   (c) $x - 2y = -6$
   (d) It could correspond to either (i) or (iv). If the three equations in two unknowns that result could now be solved uniquely for x, y and z the situation is (i). Otherwise we have three different lines of intersection that do not cross in a single point so it would be (iv).
   (e) Depending on which case we choose in (d), either answer is possible based on the information given so far. A correct justification shows that students understand the principle involved.
   (f) Again, either answer is acceptable with a correct rationale.

NOTES

(a) Eliminate z from (1) and (2)

(b) Eliminate z from (2) and (3)

(c) Eliminate z from (1) and (3)

(d) Which of the diagrams in Display 6.8 does this situation correspond to?

(e) Is this system consistent or inconsistent? Explain.

(f) Does the system have a unique solution? Explain.

4. Given the following system of equations

$$z = 2x - y + 10 \qquad (1)$$
$$z = x + y - 2 \qquad (2)$$
$$z = 2x - y + 12 \qquad (3)$$

(a) Eliminate z from (1) and (2)

(b) Eliminate z from (2) and (3)

(c) Eliminate z from (1) and (3)

(d) Which of the diagrams in Display 6.8 does this situation correspond to?

(e) Is this system consistent or inconsistent? Explain.

(f) Does the system have a unique solution? Explain.

5. Given the following system of equations

$$z = x + y - 3 \qquad (1)$$
$$z = 2x + y - 2 \qquad (2)$$
$$z = 3x + 2y - 6 \qquad (3)$$

(a) Eliminate z from (1) and (2)

(b) Eliminate z from (2) and (3)

(c) Eliminate z from (1) and (3)

(d) Which of the diagrams in Display 6.8 does this situation correspond to?

(e) Is this system consistent or inconsistent? Explain.

(f) Does the system have a unique solution? Explain.

Published by IT'S ABOUT TIME, Inc. © 2000 MATHconx, LLC

4. (a) $x - 2y = -12$
   (b) $x - 2y = -14$
   (c) This produces the statement that $0 = 2$ or $10 = 12$.
   (d) The algebra tells us that two pairs of the planes have intersection lines and two do not. The diagram representing this situation is (iii).
   (e) This makes the system inconsistent since there is nothing common to all three planes.
   (f) No.

5. (a) $x = -1$
   (b) $x + y = 4$
   (c) $2x + y = 3$
   (d) This system can now be easily solved for y and z. Hence, it conforms to diagram (i).
   (e) The system is consistent.
   (f) Yes, the one solution is $(-1, 5, 1)$.

NOTES

Chapter 6

6.3 Solving Systems of Equations

Learning Outcomes

After studying this section you will be able to:

Solve a system of two linear equations in two unknowns by the method of elimination;

Solve a system of three linear equations in three unknowns by the method of elimination;

Identify dependent systems of equations;

Demonstrate understanding of the geometric interpretation of dependent systems.

Krelling Industries has been so successful with their new expanded line of desks and tables they decide to introduce yet another product—a chair to match the desk and table. The plant manager in Maine is now trying to make sure that all of the manufacturing capacity of the plant is utilized. A chair requires 2 hours of labor time, a table 6 hours and a desk 9 hours. A chair requires 3 board feet of lumber, a table 7 and a desk 10. A chair requires 1 hour of sanding, a table requires 5 and a desk requires 4. Each week the factory has available 560 board feet of lumber, 454 labor hours and 292 hours of sanding machine time. How many chairs, tables and desks should be produced to fully utilize all of the available resources?

Problems such as this one are faced by industries all the time—usually with many more variables and factors than this simplified example. However, such an example does illustrate the basic features of more complicated problems, and it shows how systems of equations grow out of real life situations.

6.18

1. To help you set up the equations required to solve the problem stated above, copy and complete the following chart in your notebook using the information described.

	Wood (board feet)	Labor (hours)	Sanding Time (hours)
Chair *(c)*			
Table *(t)*			
Desk *(d)*			
Totals			

2. (a) What is the total amount of wood required to manufacture 3 chairs?
   (b) What is the total amount of wood required to manufacture 10 chairs?

500

Published by IT'S ABOUT TIME, Inc. © 2000 MATHconx, LLC

6.3 Solving Systems of Equations

We now examine a problem similar to one encountered in industry that involves the solution of a system of equations. This is used to motivate the need to be able to solve systems of equations that have more than just two variables. Substitution is used because of the way that it can be linked to matrices later in the chapter. Once again, although some technical proficiency is desirable, the understanding of the processes are more important than the details because the actual solution of a system can be done with matrices using a calculator. Interpreting and understanding the results, however, can only be done by a human, using judgment.

6.18

1.

	Wood (board feet)	Labor (hours)	Sanding Time (hours)
Chair *(c)*	3	2	1
Table *(t)*	7	6	5
Desk *(d)*	10	9	4
Totals Available	560	454	292

2. (a) 9   (c) 90
   (b) 30   (d) $3c$

Additional Support Materials:

Assessments	Qty
Form (A)	1
Form (B)	1

Blackline Masters	Qty
Student p.505	1

Extensions	Qty

Supplements	Qty
Solving Systems of Equations	2

Chapter 6

   (c) What is the total amount of wood required to manufacture 30 chairs?

   (d) What is the total amount of wood required to manufacture *c* chairs?

3. (a) What is the total amount of wood required to manufacture *t* tables?

   (b) What is the total amount of wood required to manufacture *d* desks?

   (c) Using the variables *c, t,* and *d,* write an equation that expresses the fact that the total amount of wood available for chairs, tables and desks is 560 board feet.

4. Write two other equations based on the information in the chart.

We are now faced with the problem of solving a system of three equations in three unknowns. Before we try to solve the problem, let's look at another method of solving two equations in two unknowns that is different from the method of substitution. We can use this new method to solve three equations with three unknowns. This method is called the *method of elimination*. It involves reducing a system down to a single equation with one variable. First, let's look at the system below that has two variables, x and y. Fortunately there is an easy way to reduce this system of two equations in two unknowns to a single equation with one variable—just add the two equations together. This will give us a single equation with one variable only.

$$
\begin{array}{r}
x + y = 7 \\
x - y = 3 \\
\hline
2x = 10 \quad \text{So, } x = 5
\end{array}
$$

Now substitute 5 for x into $x + y = 7$ to find the value of y.

$$5 + y = 7 \qquad \text{So, } y = 2$$

If we look at $x - y = 3$, we can see that the values we have found for x and y also satisfy this equation. The solution to this system of equations would be written as (5, 2) to remind us that what we have really done here is to find the point where the lines representing these equations intersect.

Published by IT'S ABOUT TIME, Inc. © 2000 MATHconx, LLC

501

3. (a) $7t$
   (b) $10d$
   (c) $3c + 7t + 10d = 560$

4. $2c + 6t + 9d = 454$
   $c + 5t + 4d = 292$

NOTES

6.19

1. Use the method of elimination to solve this system of equations. Verify your answers in both equations.

$$x + y = 15 \qquad (1)$$
$$3x - y = 1 \qquad (2)$$

2. Use the method of elimination to solve this system of equations. Verify your answers in both equations.

$$x - 2y = 4 \qquad (1)$$
$$2x + 2y = 5 \qquad (2)$$

3. Use the method of elimination to solve this system of equations. Verify your answers in both equations.

$$2x + \ y = 3 \qquad (1)$$
$$-2x + 3y = 1 \qquad (2)$$

4. Each of the systems in questions 1–3 have a common feature that made eliminating one of the variables easy. What is it?

In the equations you have been given, one variable was eliminated when you added the equations together. Now suppose you needed to solve the system.

$$x + 2y = 4 \qquad (1)$$
$$3x + 2y = 8 \qquad (2)$$

It is clear that adding equations (1) and (2) won't help, because neither the x term nor the y term will cancel out. However, if we subtracted them, our problem would be solved.

$$x + 2y = 4 \qquad (1)$$
$$\underline{3x + 2y = 8 \qquad (2)}$$
$$-2x + 0 \ = -4$$
$$x \ = 2$$

Now substitute $x = 2$ into equation (1) to find the value of y.

$$2 + 2y = 4$$
$$2 - 2 + 2y = 4 - 2$$
$$2y = 2$$
$$y = 1$$

Therefore, the solution to this system of equations is (2, 1).

Published by IT'S ABOUT TIME, Inc. © 2000 MATHconx, LLC

502

6.19

1. (4, 11)

2. (3, -0.5)

3. (1, 1)

4. In each of these systems, one variable is eliminated when the equations are added.

NOTES

Chapter 6

1. Verify that this solution also satisfies equation (2).

2. What is the geometric interpretation of the solution (2, 1)?

a
6.20

1. You wish to eliminate the variable x from the following system of equations.

6.21

   (a) Can it be done by adding the two equations? Explain.

   (b) By subtracting them? Explain.

   (c) What number could you multiply equation (2) by to make both x terms equal?

$$3x - 5y = 1 \qquad (1)$$
$$x + y = 7 \qquad (2)$$

1. If you wished to eliminate the variable x from this system of equations

$$x - 3y = 7 \qquad (1)$$
$$2x + y = 5 \qquad (2)$$

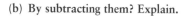

b
6.22

what number would you multiply equation (1) by?

2. If you wished to eliminate the variable x from this system of equations

$$-2x - 7y = 1 \qquad (1)$$
$$x + 4y = 9 \qquad (2)$$

what number would you multiply equation (2) by?

Using this method allows us to eliminate a variable from *any* system of equations, even linear ones such as this that have different x-coefficients and different y-coefficients.

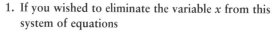

$$3x - 5y = -2 \qquad (1)$$
$$2x + 2y = 4 \qquad (2)$$

6.20

1. $3*2 + 2*1 = 6 + 2 = 8$

6.21

2. This is the intersection point of the two lines.

6.22

1. (a) No.   (b) No.   (c) by 3

1. (1) by -2

NOTES

_____

Chapter 6

Let's assume that we wanted to eliminate the variable x. The smallest multiple of 2 and 3 is 6. To eliminate x we could multiply equation (1) by 2 and equation (2) by -3.

$$2 \times (1) \rightarrow \quad 6x - 10y = -4$$
$$-3 \times (2) \rightarrow -6x - \quad 6y = -12$$

Then add these two equations to produce

$$-16y = -16$$
$$y = 1$$

Now we substitute $y = 1$ into equation (2) to find x. So $x = 1$. Therefore, the solution for this system of equations is (1, 1). Remember that we should also verify that this solution works in equation (1) this time, because we substituted in equation (2) to find the value for y.

6.23

Let's assume that we wanted to eliminate the variable y.

$$3x - 5y = -2 \qquad (1)$$
$$2x + 2y = \;\; 4 \qquad (2)$$

We need to think of some number that both -5 and 2 can be converted into.

1. What is the smallest number of which -5 and 2 are factors?

2. What numbers would you need to multiply equations (1) and (2) by to make this conversion?

6.24

1. (a) Solve the system of equations by first eliminating the variable y.

$$3x - 5y = -2 \qquad (1)$$
$$2x + 2y = \;\; 4 \qquad (2)$$

    (b) Did you get the same answer when you eliminated the variable x first? Does it make any difference which variable you eliminate first? Explain.

2. In the following table, you are given a number of systems of two equations in two unknowns. Write out how you would eliminate the given variable. The first one has been done for you. Note that although there are many correct answers for each of these questions, normally we try to use the smallest numbers we can to make the arithmetic easier.

6.23

2. (2) by 2

6.24

1. 10 or -10   2. Equation (1) by 2 and equation (2) by 5.

1. (a) (1, 1)
   (b) The answer should be exactly the same. It does not matter which variable you eliminate, pick which one seems easier.

NOTES

_____

Chapter 6

System	To Eliminate x	To Eliminate y
$3x + 3y = 2$ (1) $x + 2y = 7$ (2)	(2) × -3	(1) × -2 (2) × 3
$2x - 3y = 1$ (1) $7x + 6y = 1$ (2)		
$4x - 5y = 9$ (1) $3x + 2y = 7$ (2)		
$9x + 3y = 1$ (1) $2x + 4y = 7$ (2)		
$6x - 5y = 4$ (1) $-3x + 3y = 7$ (2)		
$7x - 6y = 2$ (1) $2x + 5y = -3$ (2)		
$3x - 5y = 6$ (1) $3x + 5y = 7$ (2)		

3. Solve the following systems of two equations in two unknowns by the method of elimination. Verify that your answers satisfy both equations.

$$2x + 3y = -4 \qquad (1)$$
$$4x - y = 6 \qquad (2)$$

$$5x - 2y = 6 \qquad (1)$$
$$3x + y = 8 \qquad (2)$$

You may be wondering why we needed to learn the method of elimination. Wasn't knowing the method of substitution enough? Let's compare the two methods when solving a complex system of two equations in two unknowns.

$$4x - 5y = 9 \qquad (1)$$
$$3x + 2y = 7 \qquad (2)$$

Using substitution, equation (1) becomes

$$y = \frac{4}{5}x - \frac{9}{5}$$

and equation (2) becomes

$$y = -\frac{3}{2}x + \frac{7}{2}$$

So,

$$\frac{4}{5}x - \frac{9}{5} = -\frac{3}{2}x + \frac{7}{2}$$

Published by IT'S ABOUT TIME, Inc. © 2000 MATHconx, LLC

2.

System	To Eliminate x	To Eliminate y
(1) $3x + 3y = 2$ (2) $x + 2y = 7$	(2) $\times$ -3	(1) $\times$ -2 (2) $\times$ $+3$
(1) $x - 3y = 1$ (2) $7x + 6y = 1$	(1) $\times$ -7	(1) $\times$ 2
(1) $4x - 5y = 9$ (2) $3x + 2y = 7$	(1) $\times$ -3 (2) $\times$ 4	(1) $\times$ 2 (2) $\times$ 5
(1) $9x + y = 1$ (2) $2x + 4y = 7$	(1) $\times$ 2 (2) $\times$ -9	(1) $\times$ -4
(1) $6x - 5y = 4$ (2) $-3x + 3y = 7$	(2) $\times$ 2	(1) $\times$ 3 (2) $\times$ 5
(1) $7x - 6y = 2$ (2) $2x + 5y = -3$	(1) $\times$ -2 (2) $\times$ 7	(1) $\times$ 5 (2) $\times$ 6
(1) $x - 5y = 6$ (2) $3x + y = 7$	(1) $\times$ -3	(2) $\times$ 5

3. (a) $(1, -2)$ (b) $(2, 2)$

Chapter 6

We won't finish solving the equation because you can see that it will require a lot more algebra to solve for x.

Using the method of elimination,

$$
\begin{array}{lll}
4x - 5y = 9 & \rightarrow & 8x - 10y = 18 \\
3x + 2y = 7 & \rightarrow & 15x + 10y = 35 \\
\hline
\text{add} & & 23x + 0 = 53 \\
& & x = \dfrac{53}{23}
\end{array}
$$

So no matter how difficult solving a system of two equations in two unknowns appears to be, the method of elimination seems more efficient than the method of substitution.

Let's use the method of elimination when solving a system of three linear equations in three unknowns. Basically we proceed in a similar fashion, eliminating a single variable from two different pairs of equations. This then gives us a system of two equations in two unknowns that we already know how to solve!

6.25

1. **In the following system, which variable do you think would be the easiest to eliminate from all three equations? Explain.**

   $x + y - 3z = -6$ (1)

   $2x - y + z = 5$ (2)

   $x + y + 2z = 4$ (3)

2. **We choose to eliminate the variable y. Here is one way to do so. Add equation (1) and equation (2) and call the resulting equation (4).**

3. **Add equation (2) and equation (3) and call the resulting equation (5).**

4. **Solve equations (4) and (5) using the method of elimination.**

5. **Now that you have a value for x and a value for z, pick one of equations (1), (2) or (3) and substitute the values of x and z that you have found. Use this equation to find y.**

6. **Write the solution of this system as an ordered triple (x, y, z).**

Published by IT'S ABOUT TIME, Inc. © 2000 MATHconx, LLC

6.25

1. Answers may vary.

2. Equation (4) is $3x - 2z = -1$.

3. Equation (5) is $3x + 3z = 9$.

4. This produces $z = 2$ and by substitution $x = 1$.

5. This produces $y = -1$.

6. $(1, -1, 2)$

NOTES

_____

7. Verify that your answer makes all three equations true statements.

8. What is the geometrical interpretation of the solution of this system of three equations in three unknowns?

At the beginning of this section you developed a series of three linear equations in three unknowns from a problem about Krelling Industries. Look back in your notebook and confirm that the three equations you developed are the same as the system below

$$7t + 10d + 3c = 560 \qquad (1)$$
$$6t + 9d + 2c = 454 \qquad (2)$$
$$5t + 4d + c = 292 \qquad (3)$$

where t represents the number of tables, d represents the number of desks and c represents the number of chairs to be produced.

1. Eliminate the variable c from equations (2) and (3). Label the resulting equation (4).

2. Eliminate the variable c from equations (1) and (3). Label the resulting equation (5).

6.26

3. Solve the system of two equations in two unknowns labeled as (4) and (5).

4. Substitute back in either (1), (2) or (3) to find the remaining variable.

5. How many tables, desks and chairs will fully utilize the available resources in this factory?

The ability to make efficient use of resources is crucial to businesses and industry, therefore problems such as these are important ones to be able to solve.

Before we finish this section let's reconsider the idea with which we started the chapter. We originally examined a problem that tried to establish when the profits at two different manufacturing plants would be the same. The equations had three variables and we only had two equations. We found that it was not possible to find a unique solution in this case.

Published by IT'S ABOUT TIME, Inc. © 2000 MATHconx, LLC

507

7. Students will verify answers.

8. The three planes intersect in a single point.

6.26

1. (4) $-4t + d = -130$

2. (5) $-8t - 2d = -316$

3. This produces $t = 36$ and $d = 14$.

4. $c = 56$

5. The resources are fully utilized when they make 56 chairs, 14 desks and 36 tables.

NOTES

Chapter 6

Let's review why. The two equations represent planes that intersect in a line. So this system of two equations does not have a unique solution. Systems that do not have a unique solution are said to be **dependent**. For example, a system with two equations in two unknowns such as

$$y = 3x + 2 \qquad (1)$$
$$2y = 6x + 4 \qquad (2)$$

is a dependent system. Equation (2) is a multiple of (1). That is, they are the same line, so there is no unique solution.

Recall what happens in two dimensions. If you graph the two equations

$$y = x + 3 \qquad (3)$$
$$y = -x + 5 \qquad (4)$$

on your graphing calculator, you will find that they intersect in a single point. A system that has exactly one solution is said to be **independent**. This is reasonable because we should be able to solve two equations in two unknowns and find a unique solution if one equation is not a multiple of the other.

This idea will come up again, and it is important because it is difficult to tell if a system is dependent or independent just by looking at it. However, when you begin to solve a system of two equations in two unknowns you will realize what is happening if you cannot find a unique solution.

Now that the world has a global economy, the ability to make efficient use of resources is crucial. In fact, problems in industry, science and financial business often involve more than three variables. Solving large systems of equations such as 100 equations in 100 unknowns is not something that we can accomplish in any reasonable amount of time. We can, however, understand the process and then program a computer to solve such problems. This is the reason that *understanding* the ideas is so important. In the rest of this chapter we will look at a way that mathematicians have developed of making this process somewhat more refined by using arrays of numbers called *matrices*. The matrices then allow the use of calculators and computers to solve much more complex problems.

508

Published by IT'S ABOUT TIME, Inc. © 2000 MATHconx, LLC

NOTES

Chapter 6

Problem Set: 6.3

1. Solve each of the following systems by the method of elimination.

(a) $\quad x - y = -1 \qquad$ (1)
$\qquad x + y = 3 \qquad$ (2)

(b) $\quad 2x - y = -7 \qquad$ (1)
$\qquad 3x + y = -3 \qquad$ (2)

(c) $\quad 2x - 3y = 3 \qquad$ (1)
$\qquad 3x + 4y = 13 \qquad$ (2)

(d) $\quad 5x - y = 15 \qquad$ (1)
$\qquad 3x - 3y = -3 \qquad$ (2)

2. What is the maximum number of variables that need to be eliminated to solve

(a) 2 equations in two unknowns?

(b) 3 equations in three unknowns?

(c) 4 equations in four unknowns?

(d) 5 equations in five unknowns?

(e) 200 equations in 200 unknowns?

(f) If you could do one elimination in 2 minutes, how long would it take (in hours) to find the value of the first variable in (e)?

3. A chemist has two solutions of sulphuric acid, one a 20% solution and the other a 50% solution. Unfortunately, his experiment requires a 40% solution. How many liters of each should he mix to produce 5 liters of 40% solution?

Published by IT'S ABOUT TIME, Inc. © 2000 MATHconx, LLC

Problem Set: 6.3

1. (a) $x = 1$; $y = 2$
    (b) $x = -2$; $y = 3$
    (c) $x = 3$; $y = 1$
    (d) $x = 4$; $y = 5$

2. This is an interesting problem that is similar to the handshake problem. The pattern makes the generalization easier to see. It also provides a good starting point for discussion about why mathematicians have worked hard at finding other ways to solve large systems of linear equations.

    (a) 1 or $\dfrac{(2)(2 - 1)}{2}$

    (b) 3 or $\dfrac{(3)(3 - 1)}{2}$

    (c) 6 or $\dfrac{(4)(4 - 1)}{2}$

    (d) 10 or $\dfrac{(5)(5 - 1)}{2}$

    (e) $\dfrac{(200)(200 - 1)}{2} = 19,900$

    (f) Approximately 663 hours.

3. He should use approximately 3.33 liters of 50% solution and 1.67 liters of 20% solution. Have students note that this makes sense—more of the 50% solution is needed because the final result is to be a 40% solution and the 50% solution is closer than the 20% solution to this target.
    $20\% * x + 50\% * y = 40\% * 5$
    $x + y = 5$

Chapter 6

4. Solve each of the following systems by the method of elimination.

(a)
$$\begin{array}{ll} x + y + z = 3 & (1) \\ x - y + z = 3 & (2) \\ 2x + y + z = 4 & (3) \end{array}$$

(b)
$$\begin{array}{ll} 2x - y + z = 1 & (1) \\ x + y - 2z = -5 & (2) \\ 3x + y - z = -6 & (3) \end{array}$$

(c)
$$\begin{array}{ll} x - y + z = 2 & (1) \\ x + 2y - z = 2 & (2) \\ 3x + 3y - z = 6 & (3) \end{array}$$

(d)
$$\begin{array}{ll} 2x + y + z = -2 & (1) \\ x - 3y + z = 3 & (2) \\ 3x - 4y - 2z = -1 & (3) \end{array}$$

5. Krelling Industries has a second furniture plant (in California) making chairs, tables and desks. The manager wants to be sure that all of the manufacturing capacity of the plant is utilized. However, the sanding equipment is different in the second plant and so he cannot use the same schedule. In the second plant, a chair requires 2 hours of labor time, a table 6 hours and a desk 9 hours. A chair requires 3 board feet of lumber, a table 7 and a desk 10. A chair requires 1 hour of sanding, a table requires 5 and a desk requires 8. Each week the factory has available 560 board feet of lumber, 454 labor hours and 348 hours of sanding machine time. How many chairs, tables and desks should be produced to fully utilize all of the available resources? Explain your result.

6. Sylvia works for a large corporation that sponsors a party for 250 children at the local hospital. She is given $2,500 to spend and she plans to buy three types of toys: stuffed toys, plastic models and assorted board games.
A local merchant has agreed to sell stuffed toys for $10 each, plastic models for $9 each and board games for $11 each. She knows that this year there are quite a few very young children in the hospital so she wants to buy 25 more stuffed toys than board games. How many of each type of toy should she buy to spend all her money?

Published by IT'S ABOUT TIME, Inc. © 2000 MATHconx, LLC

510

4. (a) (1, 0, 2)
   (b) (–1, –2, 1)
   (c) This system is not independent and does not have a unique solution. In fact $(2) \times 2 + (1)$ gives equation (3).
   (d) (–1, –1, 1)

5. Although this problem appears to be the same as the previous one, it is not. The system is not independent, so there is no unique solution. Students should realize this when they get a statement such as $0 = 0$. The system is still consistent, however, because there are in fact many solutions. Geometrically this means that the three planes represented by these equations all intersect in the same straight line.

6. She should buy 100 stuffed toys, 75 plastic models and 75 board games.

NOTES

6.4 Solving Linear Systems Using Matrices

Let's return to a problem that you have encountered before.

Krelling Industries makes chairs, tables, and desks. A chair requires 3 board feet of lumber, a table 7 feet, and a desk 10 feet. A chair requires 2 labor hours of time, a table 6 hours, and a desk 9 hours. Wood used in construction is sanded by machine. A chair requires 1 hour of sanding, a table 5 hours, and a desk 4 hours. Each week the company has available 560 board feet of lumber, 454 labor hours, and 292 hours of machine availability for sanding. How many chairs, tables and desks should be produced each week if the company plans to utilize all the resources including lumber, labor hours, and machine availability?

Learning Outcomes

After studying this section you will be able to:

Represent a system of equations as a matrix;

Explain what is meant by a row operation on a matrix;

Use Gaussian elimination to solve systems of equations;

Use a calculator to perform row operations.

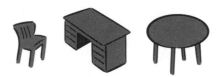

This problem led to the following system of equations.

$$3x + 7y + 10z = 560 \qquad \text{(a)}$$
$$(1) \qquad 2x + 6y + 9z = 454 \qquad \text{(b)}$$
$$1x + 5y + 4z = 292 \qquad \text{(c)}$$

where x denotes the number of chairs produced, y the number of tables produced, and z the number of desks produced. The first equation (a) represents information about the lumber used and available, the second equation (b) represents information about the labor hours used and available, and the third equation (c) represents information about the sanding machine hours used and available.

There are a number of operations that can be performed on the system in (1) without changing the solution set. First, it

6.4 Solving Linear Systems Using Matrices

The purpose of this section is to look at the process of elimination, started in the previous section, from a different point of view. In particular, the use of matrices reduces the amount of writing that one must do in solving linear systems. It is important for students to understand that using row operations on matrices is equivalent to changing equations in systems, yielding new systems which have the same solution set.

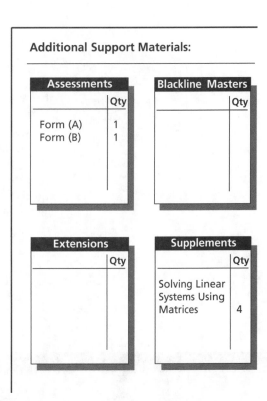

Additional Support Materials:

Assessments	Qty
Form (A)	1
Form (B)	1

Blackline Masters	Qty

Extensions	Qty

Supplements	Qty
Solving Linear Systems Using Matrices	4

Chapter 6

should be clear that if we interchange any two of the equations, the solution set will not be changed. For example, interchanging the first equation (a) and the third equation (c) in (1) will not change the solution set.

$$1x + 5y + 4z = 292 \qquad \text{(c)}$$

$$2x + 6y + 9z = 454 \qquad \text{(b)}$$

$$3x + 7y + 10z = 560 \qquad \text{(a)}$$

a

6.27

1. Write the system that results from interchanging the second equation (b) and third equation (c) in (1).

2. Write the system that results from interchanging the first equation (a) and the second equation (b) in (1).

Second, if any equation is multiplied by a nonzero constant, the solution set will not be changed. For example, multiplying the second equation (b) in (1) by 3 will not change the solution set:

$$3x + 7y + 10z = 560 \qquad \text{(a)}$$

$$(3)(2x + 6y + 9z) = (3)454 \qquad \text{(b)}$$

$$1x + 5y + 4z = 292 \qquad \text{(c)}$$

or

$$3x + 7y + 10z = 560$$

$$6x + 18y + 27z = 1362$$

$$1x + 5y + 4z = 292$$

b

6.28

1. Write the system that results from multiplying the first equation (a) in (1) by 5.

2. Write the system that results from multiplying the third equation (c) in (1) by 4.

Third, if any equation is changed by adding a constant times another equation to it, the solution set is unchanged. For example, multiplying the third equation (c) in (1) by –2

$$3x + 7y + 10z = 560 \qquad \text{(a)}$$

$$2x + 6y + 9z = 454 \qquad \text{(b)}$$

$$(-2)(1x + 5y + 4z) = (-2)292 \qquad \text{(c)}$$

Published by IT'S ABOUT TIME, Inc. © 2000 MATHconx, LLC

512

6.27

1. $3x + 7y + 10z = 560$ (a)
    $1x + 5y + 4z = 292$ (c)
    $2x + 6y + 9z = 454$ (b)

2. $2x + 6y + 9z = 454$ (b)
    $3x + 7y + 10z = 560$ (a)
    $1x + 5y + 4z = 292$ (c)

6.28

1. $15x + 35y + 50z = 2800$
    $2x + 6y + 9z = 454$
    $1x + 5y + 4z = 292$

2. $3x + 7y + 10z = 560$
    $2x + 6y + 9z = 454$
    $4x + 20y + 16z = 1168$

NOTES

Then add the altered (c), which is $-2x - 10y - 8z = -584$ to (a). This leads to the system

$$1x - 3y + 2z = -24 \qquad \text{(d)}$$
$$2x + 6y + 9z = 454 \qquad \text{(b)}$$
$$1x + 5y + 4z = 292 \qquad \text{(c)}$$

Note that equation (a) has changed to (d). Equation (c) in system (1) remains intact, and the solution set does not change.

1. Write the system of equations that results from (1) if the second equation (b) is multiplied by -3 and the resulting equation is added to the first equation (a).

6.29

2. Write the system of equations that results from (1) if the first equation (a) is multiplied by 2 and the resulting equation is added to the third equation (c).

3. Write the system of equations that results from (1) if the following two operations are performed consecutively.

(a) The second equation is multiplied by 2.

(b) The second equation is multiplied by 2 and the resulting equation is added to the third equation.

If you think about it, the x, y, and z appearing in the system

$$3x + 7y + 10z = 560$$
$$2x + 6y + 9z = 454$$
$$1x + 5y + 4z = 292$$

are simply used to separate, in order, the numbers 3, 7, 10, etc. All of the information you need about the system is given in the following *matrix*.

(2)
$$\begin{pmatrix} 3 & 7 & 10 & 560 \\ 2 & 6 & 9 & 454 \\ 1 & 5 & 4 & 292 \end{pmatrix}$$

A **matrix** is a rectangular array of numbers. If there are n rows (horizontal array) and m columns (vertical array), we say the size of the matrix is $n \times m$. The matrix in (2) is a 3×4 matrix. Note that a dotted vertical line is used in (2) to indicate the position of the equality symbols.

Published by IT'S ABOUT TIME, Inc. © 2000 MATHconx, LLC

513

6.29

1. $-3x - 11y - 17z = -802$
$2x + 6y + 9z = 454$
$1x + 5y + 4z = 292$

2. $3x + 7y + 10z = 560$
$2x + 6y + 9z = 454$
$7x + 19y + 24z = 1412$

3. $3x + 7y + 10z = 560$
$4x + 12y + 18z = 908$
$9x + 29y + 40z = 2108$

NOTES

Chapter 6

6.30

1. State the size of each of the following matrices.

(a) $\begin{pmatrix} 1 & 3 \\ 5 & -1 \\ 0 & 2 \\ 5 & 9 \end{pmatrix}$ (b) $\begin{pmatrix} 1 & 4 \\ -6 & 8 \end{pmatrix}$ (c) $\begin{pmatrix} 8 & 5 & 3 \\ -2 & 0 & 5 \\ 7 & 9 & 4 \\ 3 & 6 & 8 \\ 4 & 4 & -1 \end{pmatrix}$

2. Write a matrix to represent the following system of equations.

$$4x + 7y - 2z = 18$$

$$3x - 9y + 8z = 25$$

$$6x - 5y - 4z = 17$$

Use a dotted vertical line to represent the equality symbols.

3. Write a matrix to represent the following system of equations.

$$2x + 3y = 29$$

$$4x - 9y = 37$$

$$3x + 4y = 52$$

Since we are now able to represent systems of equations by using matrices, it should follow that certain operations are allowed on such matrices in the same way that certain operations were allowed on systems of equations. Recall that the following were allowed.

I) Interchange two equations.

II) Multiply an equation by a nonzero constant.

III) Multiply an equation by a nonzero constant and add to another equation.

We can translate these operations into operations on matrices as follows.

I) Interchange two rows.

II) Multiply a row by a nonzero constant.

III) Multiply one row by a nonzero constant and add the result to another row.

Published by IT'S ABOUT TIME, Inc. © 2000 MATHconx, LLC

6.30

1. (a) 4×2 (b) 2×2 (c) 5×3

2. $\begin{pmatrix} 4 & 7 & -2 & | & 18 \\ 3 & -9 & 8 & | & 25 \\ 6 & -5 & -4 & | & 17 \end{pmatrix}$ 3. $\begin{pmatrix} 2 & 3 & | & 29 \\ 4 & -9 & | & 37 \\ 3 & 4 & | & 52 \end{pmatrix}$

NOTES

Chapter 6

Each of these operations on a matrix is known as an **elementary row operation.** There is a notation we will use for row operations which is illustrated in the following examples.

I)   Interchange rows 2 and 3.

$$\begin{pmatrix} 1 & 3 & 5 & 7 \\ 9 & -1 & 4 & 2 \\ 3 & 0 & 5 & 4 \end{pmatrix} \xrightarrow{\;R_2 \leftrightarrow R_3\;} \begin{pmatrix} 1 & 3 & 5 & 7 \\ 3 & 0 & 5 & 4 \\ 9 & -1 & 4 & 2 \end{pmatrix}$$

II)   Multiply row 2 by 5.

$$\begin{pmatrix} 1 & 3 & 5 & 7 \\ 9 & -1 & 4 & 2 \\ 3 & 0 & 5 & 4 \end{pmatrix} \xrightarrow{\;5R_2\;} \begin{pmatrix} 1 & 3 & 5 & 7 \\ 45 & -5 & 20 & 10 \\ 3 & 0 & 5 & 4 \end{pmatrix}$$

III)   Multiply row 3 by –4 and add to row 2.
    Note that row 2 is the row that changes.

$$\begin{pmatrix} 1 & 3 & 5 & 7 \\ 9 & -1 & 4 & 2 \\ 3 & 0 & 5 & 4 \end{pmatrix} \xrightarrow{\;-4R_3 + R_2\;} \begin{pmatrix} 1 & 3 & 5 & 7 \\ -3 & -1 & -16 & -14 \\ 3 & 0 & 5 & 4 \end{pmatrix}$$

Perform the specified row operation on the given matrices.

6.31

1. $\begin{pmatrix} 3 & 2 \\ 4 & 5 \\ 9 & -2 \end{pmatrix} \xrightarrow{\;R_1 \leftrightarrow R_3\;}$ 2. $\begin{pmatrix} 1 & 3 & 7 \\ 4 & 2 & 8 \end{pmatrix} \xrightarrow{\;3R_2\;}$

3. $\begin{pmatrix} 1 & 2 \\ 4 & 5 \\ 6 & 9 \end{pmatrix} \xrightarrow{\;-6R_1 + R_3\;}$

Published by IT'S ABOUT TIME, Inc. © 2000 MATHconx, LLC

515

6.31

1. $\begin{pmatrix} 9 & -2 \\ 4 & 5 \\ 3 & 2 \end{pmatrix}$

2. $\begin{pmatrix} 1 & 3 & 7 \\ 12 & 6 & 24 \end{pmatrix}$

3. $\begin{pmatrix} 1 & 2 \\ 4 & 5 \\ 0 & -3 \end{pmatrix}$

NOTES

Chapter 6

Published by IT'S ABOUT TIME, Inc. © 2000 MATHconx, LLC

One can use these row operations as a tool for solving systems of linear equations. We illustrate a systematic approach with two equations involving two variables. Our objective is to solve the system

$$3x + 4y = 23$$

$$x + 3y = 11$$

The matrix for this system is given by

$$\begin{pmatrix} 3 & 4 & \vdots & 23 \\ 1 & 3 & \vdots & 11 \end{pmatrix}$$

Our first step is to get a 1 as the entry in the first row-first column. One way would be to multiply the first row by $\frac{1}{3}$. This process, however, introduces fractions. An easier way is simply to interchange rows 1 and 2.

$$\begin{pmatrix} 3 & 4 & \vdots & 23 \\ 1 & 3 & \vdots & 11 \end{pmatrix} \xrightarrow{\ R_1 \leftrightarrow R_2\ } \begin{pmatrix} 1 & 3 & \vdots & 11 \\ 3 & 4 & \vdots & 23 \end{pmatrix}$$

The algorithm now makes use of that 1 to get a zero below it. This is possible by multiplying row 1 by −3 and adding the result to row 2.

$$\begin{pmatrix} 1 & 3 & \vdots & 11 \\ 3 & 4 & \vdots & 23 \end{pmatrix} \xrightarrow{\ -3R_1 + R_2\ } \begin{pmatrix} 1 & 3 & \vdots & 11 \\ 0 & -5 & \vdots & -10 \end{pmatrix}$$

The next step is to get (if possible) a 1 in the entry of the second row-second column. This can be done by multiplying $-\frac{1}{5}$ times row 2.

$$\begin{pmatrix} 1 & 3 & \vdots & 11 \\ 0 & -5 & \vdots & -10 \end{pmatrix} \xrightarrow{\ \left(-\frac{1}{5}\right)R_2\ } \begin{pmatrix} 1 & 3 & \vdots & 11 \\ 0 & 1 & \vdots & 2 \end{pmatrix}$$

The solution set of the system represented by $\begin{pmatrix} 1 & 3 & \vdots & 11 \\ 0 & 1 & \vdots & 2 \end{pmatrix}$

is the same as the solution set of the original system. From the

Published by IT'S ABOUT TIME, Inc. © 2000 MATHconx, LLC

NOTES

matrix $\begin{pmatrix} 1 & 3 & \vdots & 11 \\ 0 & 1 & \vdots & 2 \end{pmatrix}$ our system of equations is now

$$x + 3y = 11$$

$$0x + y = 2 \text{ or } y = 2$$

Substituting $y = 2$ back into the first equation (backsolving) gives $x = 5$. The solution is thus given by $x = 5$ and $y = 2$.

The process we have just illustrated is known as **Gaussian elimination with backsolving.**

Use Gaussian elimination with backsolving to solve each of the following systems.

6.32

1. $2x + 3y = 10$ **2.** $5x + y = 17$

     $x + 5y = 19$ $x + 3y = 9$

In the problems considered thus far, it was easy to get a 1 into the first row-first column of the appropriate matrix. We simply interchanged rows. The following example shows another process when this is not possible. We want to solve

$$3x + 4y = 23$$

$$2x + 5y = 20$$

Consider the following operations,

$$\begin{pmatrix} 3 & 4 & \vdots & 23 \\ 2 & 5 & \vdots & 20 \end{pmatrix} \xrightarrow{-1R_2 + R_1} \begin{pmatrix} 1 & -1 & \vdots & 3 \\ 2 & 5 & \vdots & 20 \end{pmatrix}$$

$$\xrightarrow{-2R_1 + R_2} \begin{pmatrix} 1 & -1 & \vdots & 3 \\ 0 & 7 & \vdots & 14 \end{pmatrix} \xrightarrow{\left(\frac{1}{7}\right)R_2} \begin{pmatrix} 1 & -1 & \vdots & 3 \\ 0 & 1 & \vdots & 2 \end{pmatrix}$$

which leads to

$$x - y = 3$$

$$0x + y = 2 \text{ or } y = 2$$

Backsolving using $y = 2$ leads to the solution $x = 5$ and $y = 2$.

Published by IT'S ABOUT TIME, Inc. © 2000 MATHconx, LLC

6.32

1. $x = -1, y = 4$ 2. $x = 3, y = 2$

NOTES

6.33

Use Gaussian elimination with backsolving to solve each of the following.

1. $4x + 3y = 13$ 2. $7x + 3y = 29$

   $3x - 2y = 14$ $3x - 2y = 19$

The row operations we have been using can be performed on some calculators. On a TI-82 (TI-83) the operations are identified as follows.

I)   Interchange rows A and B $\leftrightarrow$
     rowSwap(*matrix, rowA, rowB*)

II) Multiply a row by a nonzero constant $\leftrightarrow$
     *row(*value, matrix, row*)

III) Multiply row A by a constant and add the result to
     B $\leftrightarrow$ *row+(*value, matrix, rowA, rowB*)

We illustrate this procedure on the TI-82 (TI-83) with our previous example.

$$3x + 4y = 23$$

$$x + 3y = 11$$

This problem was solved by using the following sequence of row operations.

$$\begin{pmatrix} 3 & 4 & \vdots & 23 \\ 1 & 3 & \vdots & 11 \end{pmatrix} \xrightarrow{R_1 \leftrightarrow R_2} \begin{pmatrix} 1 & 3 & \vdots & 11 \\ 3 & 4 & \vdots & 23 \end{pmatrix} \xrightarrow{-3R_1 + R_2}$$

$$\begin{pmatrix} 1 & 3 & \vdots & 11 \\ 0 & -5 & \vdots & -10 \end{pmatrix} \xrightarrow{\left(-\frac{1}{5}\right)R_2} \begin{pmatrix} 1 & 3 & \vdots & 11 \\ 0 & 1 & \vdots & 2 \end{pmatrix} \text{ which gave us } y = 2.$$

The final step was to use backsolving to get $x = 5$.

On a TI-82 (TI-83)

1.   Press MATRIX

2.   Use the arrow $\leftarrow$ or $\rightarrow$ key to highlight EDIT. Press ENTER

3.   You now need to enter the size 2×3. Press ENTER

4.   Insert each entry. For example,

     press 3   ENTER   press 4   ENTER   etc.

6.33

1. $x = 4, y = -1$ 2. $x = 5, y = -2$

NOTES

Chapter 6

5. Press QUIT Press MATRIX 1 ENTER You should see

   [[3 4 23]

   [1 3 11]]

6. To interchange rows 1 and 2, press MATRIX, arrow over to MATH, and arrow up or down to rowSwap(

7. Press ENTER

8. Press MATRIX 1, 1, 2) (commas are necessary) ENTER

9. You should now see

   [[1 3 11]

   [3 4 23]]

10. To multiply row 1 by ‑3 and add the result to row 2, press MATRIX, arrow to MATH, and arrow up or down to *row + (ENTER

11. Press ‑3, ANS, 1, 2) ENTER

12. You should now see

    [[1 3 11]

    [0 ‑5 ‑10]]

13. To multiply row 2 by $(-\frac{1}{5})$, press MATRIX, arrow to MATH, and arrow up or down to
    *row(ENTER

14. Press (‑1 ÷ 5), ANS, 2) ENTER

15. You should now see

    [[1 3 11]

    [0 1   2]]

At this point you are ready to backsolve for x knowing that $y = 2$.

Use a calculator, row operations, and backsolving to solve each of the following.

6.34

1. $4x + 3y = 13$ 2. $7x + 3y = 29$

   $3x - 2y = 14$ $3x - 2y = 19$

Published by IT'S ABOUT TIME, Inc. © 2000 MATHconx, LLC

519

6.34

1. $x = 4, y = -1$ 2. $x = 5, y = -2$

NOTES
_____

Chapter 6

Return to the problem of Krelling Industries. They need to solve the following system of three equations with three variables.

$$3x + 7y + 10z = 560$$

$$2x + 6y + 9z = 454$$

$$1x + 5y + 4z = 292$$

In matrix form, this system can be written as

$$\begin{pmatrix} 3 & 7 & 10 & \vdots & 560 \\ 2 & 6 & 9 & \vdots & 454 \\ 1 & 5 & 4 & \vdots & 292 \end{pmatrix}$$

Using our algorithm (Gaussian elimination with backsolving) we get a 1 as the entry in the first row-first column by interchanging rows 1 and 3.

$$\begin{pmatrix} 3 & 7 & 10 & \vdots & 560 \\ 2 & 6 & 9 & \vdots & 454 \\ 1 & 5 & 4 & \vdots & 292 \end{pmatrix} \xrightarrow{R_1 \leftrightarrow R_3} \begin{pmatrix} 1 & 5 & 4 & \vdots & 292 \\ 2 & 6 & 9 & \vdots & 454 \\ 3 & 7 & 10 & \vdots & 560 \end{pmatrix}$$

The next step involves using that 1 to get zeroes below it. The following two operations accomplish this goal.

$$\begin{pmatrix} 1 & 5 & 4 & \vdots & 292 \\ 2 & 6 & 9 & \vdots & 454 \\ 3 & 7 & 10 & \vdots & 560 \end{pmatrix} \xrightarrow[-3R_1 + R_3]{-2R_1 + R_2} \begin{pmatrix} 1 & 5 & 4 & \vdots & 292 \\ 0 & -4 & 1 & \vdots & -130 \\ 0 & -8 & -2 & \vdots & -316 \end{pmatrix}$$

The next step is to get a 1 as the first nonzero entry in the second row. This can be accomplished by multiplying row 2 by $-\frac{1}{4}$.

$$\begin{pmatrix} 1 & 5 & 4 & \vdots & 292 \\ 0 & -4 & 1 & \vdots & -130 \\ 0 & -8 & -2 & \vdots & -316 \end{pmatrix} \xrightarrow{(-\frac{1}{4})R_2} \begin{pmatrix} 1 & 5 & 4 & \vdots & 292 \\ 0 & 1 & -\frac{1}{4} & \vdots & \frac{65}{2} \\ 0 & -8 & -2 & \vdots & -316 \end{pmatrix}$$

Published by IT'S ABOUT TIME, Inc. © 2000 MATHconx, LLC

NOTES

Use that 1 to get a zero below it. This is accomplished as follows.

$$\begin{pmatrix} 1 & 5 & 4 & | & 292 \\ 0 & 1 & -\frac{1}{4} & | & \frac{65}{2} \\ 0 & -8 & -2 & | & -316 \end{pmatrix} \xrightarrow{8R_2 + R_3} \begin{pmatrix} 1 & 5 & 4 & | & 292 \\ 0 & 1 & -\frac{1}{4} & | & \frac{65}{2} \\ 0 & 0 & -4 & | & -56 \end{pmatrix}$$

Finally we get a 1 as the first nonzero entry in the third row by multiplying row 3 by $-\frac{1}{4}$.

$$\begin{pmatrix} 1 & 5 & 4 & | & 292 \\ 0 & 1 & -\frac{1}{4} & | & \frac{65}{2} \\ 0 & 0 & -4 & | & -56 \end{pmatrix} \xrightarrow{(-\frac{1}{4})R_3} \begin{pmatrix} 1 & 5 & 4 & | & 292 \\ 0 & 1 & -\frac{1}{4} & | & \frac{65}{2} \\ 0 & 0 & 1 & | & 14 \end{pmatrix}$$

The system of equations now reads

$$x + 5y + 4z = 292$$
$$y - (\tfrac{1}{4})z = \frac{65}{2}$$
$$z = 14$$

Substituting $z = 14$ into the second equation gives $y = 36$. Now substituting $z = 14$ and $y = 36$ into the first equation gives $x = 56$. That is, Krelling Industries should produce 56 chairs, 36 tables, and 14 desks.

All of the previous row operations can be performed on a TI-82 (TI-83).

Solve problems 1 and 2 using Gaussian elimination with backsolving.

6.35

1. $4x + 9y - 2z = 11$
   $x + 2y - z = 1$
   $2x + 7y + 9z = 38$

2. $3x + 10y + 6z = 47$
   $x + 3y + z = 10$
   $2x + 10y + 16z = 98$

3. **In terms of the geometry discussed in earlier sections of this chapter, what is an interpretation of the solutions found in problems 1 and 2?**

Each of the systems studied thus far in this section have had a unique solution. However, the previous study of geometry in this chapter suggests that systems we have been studying can also have no solution or an infinite number of solutions. What happens to the Gaussian elimination process in these cases? We will let you discover what happens.

Published by IT'S ABOUT TIME, Inc. © 2000 MATHconx, LLC

6.35

1. $x = 2, y = 1, z = 3$ 2. $x = {}^{-}1, y = 2, z = 5$

3. Each equation represents a plane. The point found in the solutions is the intersection of those three planes.

NOTES

6.36

1. The system

$$2x + 3y - z = 15$$
$$x + 2y - 4z = 17$$
$$3x + 5y - 5z = 12$$

has no solution. How would you explain to a friend what happens when you try to use Gaussian elimination with backsolving? Explain what you suspect is happening geometrically.

2. The system

$$2x + 3y - z = 15$$
$$x + 2y - 4z = 17$$
$$3x + 5y - 5z = 32$$

has an infinite number of solutions. How would you explain to a friend what happens when you try to use Gaussian elimination with backsolving? Explain what you suspect is happening geometrically.

Problem Set: 6.4

1. The Kopykat Toy Company makes Darbee dolls and model rockets. Each doll requires 1 quart of plastic while each rocket requires 4 quarts of plastic. A doll requires 3 hours of labor while a rocket requires 1 hour of labor. Last week the company used 230 gallons of plastic while each of 14 employees put in 40 hours of labor time. Using Gaussian elimination with backsolving, determine how many dolls and how many rockets were made last week.

2. The Kopykat Toy Company has decided to launch 2 new items called Star Peace figures. One is called 3CP1 while the other is called 3CP2. Each 3CP1 figure requires 15 minutes of labor while each 3CP2 requires 30 minutes of labor. Plastic for each 3CP1 costs $2 while plastic for each 3CP2 costs $3. Yesterday the company spent $124 on plastic and used 18 hours of labor time in making the new figures. Using Gaussian elimination with backsolving, determine how many of each figure were made yesterday.

Published by IT'S ABOUT TIME, Inc. © 2000 MATHconx, LLC

6.36

1. Using row operations, the matrix representing this system reduces to

$$\begin{pmatrix} 1 & 2 & -4 & \vdots & 7 \\ 0 & 1 & -7 & \vdots & 19 \\ 0 & 0 & 0 & \vdots & 1 \end{pmatrix}$$

The last equation is $0 = 1$ which has no solution. Thus, these planes do not have a common point of intersection.

2. Using row operations, the matrix representing this system reduces to

$$\begin{pmatrix} 1 & 2 & -4 & \vdots & 17 \\ 0 & 1 & -7 & \vdots & 19 \\ 0 & 0 & 0 & \vdots & 1 \end{pmatrix}$$

The corresponding equations are
$$x + 2y - 4z = 17$$
$$0x + y - 7z = 19$$
$$0x + 0y + 0z = 1$$

Select any value of z. Then take $y = 7z + 19$ and finally compute x from $x + 2y - 4z = 17$. Thus there are an infinite number of solutions. Geometrically, the three planes intersect in a line.

Problem Set: 6.4

1. 120 dolls and 200 rockets.

2. 32 of CPE1 and 20 of CPE2.

Chapter 6

3. A system of three linear equations with three variables is represented by a matrix. Row operations lead to the matrix

$$\begin{pmatrix} 1 & 3 & 5 & | & -4 \\ 0 & 1 & 4 & | & 2 \\ 0 & 0 & 0 & | & 1 \end{pmatrix}$$

   What can be said about the solution set of this system?

4. A system of three linear equations with three variables is represented by a matrix. Row operations lead to the matrix

$$\begin{pmatrix} 1 & 3 & 5 & | & -4 \\ 0 & 1 & 4 & | & 2 \\ 0 & 0 & 0 & | & 0 \end{pmatrix}$$

   What can be said about the solution set of this system?

5. A farmer in Delaware has 1200 acres of land which is to be planted with corn, tomatoes, and cantaloupes. When planting the crops, an acre of corn requires 3 hours of labor time, an acre of tomatoes requires 1 hour, and an acre of cantaloupes 2 hours. Labor costs $12 an hour. Seed costs are $10 an acre for corn, $15 an acre for tomatoes, and $5 an acre for cantaloupes. The farmer has decided to spend $24,000 on labor and $13,000 on seed. Using Gaussian elimination with backsolving, determine how many acres of each crop should be planted to use all the money allotted.

6. Use Gaussian elimination with backsolving to solve

$$x + y + z = 3$$

$$x - y + z = 3$$

$$2x + y + z = 4$$

Published by IT'S ABOUT TIME, Inc. © 2000 MATHconx, LLC

3. Since the last equation is $0 = 1$, there is no solution.

4. The corresponding equations are

$$x + 3y - 5z = -4$$
$$y + 4z = 2$$
$$0 = 0$$

Select any value of z. Then take $y = 2 - 4z$ and finally compute x from $x + 3y + 5z = -4$. Thus, there are an infinite number of solutions.

5. An appropriate system of equations is
$$x + y + z = 1200$$
$$3x + y + 2z = 2000$$
$$10x + 15y + 5z = 13{,}000$$

where x is the number of acres of corn planted, y is the number of acres of tomatoes planted, and z is the number of acres of cantaloupes planted. Using Gaussian elimination with backsolving gives $x = 200$, $y = 600$, and $z = 400$.

6. $x = 1$, $y = 0$, and $z = 2$

NOTES

....

....

....

....

....

....

....

....

....

Chapter 6

6.5 Matrix Operations

6.5 Matrix Operations

Learning Outcomes

After studying this section you will be able to:

Perform the addition, scalar multiplication, and multiplication of matrices;

Explain the properties of a zero matrix;

Explain the properties of an identity matrix;

Use matrices to represent information from real world situations.

In the previous section you have seen how matrices can be used to help solve systems of equations. Matrices have other uses, however. For example, matrices are a great tool for organizing, storing and manipulating information. In this section you will see how matrices can be added, subtracted, and multiplied. In many ways, these operations with matrices have many properties in common with the corresponding operations with numbers. We shall use small matrices to introduce these ideas, but you should realize that although applications in the real world frequently use larger matrices, the basic operations remain the same.

Situation 1. Newton School is a small school in Iowa.

Two homeroom classes are voting for a class president and treasurer. The three students who have been nominated are Trina, Ricardo, and Beth. There are 30 students in each of two homeroom classes, rooms A and B, who will vote. Students are given a slip of paper with the names of the nominees

_____	Trina
_____	Ricardo
_____	Beth

Published by IT'S ABOUT TIME, Inc. © 2000 MATHconx, LLC

6.5 Matrix Operations

The purpose of this section is to introduce the arithmetic (or algebra) of matrices. The addition and scalar multiplication of matrices should go very quickly, but the multiplication of matrices will require some time and effort. Students should see that the standard rules of algebra, such as the Commutative and Associative Laws, also apply to the addition and scalar multiplication operations. It is important for students to realize, however, that the multiplication of matrices is not commutative.

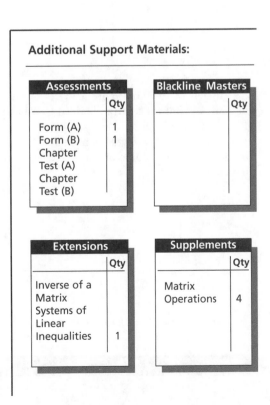

Additional Support Materials:

Assessments	Qty
Form (A)	1
Form (B)	1
Chapter Test (A)	
Chapter Test (B)	

Blackline Masters	Qty

Extensions	Qty
Inverse of a Matrix	
Systems of Linear Inequalities	1

Supplements	Qty
Matrix Operations	4

Chapter 6

When voting, students put a P on the line in front of the candidate they want to be president and T on the line in front of the candidate they want to be treasurer. After the voting, the results from the two classes were put into two matrices as follows.

$$
\begin{array}{c}
\ \ \text{Trina}\ \ \text{Ricardo}\ \ \text{Beth} \\
\begin{array}{c} P \\ T \end{array}
\left(
\begin{array}{ccc}
15 & 8 & 7 \\
10 & 9 & 11
\end{array}
\right)
\end{array}
\qquad
\begin{array}{c}
\ \ \text{Trina}\ \ \text{Ricardo}\ \ \text{Beth} \\
\begin{array}{c} P \\ T \end{array}
\left(
\begin{array}{ccc}
14 & 5 & 11 \\
8 & 7 & 15
\end{array}
\right)
\end{array}
$$

$$\text{Class } A \qquad\qquad\qquad \text{Class } B$$

1. Write a matrix where each entry gives the total number from the two classes of P's and T's obtained by each of the candidates.

6.37

2. Describe how you obtained the entries of the matrix in 1.

3. Which candidate do you think should be president?

4. Which candidate do you think should be treasurer?

If you solved the previous problems correctly, you *added* two matrices. As another example of this *addition* of matrices, observe the following sum.

$$
\left(
\begin{array}{cc}
2 & 3 \\
4 & 1 \\
-1 & 6
\end{array}
\right)
+
\left(
\begin{array}{cc}
1 & -5 \\
0 & 3 \\
8 & 4
\end{array}
\right)
=
\left(
\begin{array}{cc}
3 & -2 \\
4 & 4 \\
7 & 10
\end{array}
\right)
$$

In general, if A and B are matrices of the same size, then the **sum** $A + B$ is the matrix obtained by adding corresponding elements. For example, the entry in the first row and first column of a sum $A + B$ results from adding the entry of the first row and first column of A to the entry in the first row and first column of B.

6.37

1. $\begin{pmatrix} 29 & 13 & 18 \\ 18 & 16 & 26 \end{pmatrix}$

2. Students should indicate that they added corresponding entries of the two matrices.

3. Expect most students to name Trina.

4. Expect most students to name Beth.

NOTES

Now it's your turn.

6.38

1. Find the sum

$$\begin{pmatrix} 2 & 3 & 1 & 5 \\ 1 & 4 & 0 & -2 \end{pmatrix} + \begin{pmatrix} 3 & 2 & 5 & -2 \\ -1 & 0 & 7 & 8 \end{pmatrix}$$

2. Find the sum

$$\begin{pmatrix} 1 & 2 \\ 3 & 4 \\ 5 & 6 \\ 7 & 8 \end{pmatrix} + \begin{pmatrix} 3 & -1 \\ -1 & 5 \\ 1 & 0 \\ 0 & 2 \end{pmatrix}$$

3. Comment about the following sum

$$\begin{pmatrix} 2 & 3 & 1 \\ -1 & 4 & 0 \end{pmatrix} + \begin{pmatrix} 3 & 1 \\ 2 & 4 \end{pmatrix}$$

4. If A and B are matrices of the same size, do you think that

$$A + B = B + A?$$

Explain.

5. If A, B, and C are matrices of the same size, do you think that

$$(A + B) + C = A + (B + C)?$$

Explain.

6. Make up a situation, different from that in Situation 1, where the sum of two matrices can be used to solve a problem.

Published by IT'S ABOUT TIME, Inc. © 2000 MATHconx, LLC

6.38

1. $\begin{pmatrix} 5 & 5 & 6 & 3 \\ 0 & 4 & 7 & 6 \end{pmatrix}$

2. $\begin{pmatrix} 4 & 1 \\ 2 & 9 \\ 6 & 6 \\ 7 & 10 \end{pmatrix}$

3. The sum is not defined since the matrices have different sizes.

4. Students should conclude that $A + B = B + A$ since when adding corresponding entries, one is performing ordinary addition which is commutative.

5. Students should conclude that $(A + B) + C = A + (B + C)$ since when adding corresponding entries, one is performing ordinary addition which is associative.

6. Student answers will vary.

NOTES
_____

Chapter 6

Let O be a matrix in which every entry is the number 0. Such a matrix is called a **zero matrix.** For example,

$$O = \begin{pmatrix} 0 & 0 \\ 0 & 0 \\ 0 & 0 \end{pmatrix}$$

is a 3×2 zero matrix while

$$O = \begin{pmatrix} 0 & 0 & 0 & 0 \\ 0 & 0 & 0 & 0 \end{pmatrix}$$

is a 2×4 zero matrix.

1. If $A = \begin{pmatrix} 1 & 2 & 3 \\ 4 & 5 & 6 \end{pmatrix}$ and O is a 2×3 zero

   6.39

   matrix, what is the matrix $A + O$? What is the matrix $O + A$?

2. If $A = \begin{pmatrix} 1 & 3 \\ 5 & 7 \\ -4 & 2 \end{pmatrix}$ and O is a 3×3 zero matrix,

   what can be said about the sum $A + O$?

Let A be the matrix $\begin{pmatrix} 1 & 3 \\ 2 & -1 \\ -4 & 2 \end{pmatrix}$ If we take each entry of

A and replace that entry by the negative, or opposite, of that entry, we obtain the matrix

$$\begin{pmatrix} -1 & -3 \\ -2 & 1 \\ 4 & -2 \end{pmatrix}$$

This matrix is denoted $-A$. That is,

$$-A = \begin{pmatrix} -1 & -3 \\ -2 & 1 \\ 4 & -2 \end{pmatrix}$$

The matrix $-A$ is called the **negative** of A or the **opposite** of A.

527

6.39

1. $A + = A$ and $0 + A = A$ 2. $A + = A$

NOTES

6.40

1. If $A = \begin{pmatrix} 2 & -3 & 5 \\ 0 & -1 & 4 \end{pmatrix}$, what is the matrix $-A$?

2. If $A = \begin{pmatrix} 2 & 3 & -1 \\ -4 & -5 & 9 \end{pmatrix}$, find the matrix $A + (-A)$.

3. If A is any matrix, what do you conjecture about the matrix $A + (-A)$? Explain.

4. Let A and B be matrices of the same size. If $A + B = 0$, what conjecture would you make about the matrix B. Explain.

Situation 2. Happy Henry's has two local drugstores in northern Pennsylvania. One is the main store and the other is a branch store. Each store carries three varieties of aspirin—children's aspirin (C), regular aspirin (R), and extra strength aspirin (E).

At the start of the week beginning July 2, the inventory of each type of aspirin, in terms of the number of bottles, is given by the following matrix.

$$\begin{array}{c} \\ \text{main} \\ \text{branch} \end{array} \begin{array}{ccc} \text{C} & \text{R} & \text{E} \end{array} \\ \begin{pmatrix} 43 & 58 & 41 \\ 25 & 19 & 10 \end{pmatrix}$$

During the week of July 2, the number of bottles in each category sold by the two stores is given in the following matrix.

$$\begin{array}{c} \\ \text{main} \\ \text{branch} \end{array} \begin{array}{ccc} \text{C} & \text{R} & \text{E} \end{array} \\ \begin{pmatrix} 18 & 12 & 29 \\ 11 & 15 & 5 \end{pmatrix}$$

6.40

1. $\begin{pmatrix} -2 & 3 & -5 \\ 0 & 1 & -4 \end{pmatrix}$

2. $A + (-A) =$

3. $A + (-A) =$ ___ . When adding corresponding entries, the sum of an entry and its opposite is the number 0.

4. The conjecture should be that $B = -A$. For if one adds two entries, say a and b, and $a + b = 0$, then from ordinary arithemetic, $b = -a$. Since this is true for each entry, $B = -A$.

NOTES

Chapter 6

1. Write a matrix which gives the inventory of each type of aspirin at each store at the end of the week beginning July 2. You should assume that no new shipments of aspirin have arrived.

a
6.41

2. Describe how you obtained the entries of the matrix in 1.

If you solved the previous problems correctly then you performed the subtraction of one matrix from another. As another example of this *subtraction* process, observe the following.

$$\begin{pmatrix} 2 & 3 \\ 4 & 1 \\ -1 & 6 \end{pmatrix} - \begin{pmatrix} 1 & -5 \\ 0 & 3 \\ 8 & 4 \end{pmatrix} = \begin{pmatrix} 1 & 8 \\ 4 & -2 \\ -9 & 2 \end{pmatrix}$$

In general, if A and B are matrices of the same size, then the **subtraction** of B from A, denoted $A - B$, is the matrix obtained by subtracting the elements of B from corresponding elements of A. For example, the entry in the first row and first column of $A - B$ results from subtracting the entry of the first row and the first column of B from the entry in the first row and first column of A.

Now it's your turn.

1. Perform the subtraction

$$\begin{pmatrix} 2 & 3 & 1 & 5 \\ 1 & 4 & 0 & -2 \end{pmatrix} - \begin{pmatrix} 3 & 2 & 5 & -2 \\ -1 & 0 & 7 & 8 \end{pmatrix}$$

b
6.42

2. Perform the subtraction

$$\begin{pmatrix} 1 & 2 \\ 3 & 4 \\ 5 & 6 \\ 7 & 8 \end{pmatrix} - \begin{pmatrix} 3 & -1 \\ -1 & 5 \\ 1 & 0 \\ 0 & 2 \end{pmatrix}$$

3. What are your comments on the following subtraction?

$$\begin{pmatrix} 2 & 3 & 1 \\ -1 & 4 & 0 \end{pmatrix} - \begin{pmatrix} 3 & 1 \\ 2 & 4 \end{pmatrix}$$

Published by IT'S ABOUT TIME, Inc. © 2000 MATHconx, LLC

529

6.41

1.

$$\begin{pmatrix} 25 & 46 & 12 \\ 14 & 4 & 5 \end{pmatrix}$$

2. Students should indicate that they took each entry in the matrix

$$\begin{pmatrix} 43 & 58 & 41 \\ 25 & 19 & 10 \end{pmatrix}$$

and subtracted the corresponding entry in the matrix

$$\begin{pmatrix} 18 & 12 & 29 \\ 11 & 15 & 5 \end{pmatrix}$$

6.42

1. $\begin{pmatrix} -1 & 1 & -4 & 7 \\ 2 & 4 & -7 & -10 \end{pmatrix}$

2. $\begin{pmatrix} -2 & 3 \\ 4 & -1 \\ 4 & 6 \\ 7 & 6 \end{pmatrix}$

3. The subtraction is undefined since the matrices are of different sizes.

Chapter 6

4. If A, B, and C are matrices of the same size, do you think that
$$A + (B - C) = (A + B) - C?$$
Explain.

5. Let $A = \begin{pmatrix} 10 & 4 & 2 \\ -3 & 11 & -5 \end{pmatrix}$ and $B = \begin{pmatrix} 7 & 2 & -1 \\ 4 & -6 & 2 \end{pmatrix}$

Find the matrices $A - B$ and $A + (-B)$. Is there a difference? What general formula do you conjecture from this problem?

6. Make up a situation, different from that in Situation 2, where the subtraction of one matrix from another can be used to solve a problem.

Situation 3. The LMN automobile manufacturing company produces three types of cars; the big Limozeen (L), the sporty little Monty Carlow (M), and the midsize Neesun (N). The company has one plant in Michigan and another in Texas. The total number of cars produced at each plant last month in each category is given in the following matrix:

$$\begin{array}{c} \text{Michigan} \\ \text{Texas} \end{array} \begin{pmatrix} 2840 & 1960 & 4250 \\ 1210 & 3820 & 6120 \\ \text{L} & \text{M} & \text{N} \end{pmatrix}$$

Next month the company will be retooling for the new models and is cutting production of each type car at each plant in half.

6.43

1. Write a matrix which indicates the total number of cars that will be produced next month at each plant of each type.

2. Describe how you obtained the entries of the matrix in 1.

3. What would be your answer in problem 1 if instead of cutting production by half, the company cut production by 40%?

530

4. Students should conclude that $A + (B - C) = (A + B) - C$ since the operations involved use the ordinary arithmetic of entries and $a + (b - c) = (a + b) - c$.

5. There is no difference. Either one ($A - B$ or $A + {-B}$) gives the matrix

$$\begin{pmatrix} 3 & 2 & 3 \\ -7 & 17 & -7 \end{pmatrix}$$

6. Student answers will vary.

6.43

1. $$\begin{matrix} \text{Michigan} \\ \text{Texas} \end{matrix} \begin{pmatrix} 1420 & 980 & 2125 \\ 605 & 1910 & 3060 \end{pmatrix}$$

2. Students should indicate that they multiplied each entry of the original matrix shown below by $\frac{1}{2}$ or 0.5.

$$\begin{matrix} \text{Michigan} \\ \text{Texas} \end{matrix} \begin{pmatrix} 2840 & 1960 & 4250 \\ 1210 & 3820 & 6120 \end{pmatrix}$$

3. $$\begin{matrix} \text{Michigan} \\ \text{Texas} \end{matrix} \begin{pmatrix} 1740 & 1176 & 2550 \\ 726 & 2292 & 3672 \end{pmatrix}$$

Note that cutting production by 40% means that 60% of the original number of car types would remain.

Chapter 6

If you answered the previous questions correctly, then you multiplied a matrix by a number. In this context, a number is often called a **scalar.** As another example of this process, observe the following product of a matrix by a scalar.

$$3 \begin{pmatrix} 2 & 1 \\ 4 & 0 \\ 2 & -5 \\ -6 & 1 \end{pmatrix} = \begin{pmatrix} 6 & 3 \\ 12 & 0 \\ 6 & -15 \\ -18 & 3 \end{pmatrix}$$

In general, if A is any matrix and c is any number, then the **scalar product** of the matrix A by the scalar c, denoted cA, is the result of multiplying every entry in A by the scalar c. For example, the entry in the first row and first column of $3A$ results from multiplying the entry of the first row and first column of A by the number 3.

Now it's your turn.

1. **Compute the scalar product**

$$4 \begin{pmatrix} -1 & 0 & 3 & 5 \\ 2 & -1 & 4 & 8 \\ 3 & 1 & -2 & 9 \end{pmatrix}$$

6.44

2. **Compute the scalar product**

$$-5 \begin{pmatrix} -4 & 6 \\ 2 & -3 \\ 5 & 7 \end{pmatrix}$$

3. **If A is a matrix, c is a scalar, and d is a scalar, do you think there is a difference between $c(dA)$ and $(cd)A$? Explain.**

4. **If A and B are matrices of the same size, and c is a scalar, do you think that**

$$c(A + B) = (cA) + (cB) ?$$

Explain.

6.44

1. $\begin{pmatrix} -4 & 0 & 12 & 20 \\ 8 & -4 & 16 & 32 \\ 12 & 4 & -8 & 36 \end{pmatrix}$

2. $\begin{pmatrix} 20 & -30 \\ -10 & 15 \\ -25 & 35 \end{pmatrix}$

3. They are the same since scalar multiplication involves ordinary multiplication with entries and this multiplication is associative.

4. They are the same since ordinary arithmetic is used in operations with entries and there is a Distributive Law $c(a + b) = (ca) + (cb)$.

NOTES

Chapter 6

5. Let $A = \begin{pmatrix} 2 & 3 \\ -1 & 5 \end{pmatrix}$

(a) What is the matrix $0A$? What conjecture would you make about $0A$ for any matrix A?

(b) Find the matrices $-A$ and $(-1)A$. What conjecture would you make about $-A$ and $(-1)A$ for any matrix A?

6. Make up a situation, different from that in Situation 3, where the multiplication of a matrix by a scalar can be used to solve a problem.

Situation 4. At the grocery store, Mrs. Rodriguez buys 2 boxes of cereal (C), 5 lb. of meat (M), 3 cans of soup (S), and 1 pint of ice cream (I).

This information is put into the form of a 1×4 matrix, known as a *row matrix*.

$$(2 \quad 5 \quad 3 \quad 1)$$

A box of cereal costs $2.50, a pound of meat costs $1.50, a can of soup costs $0.75, and a pint of ice cream costs $1.80. This information is put into the following 4×1 matrix, known as a *column matrix*.

$$\begin{array}{c} C \\ M \\ S \\ I \end{array} \begin{pmatrix} 2.50 \\ 1.50 \\ 0.75 \\ 1.80 \end{pmatrix}$$

6.45

1. Find the total amount of money that Mrs. Rodriguez spent at the grocery store.

2. Describe the process you used in solving problem 1.

If you solved these problems correctly, then you performed the *multiplication* of a row matrix and a column matrix. As another example of this process, observe the following product of a row matrix (1×5) and a column matrix (5×1). Note that the row matrix has the same number of columns as the column matrix has rows.

Published by IT'S ABOUT TIME, Inc. © 2000 MATHconx, LLC

5. (a) $\begin{pmatrix} 0 & 0 \\ 0 & 0 \end{pmatrix}$ Students should conjecture that $0A = 0$.

   (b) $-A = \begin{pmatrix} -2 & -3 \\ 1 & -5 \end{pmatrix} = (-1)A$. Students should conjecture that

   for any matrix A, $-A = (-1)A$.

6. Student answers will vary.

6.45

1. \$16.55

2. Students should indicate that they multiplied the row entries of the matrix
   (2 5 3 1) times the corresponding column entries of the matrix

$$\begin{pmatrix} 2.50 \\ 1.50 \\ .75 \\ 1.80 \end{pmatrix}$$

NOTES

Chapter 6

$$(5 \ 3 \ -9 \ 7 \ 2) \begin{pmatrix} 2 \\ -1 \\ 3 \\ 4 \\ -1 \end{pmatrix} = (5 \cdot 2) + (3 \cdot (-1)) + (-9 \cdot 3) + (7 \cdot 4) + (2 \cdot (-1)) = 6$$

In general, if $A = (\ a_1 \ a_2 \ a_3 \ ... \ a_n\ )$ is a **$1 \times n$ row matrix** and

$$B = \begin{pmatrix} b_1 \\ b_2 \\ . \\ . \\ . \\ b_n \end{pmatrix}$$

is an **$n \times 1$ column matrix,** then the product AB is the number given by

$$AB = a_1b_1 + a_2b_2 + ... + a_nb_n$$

Now it's your turn.

1. **Find the product of the row matrix (-2 3 7) and**

   the column matrix $\begin{pmatrix} 5 \\ -4 \\ 2 \end{pmatrix}$

6.46

2. **Find the product of the row matrix (-1 5 10 12)**

   and the column matrix $\begin{pmatrix} 1 \\ 1 \\ 1 \\ 1 \end{pmatrix}$

3. **What can you say about the product of the row matrix (3 2 -4 9) and the column matrix?**

   $$\begin{pmatrix} 6 \\ -2 \\ 5 \end{pmatrix}$$

533

6.46

1. -8   2. 26   3. This product is undefined since the first matrix has four columns, but the second matrix has only three rows.

4. $(cA)B$ and $c(AB)$ are the same number. An entry in the sum formed in $(cA)B$ would look like $(ca)b$ which, using ordinary arithmetic, is the same as $c(ab)$ which is the corresponding entry in the sum formed in $c(AB)$.

5. Student answers will vary.

NOTES

Chapter 6

4. If A is a 1×5 row matrix, B is a 5×1 column matrix, and c is a scalar, do you think there is a difference between $(cA)B$ and $c(AB)$? Explain.

5. Make up a situation, different from Situation 4, in which the product of a row matrix and a column matrix can be used to solve a problem.

Situation 5. Julie and Richard stop in Jo Goodies music store.

Julie wants to buy 2 rock CDs, 4 classical CDs, and 3 heavy metal CDs. Richard wants to buy 3 rock CDs, 2 classical CDs, and 1 heavy metal CD. This information is put into the following matrix.

$$
\begin{array}{c}
\\
\text{Julie} \\
\text{Richard}
\end{array}
\begin{array}{ccc}
\text{Rock} & \text{Classical} & \text{Heavy} \\
& & \text{Metal} \\
\end{array}
\left(
\begin{array}{ccc}
2 & 4 & 3 \\
3 & 2 & 1
\end{array}
\right)
$$

Each CD is available as a tape cassette or an album. The cost of a rock CD is \$12 while a rock cassette tape costs \$8. A classical CD costs \$15 while a classical tape costs only \$6. A heavy metal CD costs \$11 while a heavy metal tape costs \$9. This information is put into the following matrix.

$$
\begin{array}{c}
\\
\text{Rock} \\
\text{Classical} \\
\text{Heavy Metal}
\end{array}
\begin{array}{cc}
\text{Tape} & \text{CD} \\
\end{array}
\left(
\begin{array}{cc}
8 & 12 \\
6 & 15 \\
9 & 11
\end{array}
\right)
$$

6.47

1. How much money will Julie spend if she buys all tapes?

2. How much money will Julie spend if she buys all CDs?

3. How much money will Richard spend if he buys all tapes?

534

6.47

1. $67   2. $117   3. $45   4. $77

5. Students should indicate that each number in 1–4 was obtained by

   multiplying a row of $\begin{pmatrix} 2 & 4 & 3 \\ 3 & 2 & 1 \end{pmatrix}$ regarded as a row matrix, times a

   column of $\begin{pmatrix} 8 & 12 \\ 6 & 15 \\ 9 & 11 \end{pmatrix}$ regarded as a column matrix.

6. $\begin{pmatrix} 67 & 117 \\ 45 & 77 \end{pmatrix}$

NOTES

Chapter 6

4. How much money will Richard spend if he buys all CDs?

5. Describe the process you used in answering questions 1–4.

6. Write the answers to question 1–4 as a 2×2 matrix in the form

$$\begin{pmatrix} \text{Answer 1} & \text{Answer 2} \\ \text{Answer 3} & \text{Answer 4} \end{pmatrix}$$

If you solved the previous problems correctly, then you performed the multiplication of a 2×3 matrix and a 3×2 matrix written as follows.

$$\begin{pmatrix} 2 & 4 & 3 \\ 3 & 2 & 1 \end{pmatrix} \begin{pmatrix} 8 & 12 \\ 6 & 15 \\ 9 & 11 \end{pmatrix} = \begin{pmatrix} \text{Answer 1} & \text{Answer 2} \\ \text{Answer 4} & \text{Answer 3} \end{pmatrix}$$

Note that Answer 1 is obtained from the multiplication of the first row of

$$\begin{pmatrix} 2 & 4 & 3 \\ 3 & 2 & 1 \end{pmatrix}$$

regarded as a row matrix, and the first column of

$$\begin{pmatrix} 8 & 12 \\ 6 & 15 \\ 9 & 11 \end{pmatrix}$$

regarded as a column matrix, resulting in

$$2 \cdot 8 + 4 \cdot 6 + 3 \cdot 9 = 67$$

1. Explain how Answer 2 can be regarded as the multiplication of a row matrix and a column matrix.

6.48

2. Explain how Answer 4 can be regarded as the multiplication of a row matrix and a column matrix.

Published by IT'S ABOUT TIME, Inc. © 2000 MATHconx, LLC

6.48

1. *Answer* 2 is the result of multiplying the first row of $\begin{pmatrix} 2 & 4 & 3 \\ 3 & 2 & 1 \end{pmatrix}$ regarded as a row matrix, times the second column of $\begin{pmatrix} 8 & 12 \\ 6 & 15 \\ 9 & 11 \end{pmatrix}$ regarded as a column matrix.

2. *Answer* 4 is the result of multiplying the second row of $\begin{pmatrix} 2 & 4 & 3 \\ 3 & 2 & 1 \end{pmatrix}$ regarded as a row matrix, times the second column of $\begin{pmatrix} 8 & 12 \\ 6 & 15 \\ 9 & 11 \end{pmatrix}$ regarded as a column matrix.

NOTES

Chapter 6

In general, if A and B are matrices where A has the same number of columns that B has rows, then one can find a **product matrix** AB which involves successively multiplying a row of A, regarded as a row matrix, and a column of B, regarded as a column matrix. If you select row i of matrix A and row j of matrix B, then the product of the ith row of A and the jth column of B goes into the ith row and jth column of AB. This process is known as **matrix multiplication.**

Observe how the process works in the following example.

$$\begin{pmatrix} 2 & 1 \\ 3 & -2 \\ 8 & 5 \end{pmatrix} \begin{pmatrix} 3 & 1 & 2 & 0 \\ -1 & 4 & 6 & 5 \end{pmatrix} = \begin{pmatrix} 5 & 6 & 10 & 5 \\ 11 & -5 & -6 & -10 \\ 19 & 28 & 46 & 25 \end{pmatrix}$$

For example, the first row 5 6 10 5 of the matrix product was obtained from

$$2 \cdot 3 + 1 \cdot (-1) = 5$$
$$2 \cdot 1 + 1 \cdot 4 = 6$$
$$2 \cdot 2 + 1 \cdot 6 = 10 \text{ and}$$
$$2 \cdot 0 + 1 \cdot 5 = 5$$

Note that the matrix A has 2 columns,

$$A = \begin{pmatrix} 2 & 1 \\ 3 & -2 \\ 8 & 5 \end{pmatrix}$$

while the matrix B has 2 rows.

$$B = \begin{pmatrix} 3 & 1 & 2 & 0 \\ -1 & 4 & 6 & 5 \end{pmatrix}$$

Also note that the result of multiplying a 3×2 matrix and a 2×4 matrix is a 3×4 matrix. In general, if A is an $n \times m$ matrix and B is an $m \times r$ matrix, then AB is an $n \times r$ matrix.

In the previous example we found the product

$$\begin{pmatrix} 2 & 1 \\ 3 & -2 \\ 8 & 5 \end{pmatrix} \begin{pmatrix} 3 & 1 & 2 & 0 \\ -1 & 4 & 6 & 5 \end{pmatrix}$$

Published by IT'S ABOUT TIME, Inc. © 2000 MATHconx, LLC

NOTES

Note that the product $\begin{pmatrix} 3 & 1 & 2 & 0 \\ -1 & 4 & 6 & 5 \end{pmatrix} \begin{pmatrix} 2 & 1 \\ 3 & -2 \\ 8 & 5 \end{pmatrix}$

is *not defined* since the number of columns in the first matrix is not the same as the number of rows in the second matrix. The order in which matrices appear in a product is very important.

Let's see if you have the idea of matrix multiplication.

1. Find the product. $\begin{pmatrix} 2 & 3 \\ -1 & 4 \end{pmatrix} \begin{pmatrix} 3 & -1 & 5 \\ 4 & 2 & 3 \end{pmatrix}$

6.49

2. (a) Find the product. $\begin{pmatrix} 5 & 2 \\ 3 & 4 \end{pmatrix} \begin{pmatrix} 4 & 3 \\ 2 & 2 \end{pmatrix}$

   (b) Find the product. $\begin{pmatrix} 4 & 3 \\ 2 & 2 \end{pmatrix} \begin{pmatrix} 5 & 2 \\ 3 & 4 \end{pmatrix}$

   (c) What do (a) and (b) tell you about the products *AB* and *BA* of two matrices, even when both products are defined?

3. Find the product.

$$\begin{pmatrix} 3 & 1 & 2 & 0 \\ 4 & 5 & 9 & 2 \end{pmatrix} \begin{pmatrix} 1 & 5 & -1 \\ 2 & 6 & -2 \\ 3 & 7 & 0 \\ 4 & 8 & 3 \end{pmatrix}$$

4. What can you say about the product?

$$\begin{pmatrix} 3 & 1 & 2 \\ 4 & 5 & 9 \end{pmatrix} \begin{pmatrix} 1 & 5 \\ 2 & 6 \\ 3 & 7 \\ 4 & 8 \end{pmatrix}$$

5. Make up a situation, different from that in Situation 5, where the product of two matrices can be used to solve a problem.

6.49

1. $\begin{pmatrix} 18 & 4 & 19 \\ 13 & 9 & 7 \end{pmatrix}$

2. (a) $\begin{pmatrix} 24 & 19 \\ 20 & 17 \end{pmatrix}$ (b) $\begin{pmatrix} 29 & 20 \\ 16 & 12 \end{pmatrix}$

    (c) The product AB need not equal the product BA.

3. $\begin{pmatrix} 11 & 35 & -5 \\ 49 & 129 & -8 \end{pmatrix}$

4. The product is undefined since the number of columns in the first matrix is not the same as the number of rows in the second matrix.

5. Student answers will vary.

NOTES

Chapter 6

If A is a square matrix (2×2, 3×3, etc.), then the entries in the first row-first column, second row-second column, third row-third column, etc., form the **main diagonal** of the matrix A. For example, in the 3×3 matrix

$$\begin{pmatrix} 2 & 3 & -1 \\ 5 & 4 & 0 \\ 6 & 7 & -8 \end{pmatrix}$$

the entries 2, 4, and -8 comprise the main diagonal.

If the main diagonal of a square matrix consists of all 1's, and all other entries in the matrix are 0, the matrix is called an **identity matrix**. For example,

$$\begin{pmatrix} 1 & 0 \\ 0 & 1 \end{pmatrix}, \begin{pmatrix} 1 & 0 & 0 \\ 0 & 1 & 0 \\ 0 & 0 & 1 \end{pmatrix}, \text{ and } \begin{pmatrix} 1 & 0 & 0 & 0 \\ 0 & 1 & 0 & 0 \\ 0 & 0 & 1 & 0 \\ 0 & 0 & 0 & 1 \end{pmatrix}$$

are identity matrices. The following problems should give you some idea of why such matrices are called identity matrices.

6.50

1. Find the product

$$\begin{pmatrix} 1 & 0 \\ 0 & 1 \end{pmatrix} \begin{pmatrix} 1 & 2 & 3 \\ 4 & 5 & 6 \end{pmatrix}$$

2. Find the product

$$\begin{pmatrix} 1 & 2 & 3 \\ 4 & 5 & 6 \end{pmatrix} \begin{pmatrix} 1 & 0 & 0 \\ 0 & 1 & 0 \\ 0 & 0 & 1 \end{pmatrix}$$

3. Based on the results of problems 1. and 2., what conjecture would you make about the matrices IA and BI, where I is an identity matrix and the products IA and BI are defined?

Published by IT'S ABOUT TIME, Inc. © 2000 MATHconx, LLC

6.50

1. $\begin{pmatrix} 1 & 2 & 3 \\ 4 & 5 & 6 \end{pmatrix}$ 2. $\begin{pmatrix} 1 & 2 & 3 \\ 4 & 5 & 6 \end{pmatrix}$

3. Conjecture should be $1A = A$ and $B1+ = B$

NOTES

Chapter 6

REFLECT

This chapter began with a study of planes in three dimensional space. That study led to an examination of the intersection of planes, which in turn led to systems of linear equations and their solutions. A systematic way of solving linear systems involves a matrix representation and Gaussian elimination with backsolving. Finally, the concept of a matrix was shown to be a tool for organizing and manipulating data. All together this is quite a chunk of information, and we hope the reader has appreciated the many *Connections* that appear in this chapter.

Problem Set: 6.5

1. A small company produces three video games; Star Blaster (B), Chase (C), and Private Saygar (S). The sales, in thousands of dollars, for January–June and July–December of last year are given in the matrix

$$
\begin{array}{c}
\\
\text{Jan.–June} \\
\text{July–Dec.}
\end{array}
\begin{array}{ccc}
B & C & S \\
\end{array}
\left(
\begin{array}{ccc}
3 & 2 & 4 \\
4 & 1 & 5
\end{array}
\right)
$$

   This year, the company predicts that during each 6 month period, the sales of each item will increase by 50%.

   (a) Write a matrix which indicates the predicted sales for each 6 month period of this year.

   (b) What matrix operation is involved in (a)?

2. A steel company makes 2 types of beams. One is an I beam while the other is a T beam. During 1996, the sales, in thousands of tons, for each quarter (3 month period) are given in the matrix

$$
\text{Quarter}
\begin{array}{c}
1 \\
2 \\
3 \\
4
\end{array}
\begin{array}{cc}
I & T \\
\end{array}
\left(
\begin{array}{cc}
4.9 & 5.2 \\
8.7 & 7.4 \\
10.4 & 9.3 \\
7.1 & 4.2
\end{array}
\right)
$$

Problem Set: 6.5

1. (a)

$$\begin{array}{c} \\ \text{Jan.–June} \\ \text{July–Dec.} \end{array} \begin{array}{ccc} B & C & S \\ \left(\begin{array}{ccc} 4.5 & 3 & 6 \\ 6 & 1.5 & 7.5 \end{array}\right) \end{array}$$

(b) Scalar multiplication

2. (a)

$$\text{Quarter} \begin{array}{c} 1 \\ 2 \\ 3 \\ 4 \end{array} \begin{array}{cc} I & T \\ \left(\begin{array}{cc} 10.6 & 13.5 \\ 15.9 & 13.8 \\ 18.7 & 14.5 \\ 13.9 & 12.3 \end{array}\right) \end{array}$$

(b) Matrix addition

NOTES

Chapter 6

During 1997, the sales for each quarter are given in the matrix

$$
\text{Quarter} \begin{array}{c} \\ 1 \\ 2 \\ 3 \\ 4 \end{array} \begin{array}{c} \text{I} \quad\;\; \text{T} \\ \left(\begin{array}{cc} 5.7 & 8.3 \\ 7.2 & 6.4 \\ 8.3 & 5.2 \\ 6.8 & 8.1 \end{array} \right) \end{array}
$$

(a) Write a matrix representing the total sales in each quarter during the 2 year period 1996–1997.

(b) What matrix operation is involved in (a)?

3. Happy Henry's has two local drugstores in northern Pennsylvania. One is the main store and the other is a branch store. Each store carries three varieties of aspirin— children's aspirin (C), regular aspirin (R), and extra strength aspirin (E).

During January, February, and March of 1995, the sales at the main store of each type of aspirin (number of bottles) are given by the matrix

$$
1995 \begin{array}{c} \\ \text{Jan.} \\ \text{Feb.} \\ \text{Mar.} \end{array} \begin{array}{c} \text{C} \quad\; \text{R} \quad\; \text{E} \\ \left(\begin{array}{ccc} 43 & 29 & 18 \\ 54 & 32 & 15 \\ 61 & 28 & 22 \end{array} \right) \end{array}
$$

The sales at the branch store are given by the matrix

$$
1995 \begin{array}{c} \\ \text{Jan.} \\ \text{Feb.} \\ \text{Mar.} \end{array} \begin{array}{c} \text{C} \quad\; \text{R} \quad\; \text{E} \\ \left(\begin{array}{ccc} 38 & 24 & 15 \\ 49 & 30 & 12 \\ 56 & 25 & 18 \end{array} \right) \end{array}
$$

540

Published by IT'S ABOUT TIME, Inc. © 2000 MATHconx, LLC

3. (a)

$$\begin{array}{c} & \begin{array}{ccc} C & R & E \end{array} \\ \begin{array}{c} \text{Jan.} \\ 1995 \quad \text{Feb.} \\ \text{Mar.} \end{array} & \begin{pmatrix} 205 & 135 & 84 \\ 260 & 156 & 69 \\ 295 & 134 & 102 \end{pmatrix} \end{array}$$

(b) Scalar multiplication and matrix addition.

NOTES

The year 1996 brought cold and wet weather to the East Coast topped by the famous "Blizzard of '96." As a result, sales for each type of aspirin during each of the 3 months tripled at the main store and doubled at the branch store.

(a) Write a matrix which indicates the total sales at the 2 stores of each type of aspirin during each of the first 3 months of 1996.

(b) What matrix operations were involved in (a)?

4. (a) If A is a 2×3 matrix and B is a 3×4 matrix, how many multiplications are needed to compute AB?

(b) If A is a 3×4 matrix and B is a 4×5 matrix, how many multiplications are needed to compute AB?

(c) If A is an $n \times m$ matrix and B is an $m \times r$ matrix, conjecture how many multiplications are needed to compute AB. Explain.

5. (a) If A is a 2×3 matrix and B is a 3×4 matrix, how many arithmetic operations (additions and multiplications) are needed to compute AB?

(b) If A is a 3×4 matrix and B is a 4×5 matrix, how many arithmetic operations are needed to compute AB?

(c) If A is an $n \times m$ matrix and B is an $m \times r$ matrix, conjecture how many arithmetic operations are needed to compute AB. Explain.

4. (a) 24 Select a row of *A* and a column of *B*. The product of that row and column requires 3 multiplications. For each row of *A* there are $4(3) = 12$ multiplications. Since *A* has two rows, the total is 24 multiplications.

   (b) 60 Select a row of *A* and a column of *B*. The product of that row and column requires 4 multiplications. For each row of *A* there are $4(5) = 20$ multiplications. Since *A* has three rows, the total is 60 multiplications.

   (c) *nmr* Select a row of *A* and a column of *B*. The product of that row and column requires *m* multiplications. For each row of *A* there are $m(r)$ multiplications. Since *A* has *n* rows, the total is *nmr* multiplications.

5. (a) 40 Select a row of *A* and a column of *B*. The product of that row and column requires 2 additions and 3 multiplications or 5 operations. For each row of *A* there are $5(4) = 20$ such operations. Since *A* has two rows, the total is 40 operations.

   (b) 40 Select a row of *A* and a column of *B*. The product of that row and column requires 3 additions and 4 multiplications or 7 operations. For each row of *A* there are $7(5) = 35$ such operations. Since *A* has three rows, the total is 105 operations.

   (c) Select a row of *A* and a column of *B*. The product of that row and column requires $m - 1$ additions and *m* multiplications or $2m - 1$ operations. For each row of *A* there are $r(2m - 1)$ such operations. Since *A* has *n* rows, the total is $nr(2m - 1)$ operations.

Chapter 6

6. It is possible to add and multiply matrices on some calculators. Scalar multiplication is also possible. On a TI-82 (TI-83), enter the matrix

$$\begin{pmatrix} 1 & 2 & 3 \\ 4 & 5 & 6 \end{pmatrix}$$

as the [A] (or 1) matrix,

$$\begin{pmatrix} 2 & 5 & 9 \\ 3 & 8 & 4 \end{pmatrix}$$

as the [B] (or 2) matrix, and

$$\begin{pmatrix} -1 & 2 & 4 \\ 3 & 3 & 2 \\ -2 & -1 & 0 \end{pmatrix}$$

as the [C] (or 3) matrix.

Now perform the following.

(a) Press MATRIX 1 + MATRIX 2 ENTER
What is the result?

(b) Press MATRIX 1 × MATRIX 3 ENTER
What is the result?

(c) Press 2 × MATRIX 1 ENTER
What is the result?

Published by IT'S ABOUT TIME, Inc. © 2000 MATHconx, LLC

6. Students should have learned how to enter matrices into a TI-82 (TI-83) in the previous section.

(a) $\begin{pmatrix} 3 & 7 & 12 \\ 7 & 13 & 10 \end{pmatrix}$ (b) $\begin{pmatrix} -1 & 5 & 8 \\ -1 & 17 & 26 \end{pmatrix}$ (c) $\begin{pmatrix} 2 & 4 & 6 \\ 8 & 10 & 12 \end{pmatrix}$

NOTES

Appendix A: Using a TI-82 (TI-83) Graphing Calculator

A graphing calculator is a useful tool for doing many different mathematical things. Once you begin to use it, you'll find that it is powerful, fast, and friendly. In fact, your biggest difficulty may be just getting started for the first time! Because this machine can do a lot, it has lots of complicated looking buttons. But you don't have to know about *all* of them before you start to use *any* of them! The sooner you make friends with your electronic assistant, the more it will be able to help you. Let us introduce you to each other by trying a few simple things.

The Cover

The face of the calculator is protected from dirt and scratches by a cover that slides on and off from the top. When you're using the calculator, this cover slips on the back so that you won't lose it. Always put the cover back over the face of the calculator when you finish using it.

On, Off, 2nd , and Clear

To get the calculator's attention, just press ON (at the lower left corner of the calculator). What happens? Do you see a dark block blinking in the upper left corner of the screen? That's the **cursor**, which tells you where you are on the screen. The cursor is always at the spot that will be affected by the next button you push.

Notice that the word OFF is printed in color above the ON button, a little to the left of its center. Notice also that there is one key of the same color. It is the key marked 2nd at the left end of the second row.

When you push 2nd , it makes the next key that you push behave like what is marked above it on the left.

Try it: Push 2nd . What has happened to the cursor? Do you see an up arrow inside it as it blinks? That's to remind you that 2nd key has been pushed and will affect the next key

Published by IT'S ABOUT TIME, Inc. © 2000 MATHconx, LLC

Using a TI-82 (TI-83) Graphing Calculator

This appendix is *not* intended to replace the manual for the TI-82 (TI-83) graphing calculators. Rather, it has two purposes.

- It serves as a gentle introduction to the machine by way of some simple calculations, and

- It provides a convenient reference for some of the more commonly used elementary procedures.

Before using the graphing calculators with your class for the first time, **please check each one to see that it is reset to its factory settings and that it actually turns on and off.** It will be easier to answer your students' questions if all the calculators have working batteries and behave the same when they are first turned on.

NOTES

you choose. Now push [ON]. What happens? Did the cursor disappear? You should have a blank screen; the calculator should be off.

It's always a good idea to turn your calculator off when you finish using it. If you forget, the calculator will turn itself off after a few minutes to save its batteries. Sometimes when you are using it, you may put it aside and do something else for a little while. If it is off when you pick it up again, don't worry; just press [ON]. The screen will show what was there before it shut down.

Pressing [CLEAR] gives you a blank screen that is ready for new work. But the last thing you did is still stored. Press [2nd] then [ENTER] to bring it back.

Basic Arithmetic

Doing arithmetic on a graphing calculator is no harder than on a simpler calculator. In fact, it's easier. This calculator has a screen that lets you keep track of the problem as you enter it. Let's try a few simple exercises. Turn your calculator on.

- Pick two 3-digit numbers and add them. To do this, just key in the first number, press [+] and then key in the second number. Your addition problem will appear on the screen. Press [ENTER] to get the answer.

If you make a mistake when entering a number, you can go back and fix it. The [◁] key lets you move back (left) one space at a time. When you get to your mistake, just key in the correct number over the wrong one. Then move forward (right) to the end of the line by using the [▷] key.

For instance, to add 123 and 456, press 1 2 3 [+] 4 5 6 [ENTER]. The screen will show your question on the first line and the answer at the right side of the second line, as in Display A.1.

```
123+456
             579
```

Display A.1

NOTES

- Now let's try the other three basic arithmetic operations. To clear the screen, press CLEAR. Then try subtracting, multiplying, and dividing your two 3-digit numbers. For instance, if your numbers are 123 and 456, press

  123 − 456 ENTER

  123 × 456 ENTER

  123 ÷ 456 ENTER

Your screen should look like Display A.2.

Notice that the display uses ∗ for multiplication (so that it is not confused with the letter x) and / for division.

```
123−456
              -333
123*456
             56088
123/456
        .2697368421
```

Display A.2

- Here are two, button pushing shortcuts.

If you don't want to redo a problem with just a small change in it, you don't have to reenter the whole thing. 2nd ENTER will bring back the last problem you entered. Just move to the place you want to change, key in the change, and press ENTER. For instance, add 54321 and 12345, as in Display A.3.

```
54321+12345
             66666
```

Display A.3

Published by IT'S ABOUT TIME, Inc. © 2000 MATHconx, LLC

A-3

NOTES

Now, to subtract 12345 from 54321, press [2nd] [ENTER]; the next line will show 54321 + 12345. Move your cursor back to the + sign (using [◁]) and press [−]; then press [ENTER]. Did you try it? Your screen should look like Display A.4.

```
54321+12345
            66666
54321−12345
            41976
```

Display A.4

Let's check to see that 41976 is the correct answer by adding 12345 to it and seeing if we get the first number back again. Since you want to do something to the last answer, *you don't have to reenter it*. Press [+]. Does your calculator show Ans+ and the cursor? It should. If you press an operation key right after doing a calculation, the machine assumes that you want to perform this operation on the last answer. It shows that last answer as Ans. Now key in 12345 and press [ENTER]. You should get back the first number, 54321.

A.1

1. Pick two seven-digit numbers and add them. What do you get?

2. Now subtract the second number from the first. Can you do it without rekeying the numbers? What do you get?

3. Now multiply your two seven-digit numbers. What do you get? What does the E mean?

4. Check the last answer by dividing the second of your seven-digit numbers into it. (Do it without rekeying the last answer.) Do you get your first number back again?

Appendix A: Using a TI-82 (TI-83) Graphing Calculator

A-4

A.1

Basic Arithmetic

These questions are largely for routine practice. The use of seven-digit numbers helps to encourage the students to use the shortcuts, rather than just rekeying the entries. In parts 3 and 4, it serves the additional purpose of requiring them to deal with an answer displayed in scientific notation.

1. This is straightforward.

2. To do this without rekeying, use the entry key $\boxed{\text{2nd}}$ $\boxed{\text{ENTER}}$ to get back the addition line, move back to the $+$ sign using the $\boxed{\triangleleft}$ key, press $\boxed{-}$, then press $\boxed{\text{ENTER}}$.

3. The fact that the numbers chosen have seven digits guarantees that their product appears in scientific E notation. For example, the product of 1234567 and 2345678 appears as 2.895896651E12. This means that the product is actually

    $$2.895896651 \times 10^{12} \text{ or } 2,895,896,651,000$$

4. To check without rekeying the answer, just press $\boxed{\div}$, then key in the second seven-digit number. For our example, the display will show Ans/2345678. When you press $\boxed{\text{ENTER}}$ the first seven-digit number will appear.

NOTES

Multiply 98765432 by 123456. Now check the product in two ways.

a
A.2

- Divide by pressing [÷] then entering 123456. Does it check?

- First reenter the product; then divide it by 123456. (The product is in scientific notation. To enter it as a regular number, remember that the positive number after the E tells you to move the decimal that many places to the right.) Does it check?

1. Divide 97533 by 525 and by 625. One of the answers you get will be exactly right, and the other one will be a very close approximation.

b
A.3

- Which is which?

- How can you tell?

- If you hadn't been told that one of the answers is an approximation, how could you know?

2. When an answer is too long to be displayed with ten digits, the calculator shows a ten-digit approximation. Does it do this by just chopping off (truncating) the rest of the digits, or by rounding off? What test would you give your calculator to tell which way it does this?

1. Pick any three-digit number and note it down.

c
A.4

2. Repeat its digits in the same order to form a six-digit number (like 123123, for example). Key this number into your calculator.

3. Divide your number by 7.

4. Divide your answer by 11. (How do you do this without reentering the answer?)

5. Divide the last number by 13. What do you notice about the result? Do you think that it is just a coincidence?

6. Pick another three digit number and repeat steps 2–5.

7. Try to beat the system; see if you can pick a three-digit number that doesn't work this way. What might you try? Why?

8. Can you actually prove that the pattern you see *works every time*? How might you try to do this?

Published by IT'S ABOUT TIME, Inc. © 2000 MATHconx, LLC

A-5

A.2

This exercise illustrates the fact that some answers, particularly those expressed in scientific notation, are approximations. It also gives students a chance to deal with two other matters,

- rewriting a number from scientific notation to standard form; and
- experiencing the wraparound feature of the calculator's screen display.

To do the second check, students must convert the product 1.219318517E13 to the form 12,193,185,170,000 and key in

12193185170000 ÷ 123456 ENTER

This is too long to fit on a single line of the calculator display, so it is *automatically* wrapped around to the second line. There is no need to press any sort of "carriage return" key. In fact, trying to do so probably will result in an error. The screen should look like Display A.1T.

> 12193185170000/1
> 23456
>   98765431.98

Display A.1T

A.3

These questions relate to the fact that answers are displayed with a maximum of 10 digits.

1. The easiest way to identify the exact answer is by observing that one of these answers has fewer than 10 digits; that one is exact. The next question actually tells the student this fact. The other is not, but you can't be *sure* just because it has 10 digits. The decimal form of the answer does not terminate because the divisor has prime factors other than 2 or 5.

   Multiplying back to check your answer by using Ans ✕ will give exactly the original dividend *both* times; the calculator holds a more exact approximation than it shows. However, if you rekey each answer and multiply it by its divisor, you will get an inexact original dividend in one of these cases.

2. An easy test is to divide 2 by 3. The display .6666666667 clearly shows that the calculator rounds, rather than truncates.

A.4

This question illustrates how the efficiency of a calculator permits students to focus on emerging patterns without getting tangled in computation. Steps 2 – 5 result in the original three-digit number, regardless of what was chosen to begin with. Step 7 is intended to encourage students to think about finding exceptional cases. For instance, most students will choose three different digits for their first three-digit number, and very few will use 0. They might reasonably guess that a number with repeated digits or

The Two Minus Signs

The calculator has two minus signs. The one on the blue key looks like $\boxed{-}$ and the one on the gray key looks like $\boxed{(-)}$. The blue one, on the right, is for subtraction. It is grouped with the keys for the other arithmetic operations. To subtract 3764 from 8902, for example, you would key in

$$8 \ 9 \ 0 \ 2 \ \boxed{-} \ 3 \ 7 \ 6 \ 4 \ \boxed{\text{ENTER}}$$

(Go ahead; do it. Do you get 5138?)

The gray minus key, next to the $\boxed{\text{ENTER}}$ key at the bottom, is for making a number negative. It is grouped with the digit keys and the decimal point. To add the numbers -273, 5280, and -2116, for example, you would key in

$$\boxed{(-)} \ 2 \ 7 \ 3 \ \boxed{+} \ 5 \ 2 \ 8 \ 0 \ \boxed{+} \ \boxed{(-)} \ 2 \ 1 \ 1 \ 6 \ \boxed{\text{ENTER}}$$

(Try it.) Notice that the display shows these negative signs without the parentheses, but they are smaller and raised a little. To see the difference between this negative sign and the subtraction sign, try subtracting the negative number -567 from 1234. Here are the keystrokes.

$$1234 \ \boxed{-} \ \boxed{(-)} \ 567$$

The display should look like this:

$$1234 - {}^-567$$

Raising to a Power

To raise a number to a power, press $\boxed{\wedge}$ just before entering the exponent. Thus, to compute 738^5, press

$$7 \ 3 \ 8 \ \boxed{\wedge} \ 5 \ \boxed{\text{ENTER}}$$

The screen should look like Display A.5.

```
738^5
        2.1891817E14
```

Display A.5

Published by IT'S ABOUT TIME, Inc. © 2000 MATHconx, LLC

with zeros in it would not work in the same way. Even though that turns out not to be the case, such conjectures are evidence that students are developing good thinking and exploration skills.

Proving that this "trick" works all the time is well within the reach of many students at this level. It rests on the fact that multiplication undoes what division does (and vice versa). Multiplying the three divisors, $7 \times 11 \times 13$, produces the number 1001, and multiplying a three-digit number by 1001 has the effect of repeating its digits. Thus, step 2 multiplied the original number by 1001, and steps 3, 4, and 5 divided by 1001 in three stages.

NOTES

The Menu Keys

Many keys bring a menu to the screen. A menu is a list of functions—things that the calculator is ready to do for you. For instance, press each of the keys across the row that starts with $\boxed{\text{MATH}}$. Don't worry about what all those lists say; just pick one out and look at it as you read the rest of this paragraph. Notice that it is actually a double menu. There are two cursors on it, shown as dark blocks. The one in the top left corner can be moved along the top line by using the $\boxed{\triangleleft}$ and $\boxed{\triangleright}$ keys. Each time you move it to a new place on the top line, the menu below changes. The items in each lower menu are reached by using the other cursor, which can be moved up and down along the left side of the screen by using the $\boxed{\triangle}$ and $\boxed{\triangledown}$ keys.

Once you have put the cursor on the choice you want, you actually make the choice by pressing $\boxed{\text{ENTER}}$. This makes the calculator go back to its "home" screen and display your choice. To make the calculator do what you have chosen, press $\boxed{\text{ENTER}}$ again.

1. **How many separate calculator functions can be reached through the menus of the $\boxed{\text{MATH}}$ key?**

2. **How many separate calculator functions can be reached through the menus of the $\boxed{\text{MATRX}}$ key?**

A.5

Entering Data in a List

The data handling tools are found through the statistics menu.

• Turn your calculator on and press $\boxed{\text{STAT}}$. You'll see a menu that looks like Display A.6.

```
EDIT CALC
1:Edit…
2:SortA(
3:SortD(
4:ClrList



     TI-82
```

```
EDIT CALC TESTS
1:Edit…
2:SortA(
3:SortD(
4:ClrList
5:SetUpEditor


     TI-83
```

Display A.6

The Menu Keys

A.5

These questions serve the purpose of getting the students to focus on the dual nature of the menus. They also emphasize the wide ranging power of these machines. Finally, they provide a simple exercise in counting possibilities in a somewhat novel setting.

1. On the TI-82: $10 + 6 + 6 + 4 = 26$; on the TI-83: $10 + 9 + 7 + 7 = 33$.

2. On the TI-82: $5 + 11 + 5 = 21$; on the TI-83: $10 + 16 + 10 = 36$.

Entering Data in a List

This part is self explanatory. There are no differences between the TI-82 (TI-83) for this process.

Summaries of 1-Variable Data

This is another self explanatory part with no differences between the TI-82 (TI-83) calculators. You may need to remind students that the downward pointing arrow at the beginning of the last line of the display indicates that there are more lines of information below, which can be read by pressing ▽ .

NOTES

_____

- To enter data, make sure that the top cursor is on EDIT and the left cursor is on 1: . Then press [ENTER] . Your screen display should look like Display A.7, with the cursor right under L1.

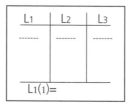

Display A.7

Note: If the display shows numbers in the L1 column, you'll have to clear the data memory. There are two ways to do this.

Get out of this display (by pressing [2nd] [QUIT]) and go back to the [STAT] menu. Press [4] . When ClrList appears, press [2nd] [1] then [ENTER] ; Done will appear. Now go back to the [STAT] screen and choose 1:Edit .

or

Without leaving this display, use the [△] and [◁] keys to move your cursor to the top of the column and highlight L1. Press [CLEAR] and then the [▽] key. List L1 should be cleared.

- Now it's time to enter the data. The calculator stores in its memory each data number you enter, along with an L1 label for that entry. The first number is called L1(1), and the second is called L1(2), and so on. (We'll ignore the L2 and L3 labels for now.) Key in the first data number, then press [ENTER] . Notice that L1(2) now appears at the bottom of the screen. Key in the second data number and press [ENTER] ; and so on, until you have put in all the data. If you make a mistake, just use the arrow keys to move the cursor to your error, type over it correctly, then move back to where you were.

At this point, the calculator has all your data stored in a way that is easy to use, and the data will stay stored even after the calculator is turned off.

A-8

NOTES

Summaries of 1-Variable Data

It is easy to get summary information about data that is stored in a single list.

- Bring up the STAT menu.

- Move the top cursor to CALC. The side cursor should be on 1:1-Var Stats. Press ENTER .

- 1:1-Var Stats will appear on your screen. Enter the list you want the calculator to summarize. For instance, if you want a summary of the data in list L1, press 2nd 1 ; then press ENTER .

That's all there is to it! A screenful of information will appear. Sections 1.3–1.7 in Chapter 1 of Book 1 explain how to interpret that information.

Putting Data in Size Order

The TI-82 (TI-83) have built-in programs that will put your data in size order automatically. Do you have any data stored in L1? If not, enter ten or a dozen numbers at random, so that you can see how the following steps work.

1. Turn the calculator on and go to the STAT menu.

2. To see what is stored in L1, ENTER 1:Edit . Then move the cursor left to L1, if it's not there already. Make sure you have some data in this list.

3. Go back to the STAT menu and choose 2:Sort A(then press ENTER . Tell the calculator to sort the L1 list by pressing 2nd 1 , then ENTER . Your calculator screen should now say Done. To see what it has done, reopen List 1 (using STAT 1:Edit) . Your data should now be listed in ascending order — that is, from smallest to largest as you read down the list. The A in SortA(stands for ascending order.

Now go back to the STAT menu and choose 2:SortD(then press ENTER . Tell the calculator to sort the L1 list again. (Press 2nd 1 ENTER .)

1. When the screen says Done, what has your calculator done? (Look at L1 again to help you answer this question.)

A.6

2. What does the "D" in SortD(stand for?

Published by IT'S ABOUT TIME, Inc. © 2000 MATHconx, LLC

Putting Data in Size Order

A.6

This should be a very quick exercise.

1. It sorts the numbers from largest to smallest as you read down the list.
2. The D stands for descending order.

If you haven't covered median yet, you can ignore the rest of this part until you get to Section 1.4 of Year 1.

NOTES
_____

Once the data are in size order, it is easy to find the median. For example, if you have 21 data items in all, the median is just the 11th one in the sorted list. Scroll through the data (using the $\boxed{\triangle}$ key) until you find L1(11). Its value is the median. If you have 20 data items, the median is halfway between the 10th and 11th items in the sorted list. Scroll through the data until you find L1(10) and L1(11). Then calculate the number halfway between them.

Finding the mode is just as easy. Count repeated items in this list. The one that is repeated the most times is the mode.

The Graph Window

This kind of calculator is called a graphing calculator because it can *draw graphs*. The screen on a graphing calculator can show line drawings of mathematical relationships. It does this with two kinds of coordinate systems—*rectangular coordinates* or *polar coordinates*. In this part we shall use only rectangular coordinates; polar coordinates will appear much later. (If you are not familiar with the idea of a rectangular coordinate system, you should review the first section of Chapter 3 in Book 1 now.)

Your calculator leaves the factory with standard coordinate axes built in. To see what they look like, turn on your calculator and press $\boxed{\text{GRAPH}}$ (in the upper right corner). You should see a horizontal and a vertical axis crossing the middle of the screen. The horizontal axis is called the **x-axis**, and the vertical axis is called the **y-axis**. (If your screen doesn't show this, press $\boxed{\text{ZOOM}}$ and choose 6:ZStandard .) Examine this display carefully; then answer the following questions.

A.7

1. Assuming that the dots along each axis mark the integer points, what is the largest possible value on the *x*-axis? On the *y*-axis?

2. What is the smallest possible value on the *x*-axis? On the *y*-axis?

3. Does it look as if the same unit of measure is being used on both axes?

4. Why do you suppose the spacing between the units is not exactly the same everywhere on an axis? Do you think that this might cause a problem?

The Graph Window

A.7

These questions are intended to get students to understand the purpose for some of the display adjustments that are available from various menus. These adjustment options will be described shortly.

1. 10 for both
2. -10 for both
3. No
4. Because the numbers of dots across and down are not multiples of 10. Both this and the apparent difference in unit length on the axes could be misleading later if not understood properly. The students probably can only guess at this now.

NOTES

The standard coordinate axis setting can be changed in several ways. This is done using the menu that appears when you press [WINDOW] . Try that now. You should get Display A.8.

```
WINDOW FORMAT
   Xmin = -10
   Xmax = 10
   Xscl = 1
   Ymin = -10
   Ymax = 10
   Yscl = 1
```
TI-82

```
WINDOW
   Xmin = -10
   Xmax = 10
   Xscl = 1
   Ymin = -10
   Ymax = 10
   Yscl = 1
   Xres = 1
```
TI-83

Display A.8

Xmin and Xmax are the smallest and largest values on the x-axis (the horizontal axis); Ymin and Ymax are the smallest and largest values on the y-axis (the vertical axis).

Xscl and Yscl are the scales for marking off points on the axes. The setting 1 means that each single integer value on the axis is marked. To see how the scale value works, change Xscl to 2. (Move the cursor down, using [▽] , then just key in 2 in place of 1.) Now press [GRAPH] . What change do you notice? Now go back to the WINDOW menu (press [WINDOW]) and change Yscl to 5. Return to the graph (press [GRAPH]). What has changed?

You can ignore the Xres = 1 line on the TI-83 for now. (If you're really curious, see p. 3–11 of the TI-83 Guidebook.)

Change the WINDOW settings so that they look like Display A.9. Then look at the graph and answer these questions.

A.8

1. Where on the screen is the origin of the coordinate system?

2. Does it look as if the same unit of measure is being used on both axes?

3. Does it look as if the spacing between the units is the same everywhere on an axis?

4. What happens when you press [△] then [▽] ?

Published by IT'S ABOUT TIME, Inc. © 2000 MATHconx, LLC

A-11

A.8

1. These settings provide a coordinate system with the origin in the lower left corner of the screen.
2. The integer points are marked on the axes, using the same distance as the unit of measure on both.
3. The units are uniformly spaced on each axis because the array of dots in the screen display is 95 by 63. The scale in this case is actually 10 dots to the unit in each coordinate direction.
4. When the two arrow keys are pressed, a cursor in the form of a cross appears exactly in the center of the screen, and the coordinates of its location are shown at the bottom: $x = 4.7$ and $y = 3.1$. This is discussed further in the next couple of paragraphs.

NOTES
_____

```
WINDOW FORMAT
  Xmin=0
  Xmax=9.4
  Xscl=1
  Ymin=0
  Ymax=6.2
  Yscl=1
```
TI-82

```
WINDOW
  Xmin=0
  Xmax=9.4
  Xscl=1
  Ymin=0
  Ymax=6.2
  Yscl=1
  Xres=1
```
TI-83

Display A.9

If you have worked through the previous questions, you found that pressing ⬆ , ⬇ puts a cross exactly in the middle of your screen and two numbers at the bottom. The cross is the cursor for the graphing screen, and the numbers are the coordinates of the point at its center. In this case, the cursor is at (4.7, 3.1). It can be moved to any point on the graph by using the four arrow keys ⬆ , ⬇ , ◁ , ▷ at the upper right of the keypad.

A.9   **Move the cursor to the point (4, 3). How far does the cursor move each time you press ◁ or ▷ ? How far does it move each time you press ⬆ or ⬇ ? Now move the cursor directly down to the bottom of the screen. What are the coordinates of the lowest point you can reach?**

These new WINDOW settings are better than the standard one in some ways, and worse in others. Let's look again at the Standard coordinate system and compare it with the one we just saw. To get back to the standard settings, press ZOOM , then press 6 to choose ZStandard. The Standard coordinate axes should appear immediately.

A.10   These questions refer to the Standard coordinate axes.

1. Where is the cursor to begin with? How do you find it if you can't remember?

2. Try to move the cursor to the point (4,3). How close can you get to it?

3. How far does the cursor move each time you press ◁ or ▷ ?

4. How far does it move each time you press ⬆ or ⬇ ?

This is a simple exercise in understanding cursor movement. It also prepares the student for an investigation of scale changes. In this case, the cursor moves 0.1 unit each time you press any arrow key. This can be seen from the change in coordinates at the bottom of the screen. The lowest point directly beneath (4, 3) on this screen is (4, 0).

A.9

This is an important Exploration for understanding how the calculator deals with coordinate systems.

A.10

1. The cursor is at (0, 0) to begin with; it can be found by pressing [△] [▽] or [◁] [▷] .

2. (4.0425532, 2.9032258)

3. 0.212766 of a unit, with some minor variations.

4. 0.3225807 of a unit, with some minor variations.

5. (4.0425532, -10)

6. There are many ways of answering these questions. A major advantage of the standard system is that it allows for negative coordinate values. It also has the same maximum and minimum values on each axis. The fact that the cursor does not move in "nice" increments is a nuisance, as is the fact that the horizontal single step amounts differ from the vertical ones. Students might also find it annoying that points with integer coordinates can't always be reached exactly.

7. This might be asking a lot of students who are unfamiliar with the calculator, but class discussion could be productive. To fix the drawbacks listed above while retaining the advantages, change the WINDOW settings so that the x-axis goes from −4.7 to 4.7 and the y-axis goes from −3.1 to 3.1. Remember to use [(−)] , instead of [−] , when entering the negative values. This will give you a coordinate system with (0, 0) in the middle of the screen, so that it will handle both positive and negative values. The cursor will move in increments of 0.1 in either direction, and hence all the points with integer coordinates can be reached exactly. The disadvantage to this system is its limited range. The exercises give students a chance to explore other options.

5. Move the cursor directly down to the bottom of the screen. What are the coordinates of the lowest point you can reach?

6. In what ways is this coordinate system better than the one we set up for the previous set of questions? In what ways is it worse?

7. How might we fix the bad features of this system without losing the good ones?

Another useful WINDOW setting is 8:ZInteger in the ZOOM menu. When you press ⑧, coordinate axes appear, but they are still the standard ones. Press ENTER to get the Integer settings.

These questions refer to the Integer coordinate axes.

1. Try to move the cursor to the point (4, 3). How close can you get to it?

A.11

2. How far does the cursor move each time you press ◁ or ▷ ?

3. How far does it move each time you press △ or ▽ ?

4. Why is this setting named Integer?

5. In what ways is this coordinate system better than the one we set up for the previous set of questions? In what ways is it worse?

6. How might we fix the bad features of this system without losing the good ones?

To plot a point (mark its location) on the graphing screen, go to the point-drawing part of the DRAW menu, like this.

Press 2nd PRGM and move the top cursor to POINTS .

Choose 1 to make the ENTER key mark cursor locations. If you want to mark some points and erase others, choose 3. This lets the ENTER key change the state of any point the cursor is on; it will mark one that isn't already marked, and will unmark one that is. (*Hint:* If you have plotted too many points and you want to start over, you can go to ZOOM menu and press ⑥ . This will wipe out everything you have plotted and return to the Standard coordinate settings. If you were using different coordinate settings, you will have to redo them in the WINDOW menu.) If you want to erase some points, see Drawing Points on a graph section in the TI Guidebook.

Published by IT'S ABOUT TIME, Inc. © 2000 MATHconx, LLC

A.11

1. This Exploration should be easier than the previous one. The cursor can be moved exactly to (4, 3).

2. & 3. It moves exactly 1 unit each time any arrow key is pressed.

4. That is, all the points that can be plotted exactly must have integer coordinates; hence, it is called the Integer setting.

5. & 6. The last two parts are open-ended, as in the previous Exploration.

NOTES
_____

Problem Set: Appendix A

1. What WINDOW settings do you need in order to put the origin at the upper right corner of your screen? What can you say about the coordinates of the points that can be plotted on this screen?

2. What WINDOW settings do you need in order to put the origin at the upper left corner of your screen? What can you say about the coordinates of the points that can be plotted on this screen?

3. Choose the Integer setting for the coordinate axes and plot the points (30, 14), (-5, 20), (-26, -11), and (6, -30). Then write the coordinates of two points that lie within the area of the graph window but cannot be plotted exactly with this setting.

4. Find WINDOW settings to form a coordinate system such that the points (120, 80) and (-60, -40) are within the window frame.

   (a) How far does the cursor move each time you press ◁ or ▷ ?

   (b) How far does it move each time you press △ or ▽ ?

   (c) Can you put the cursor exactly on (120, 80)? If not, how close can you come? Plot this point as closely as you can.

   (d) Can you put the cursor exactly on (-60, -40)? If not, how close can you come? Plot this point as closely as you can.

   (e) Can you put the cursor exactly on (0, 0)? If not, how close can you come?

5. Find WINDOW settings to form a coordinate system such that the cursor can be put exactly on the points (20, 24.5) and (-17.3, -14).

   (a) What is the initial position of the cursor?

   (b) How far does it move each time you press ◁ or ▷ ?

   (c) How far does it move each time you press △ or ▽ ?

   (d) Can you put the cursor exactly on (0, 0)? If not, how close can you come?

Published by IT'S ABOUT TIME, Inc. © 2000 MATHconx, LLC

Problem Set: Appendix A

1. Set Xmax and Ymax to 0. All points that can be plotted must have both coordinates negative (or zero).

2. Set Xmin and Ymax to 0. All points that can be plotted must have a nonnegative x-coordinate and a nonpositive y-coordinate.

3. Any point within the axes ranges that does not have integers for both coordinates cannot be plotted exactly.

4. This can be done in many different ways. Answers to the specific questions depend on the choice of coordinate extremes.

5. Perhaps the easiest, but not the only way to do this is to set

$$\text{Xmin} = -17.3, \ \text{Xmax} = 20, \ \text{Ymin} = -14, \text{ and Ymax} = 20$$

If these settings are used, then the rest of the answers are

   (a) (.95319149, 5.25)
   (b) approx. .3968 (with slight variations)
   (c) .6209677 (with minor variations in the last digit)
   (d) (.15957447, .28225806)

NOTES

Drawing Histograms

Drawing a histogram is very easy. All you have to do is choose a few numbers to tell the calculator how wide and how tall to make the bars, as follows. Turn your calculator on and press WINDOW . The screen should look like Display A.10, maybe with different numbers.

```
WINDOW FORMAT
  Xmin = -10
  Xmax = 10
  Xscl = 1
  Ymin = -10
  Ymax = 10
  Yscl = 1
```
TI-82

```
WINDOW
  Xmin = -10
  Xmax = 10
  Xscl = 1
  Ymin = -10
  Ymax = 10
  Yscl = 1
  Xres = 1
```
TI-83

Display A.10

The numbers in this WINDOW list tell the calculator how to set the horizontal (X) and vertical (Y) scales.

- Xmin, an abbreviation of X *minimum*, is the smallest data value the picture will show. You should set it at some convenient value less than or equal to the smallest value in your data set.

- Xmax, an abbreviation of X *maximum*, is the largest data value the picture will show. Set it at some convenient value greater than or equal to the largest value in your data set.

- Xscl, an abbreviation of X *scale*, says how to group the data. It is the size of the base interval at the bottom of each bar of the histogram. For instance, Xscl = 10 will group the data by 10s, starting from the value of Xmin that you chose.

- Ymin is the smallest frequency of any data group. It is never less than 0, which usually is a good choice for it.

- Ymax represents the length of the longest bar. Choose a convenient number that is not less than the largest frequency of any data group, but not much larger.

- Yscl determines the size of the steps to be marked on the vertical (frequency) scale. For small data sets, set it to 1. If your setting for Ymax is much larger than 10, you might want to set Yscl larger than 1. A little experimenting will show you how to choose a helpful setting.

Published by IT'S ABOUT TIME, Inc. © 2000 MATHconx, LLC

Drawing Histograms

This is a self explanatory section. No exercises are included because there is ample opportunity to practice this process while working on problems in the text, particularly in Section 1.3 of Year 1.

NOTES

- If you have a TI-83, the last line at the bottom of this display is Xres = 1. It's a pixel resolution setting for graphing functions; ignore it for now.

Now your calculator is ready to draw a histogram.

- Press [STAT PLOT] (actually, [2nd] [Y=]), choose 1: and press [ENTER] .

- Choose these settings from each row by moving the cursor to them with the arrow keys and pressing [ENTER] each time.

  – Highlight On.

  – Highlight the histogram picture.

  – Set Xlist to the list containing your data (L1, L2, etc.).

  – Set Freq:1 .

Now press [GRAPH] — and there it is!

Drawing Boxplots

The TI-82 (TI-83) calculators can draw boxplots. All they need are the data and a few sizing instructions. Here's how to do it.

- Turn the calculator on, press [STAT] and choose 1:Edit... from the EDIT menu. Check which list contains the data you want to use. Let's assume it's in L1.

- Press [WINDOW] and set the horizontal (X) and vertical (Y) scales. (If you have forgotten how to set your WINDOW, refer to "The Graph Window" section.) Choose convenient numbers for the X range—Xmin less than your smallest data value and Xmax greater than your largest data value, but not too small or too large. (You don't want the picture to get squeezed into something you can't see well!) Also set Xscl to some convenient size.

- The Y settings don't matter as much. However, for the TI-82, if you set Ymin to –1 (Be sure to use the [(–)] key!) and Ymax to 4, the X scale will appear nicely in a readable

Published by IT'S ABOUT TIME, Inc. © 2000 MATHconx, LLC

Drawing Boxplots

This is another self explanatory section. No exercises are included because there is ample opportunity to practice this process while working on problems in the text, particularly in Section 1.5 of Year 1.

NOTES

location under the boxplot. For the TI-83, Ymin = -2 and Ymax = 2 work a little better.

- Press [STAT PLOT] (actually, [2nd] [Y=]), choose 1: and press [ENTER] . Select these settings from each row by moving the cursor to them with the arrow keys and pressing [ENTER] each time.

    On; the boxplot picture; L1 from the Xlist; 1 from Freq

- Now press [GRAPH] —and there it is!

- To read the five-number summary, press [TRACE] and use the [◁] and [▷] to display the five numbers one at a time.

Graphing and Tracing Lines

If you want the calculator to graph a line or a curve, you must first be able to describe the line or curve by an algebraic equation. Once you have the equation for what you want to draw, you must put it in the form

$$y = [\text{something}]$$

For a straight line, that's not a problem; we often put the equation in this form, anyway. For some other kinds of curves, putting them in this form can be a little messy. In this section we shall deal only with straight lines.

All graphing begins with the [Y=] key. When you press this key for the first time, you get the screen in Display A.11.

| $Y_1 =$ |
| $Y_2 =$ |
| $Y_3 =$ |
| $Y_4 =$ |
| $Y_5 =$ |
| $Y_6 =$ |
| $Y_7 =$ |

TI-82

| Plot1 | Plot2 | Plot3 |
| $\backslash Y_1 =$ |
| $\backslash Y_2 =$ |
| $\backslash Y_3 =$ |
| $\backslash Y_4 =$ |
| $\backslash Y_5 =$ |
| $\backslash Y_6 =$ |
| $\backslash Y_7 =$ |

TI-83

Display A.11

A-17

Graphing and Tracing Lines

This section should be helpful for students when they get to Chapter 3 of Year 1.

NOTES

These lines allow you to put in as many as ten different algebraic equations for things you want drawn. (The subscript number gives you a way to keep track of which equation goes with which picture on the graph.) To see how the process works, we'll make the first example simple—two straight lines through the origin.

Key in -.5X on the $Y_1=$ line, *using the* $\boxed{\text{X,T,}\theta}$ *key to make the* X; then press $\boxed{\text{ENTER}}$.

(It is important to use $\boxed{\text{X,T,}\theta}$ for X because that's how the calculator knows that you are referring to the horizontal axis.)

Key in -.25X on the $Y_2=$ line and $\boxed{\text{ENTER}}$ it.

Be sure to use the $\boxed{\text{(−)}}$ key for the negative sign. If you don't, you'll get an error message when you ask for the graph. If you want to wipe out one of these equations and redo it, just move the cursor back to the equation and press $\boxed{\text{CLEAR}}$.

Now your work is done. Press $\boxed{\text{GRAPH}}$ and just watch as the calculator draws the lines. If you forget which line goes with which equation, or if you want to see the coordinates of the points along your lines, press $\boxed{\text{TRACE}}$ and then move the cursor with the $\boxed{\triangleleft}$ and $\boxed{\triangleright}$ keys. When you do this, the coordinates of the cursor's position appear at the bottom of the screen. For the TI-82, a number appears in the upper right corner to tell you which equation you're tracing. For the TI-83, the equation appears in the upper left hand corner. In this example, when you press $\boxed{\text{TRACE}}$ you will be on $Y_1=$ -.5X, the first of the two lines we entered. Try it. Now move back and forth along this line.

To switch from one line to another, use the $\boxed{\triangle}$ and $\boxed{\triangledown}$ keys. Notice that, in this case, either of these keys gets you to the other line. That's because we are only graphing two equations. If we were graphing more than two, these keys would move up and down the *list of equations*, regardless of where the graphs appeared on the screen.

There is a way to remove the graph of an equation from the screen without erasing the equation from your list. For example, let us remove the line $Y_1=$ -.5X from the picture. Go back to the $\boxed{\text{Y=}}$ list. Notice that the = sign of each equation appears in a dark block. This shows that the graph of

NOTES

this equation is turned <u>on</u>. To turn it off, move the cursor to the = sign and press ENTER . The dark block will disappear. To turn it back on, put the cursor back on = and press ENTER again.

Approximating Data by a Line

This section refers to a situation that commonly arises in the analysis of two variable data. Such data can be represented as points on a coordinate plane, and it is often useful to know if the pattern of points can be approximated by a straight line. A common way of doing this is called *least-squares approximation*. An explanation of this process and its use appears in Chapter 4 of Year 1. This calculator section provides a simple example of how to get the TI-82 (TI-83) to give you a least-squares approximation of a set of data.

Let's look at a very small, simple data set. (The process is exactly the same for bigger, more complicated data sets.) Here are four points of two variable data.

$$(1, 2) \quad (2, 3) \quad (3, 5) \quad (4, 6)$$

If you plot these points on a coordinate plane, you will see that they don't all lie on the same line. (Don't just take our word for it; make a sketch!) The calculator uses the least-squares method to find automatically the line of "best fit." Section 4.3 (in Chapter 4) describes how this method works and what best fit means. These are the instructions for getting the calculator to do all the tedious work for you.

First of all, you need to have the data entered in two *separate data storage lists. You get to* these lists by pressing STAT and choosing 1:EDIT… from the EDIT menu. When you press ENTER , you should get Display A.12.

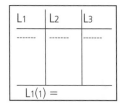

Display A.12

Published by IT'S ABOUT TIME, Inc. © 2000 MATHconx, LLC

Approximating Data by a Line

This simple example relies on some of the calculator's default choices. For instance, the data are entered in lists L1 and L2, which are the default comparison lists for LinReg. To compare other data lists with this function, you must specify your list choices. See the calculator's instruction manual for details.

Note that when you go through the linear regression process described here, the correlation coefficient r shows up automatically on the TI-82, but not on the TI-83. To get it on the TI-83, go to the CATALOG menu, choose DiagnosticOn, and $\boxed{\text{ENTER}}$ it. Then you will get both r^2 and r in the linear regression display.

NOTES

If the columns already contain data that you don't want, you can clear them out in either of two ways.

- Press [STAT] and choose 4:ClrList from the menu that appears. When the message ClrList appears, enter the name of the list you want to clear. (Press [2nd] [1] for L1, [2nd] [2] for L2, etc.) Then press [ENTER]; the screen will say Done. Now press [STAT] to return to the process of entering data.

- Go to the MEM screen (press [2nd] [+]) and choose 2:Delete... (press [2]). Choose 3:List... from the menu that appears. The screen will show the name of each data storage list that contains data. Use the arrow keys to pick the ones that you want to clear out; press [ENTER] for each one. Now press [STAT] to return to the process of entering data.

Enter the first coordinate of each data point into list L1; put its second coordinate in list L2. The four data points of our example should appear as shown in Display A.13.

L1	L2	L3
1	2	-------
2	3	
3	5	
4	6	
-------	-------	
L2(5) =		

Display A.13

Now we are almost done. Press [STAT] and go to the CALC menu. Choose LinReg(ax + b) . When you press [ENTER] , the screen will display an algebraic description of the line of best fit. For our example, it looks like Display A.14.

- The second line, y = ax + b, just tells you that the information is for slope-intercept form. (Notice that the TI-82 (TI-83) use a, not m, for the slope here.)

- The third line says that the slope is 1.4.

- The fourth line says that the y-intercept is .5.

Note: On the TI-82, the last line shows the **correlation coefficient,** a measure of how good the fit is. The correlation coefficient is not discussed in your textbook. A detailed explanation of how it works will have to wait until you study

NOTES

statistics in more depth. But, in case you are curious about it, here is a little more information. The correlation coefficient is always a number between –1 and 1, inclusive. 1 and –1 stand for a perfect fit, with all points exactly on the line. (1 is for lines with positive slope; –1 is for lines with negative slope.) The closer r is to 0, the worse the fit.

LinReg	LinReg
y = ax + b	y = ax + b
a = 1.4	a = 1.4
b = .5	b = .5
r = .9899494937	
TI-82	TI-83

Display A.14

Putting together this information about our example, we see that the least-squares line is described by the equation

$$y = 1.4x + .5$$

Graph the line $y = 1.4x + .5$. Are any of the four data points on it? How can you be sure?

A.12

Using Formulas to Make Lists

Sometimes it is useful to make a new list of data from an old one by doing the same thing to each data value. For instance, you might want to add a fixed number to each value, square each value, or find the distance of each value from some particular number. Instead of computing the new list one entry at a time, you can do it all at once if you can express your process as a formula.

Here's how the process works.

- Go to the STAT menu. Enter a list of data in L1, and then clear L2 and L3.

- To add 5 to each entry in L1, move the cursor over to the second column, then up to the heading, L2. The bottom line of your display should read L2= (without any number in parentheses).

A.12

There are two ways to check whether or not the points are exactly on the line. One is to substitute each one into the equation and see if the result is true. The other is to have the calculator plot the line with the Integer setting and move the [TRACE] cursor to the x-values 1, 2, 3, and 4. Either way, the students can see that none of the points are exactly on the line. By using the calculator, they will see the actual y-value of each point on the line with the same x-value as the data point they are checking, thus seeing how close the line comes to each data point.

Using Formulas to Make Lists

This section probably will be more useful to students after they have covered functions expressed as formulas. It is not essential to their basic understanding of the calculators. The process explained here is very much like that used in working with electronic spreadsheets. However, there is one crucial difference: The data must be in the "domain" list *before* the formula is entered at the top of the output column. Unlike spreadsheets, the TI-82 (TI-83) calculators do not store the formula. They simply use the formula to calculate the column entries, and then store only the entries. If you change a domain entry later, the corresponding entry in the output column will *not* change.

NOTES

- The trick here is to let the symbol L1 stand for each element of the list L1. That is, we make L1 *a variable.*
  Key in L1 + 5 ; the bottom of your screen should read
  L2 = L1 + 5 .

- Now press ENTER and watch the entire column for L2 fill out automatically!

- To list in L2, the square of each entry in L1, put the cursor on L2 (at the top of the column). Then enter L1^2
  (or L1 ∗ L1).

- Now let us list in L3 the midpoint between the L1 entry and the L2 entry. Put the cursor back on L3 at the top of the column and press CLEAR . This removes the old formula. Now key in (L1 + L2)/2 and press ENTER .

1. List at least ten data values in L1.

2. Write a formula to list in L2 the distance between 17 and each entry in L1. Remember: Distances are never negative numbers. Then use it.

A.13

3. Write a formula to list in L3 the square of the difference (which may be negative) between each entry in L_1 and 17. Then use it.

4. Write a formula to list in L4 the square root of each entry in L3. Then use it.

5. How are columns L2 and L4 related? Explain.

Drawing Circles

To draw circles directly on a graph, use 9:Circle(in the DRAW menu. (The DRAW menu appears when you press
2nd DRAW .) 2:Line(can be used to draw segments, which lets you add radii, diameters, and other segments to your drawings of circles.

Before beginning, make sure that all the functions on your Y= screen are turned off. If they are not, their graphs will appear when you draw circles and segments. Also make sure that all STAT PLOTS are turned off.

Follow these instructions to draw a circle directly on a graph.

1. From the ZOOM menu, choose ZStandard (to clear any unusual WINDOW settings). Then choose ZSquare or ZInteger, which displays the graph window.

A.13

1. This is mostly routine practice, but it also illustrates how to get absolute value without using the abs function.

2. L2 = abs $(17 - L1)$ or L2 = abs $(L1 - 17)$

3. L3 = $(L1 - 17)^2$ or L3 = $(L1 - 17) * (L1 - 17)$

4. L4 = $\sqrt{L3}$

5. They are the same. Squaring makes a number positive and the calculator's value for square root is the positive root. The net effect of these two steps is to make positive whatever number you start with; that is, to give you its absolute value.

NOTES

_____

_____

_____

_____

_____

_____

_____

_____

_____

_____

_____

_____

2. From the DRAW menu, choose 9:Circle(.

3. Choose a point for the center by moving the cursor to this point and pressing ENTER .

4. Choose the radius for your circle by moving the cursor this many units away from the center and pressing ENTER .

You can continue to draw circles by repeating the last two steps. To clear the screen before drawing a new circle, use :ClrDraw in the DRAW menu. If you want to stop drawing circles, press CLEAR .

Follow the steps above to draw each of these items.

1. a circle with center (0, 10) and radius 5

2. a circle with center (12, −7) and radius 15

3. four circles with center (0, 0)

a

A.14

You can also draw a circle from the Home Screen (the calculator's primary display WINDOW) by following these instructions. (You can use this same method to draw circles from a program.)

1. From the Home Screen, choose Circle(from the DRAW menu.

2. Input the coordinates of the center, followed by the radius; then press ENTER . For example, if you enter (0, 10, 5), the calculator will draw a circle with center at (0, 10) and radius 5, using whatever ZOOM WINDOW setting is current.

3. To return to the Home Screen, press CLEAR .

1. Draw a circle with center (3, 2) and radius 7 directly from the Home Screen. (If your graph does not look like a circle, how can you adjust the graph WINDOW so that it does?)

b

A.15

2. Draw four concentric circles around (0, 0) directly from the Home Screen. Earlier you were asked to draw this figure directly on a graph. Which method is easier for you? Why?

Drawing Circles

This material probably will be most useful for Year 2, particularly in Chapter 4.

These questions are for practicing the calculator skill just described. (a) and (b) are straightforward. For (c), students can choose any radii for the four circles; they have to enter the center point in each case.

A.14

This is another skill reinforcement exercise.

A.15

1. Circles will look oval in Standard WINDOW setting and in any setting where the x and y axes are of equal length in calculator units. To get a circle that looks round, choose the ZSquare WINDOW setting in the ZOOM menu.

2. This is straightforward. Answers as to which method is easier are likely to vary, depending on the learning style preferences of the individual students.

NOTES

Using a Spreadsheet

Computers give us many different tools for doing and using mathematics. One of these tools is called a **spreadsheet**. These days, a spreadsheet is an easy-to-use and very powerful computer program, but the idea of a spreadsheet is really much simpler and older than computers. Originally, a spreadsheet was just an oversized piece of paper, with lines and columns that made it easier for accountants and bookkeepers to keep their work in order.

You can make a spreadsheet on a lined piece of paper.

- Make a narrow border across the top and down the left side of the sheet.

- Divide the rest of the paper into columns from top to bottom. Six columns of about equal width will do for now.

- In the left margin, number the lines, beginning with 1, to the bottom of the page.

- Across the top margin, name each column with a letter from A to F in alphabetical order.

Your paper should look something like Display B.1.

	A	B	C	D	E	F
1						
2						
3						
4						
5						
6						
⋮						

Display B.1

Published by IT'S ABOUT TIME, Inc. © 2000 MATHconx, LLC

B-1

Using a Spreadsheet

This appendix is *not* intended to replace a manual for your electronic spreadsheet program. Rather, it has two purposes.

- It serves as a gentle introduction to electronic spreadsheets by way of some simple exercises.

- It provides a convenient reference for some of the more commonly used elementary procedures.

The instructions provided here are fairly generic. They have been crafted to apply equally well to Microsoft Excel and to Lotus 1-2-3, with appropriate comments about specific differences. You should be able to use this Appendix with virtually any electronic spreadsheet program. However, if you are using an older and/or less common spreadsheet, it probably would be a good idea to work through these instructions with it in some detail. In any event, your spreadsheet manual should be regarded as the definitive reference source.

NOTES

The Cell Names

Each box in this grid has its own address — the letter of its column followed by the number of its row. For instance, C4 refers to the box, third column (column C), on the fourth line (row). The electronic spreadsheets that computers handle look just like this, and each position in them is addressed in just the same way. Electronic spreadsheet manuals often call the boxes **cells**. We'll do the same thing, so that you become used to the term.

B.1

Here are a couple of questions to get you comfortable with the way cells are addressed.

- Make a copy of Display B.1 and shade in these cells: A2, B3, C4, D5, E6, A6, B5, D3, E2. What shape do you get?

- If you wanted Display B.1 to be shaded in a checkerboard pattern, with alternating cells filled in, which cells would you shade? Write out all their addresses. (There's more than one way to do this.)

The advantage of electronic spreadsheets over handmade ones is that the electronic ones do the computations for you, *IF* you ask them properly. If you know how to speak the language of your spreadsheet program, you can get it to do all the hard work very quickly. The main idea to remember is

> A spreadsheet is powerful because it can find and work with numbers that appear anywhere on it by using the cell names.

Therefore,

> When working with a spreadsheet, always try to build what you want, step by step, from the first data you enter. The fewer numbers you have to enter, the easier it is for the spreadsheet to do your work.

Published by IT'S ABOUT TIME, Inc. © 2000 MATHconx, LLC

The Cell Names

B.1

The first question makes an X centered at C4. The second question can be done in two ways, depending on whether or not Cell A1 is to be shaded. If it is, then the cells to be shaded are

A1, C1, E1, B2, D2, F2, A3, C3, E3, B4, D4, F4, A5, C5, E5, B6, D6, F6

Otherwise, the shaded cells are

B1, D1, F2, A2, C2, E2, B3, D3, F3, A4, C4, E4, B5, D5, F5, A6, C6, E6

NOTES

The rest of this appendix shows you how to get an electronic spreadsheet to work for you. For practice, each new process will be introduced by using it to deal with this problem.

> You are sent to the local supermarket to buy at least 2 pounds of potato chips for a club picnic. The club treasurer tells you to spend as little money as possible.

Now, there are many different brands of potato chips, and each brand comes in several different size bags. How can you compare prices in a useful way? Well, the bag sizes are measured in ounces. If you divide the price of the bag by the number of ounces, you'll get the price per ounce (this approach is called *unit* pricing). We'll set up a spreadsheet to tell you the price per ounce of every kind of potato chip bag your market sells.

> *Don't just read the rest of this appendix*: **DO IT! Work along with the instructions using your own spreadsheet.**   **B.2**

Entering Numbers and Text   **B.3**

There are three different kinds of things you can put in a cell — numbers, text, and formulas. Most spreadsheets distinguish between numbers and text automatically.

1. If you enter numerical symbols only, the entry is treated as a number.

2. If you begin an entry with letters or other symbols not related to numbers (even if numbers are entered along with them), the entry is treated as text.

[Note: If you want a number (such as a date or a year) or a number related symbol (such as $) to be treated as a text entry, you have to tell the machine somehow. Check your user's manual for the way your spreadsheet program does it.]

Display B.2 lists the prices of different brands and sizes of potato chips, including the special sale prices for the day. These are actual data from a supermarket. To enter these data in

Published by IT'S ABOUT TIME, Inc. © 2000 MATHconx, LLC

B.2 It is important to have the students actually work through this example, step by step, on a computer, if possible. If you have enough access to computers to allow for small group work, that probably would be best. If you only have access to one computer for demonstration purposes, have students take turns working through the steps. Either way, check their understanding at each stage and discuss difficulties as they arise.

NOTES

their most useful form, you should use *three* columns—one for the brand, one for the weights (in ounces), and one for the prices. Put the information of Display B.2 into columns A, B, and C now.

Brand of Chip	No. of Ounces	$Cost of Bag
Cape Cod	11 oz.	2.49
Eagle Thins	9.5 oz.	1.99
Humpty Dumpty	6 oz.	1.19
Humpty Dumpty	10 oz.	1.68
Lay's	6 oz.	0.95
Lay's	14 oz.	2.79
O'Boisies	14.5 oz.	2.79
Ruffles	6 oz.	1.39
Ruffles	14 oz.	2.79
Tom's	6 oz.	1.39
Tom's	11 oz.	1.69
Wise	6 oz.	1.39
Wise	10 oz.	1.48

Display B.2

The standard column width of your spreadsheet probably is not big enough to handle some of the brand names. Find the Column Width command and adjust the width of column A to 15 spaces. While you're at it, you might as well adjust the width of column B (the ounces) and column C (the price) each seven spaces wide. This will make the display look a little neater.

B-4

NOTES

Entering Formulas

If you want the spreadsheet to calculate an entry from other data, you have to give it a formula to use. You also have to begin with a special symbol to let it know that a formula is about to be entered. The special symbol depends on the type of spreadsheet you have. Excel uses the symbol = ; Lotus 1-2-3 uses the symbol + ; your software might use something else.

Calculate the price per ounce of Cape Cod chips by entering the formula C1/B1 into cell D1. As soon as you enter it, the number 0.226363 should appear. This is correct, but more accurate than we need. Three decimal places should be enough. Find the spreadsheet command that fixes the number of decimal places and use it to set the column D display to 3 places.

Copying Formulas

To get the price per ounce of Eagle Thins, all you have to do is copy the formula from cell D1 to cell D2. Do that. (Check the spreadsheet manual to see how to copy from one cell to another.) As soon as you do it, the number 0.209 will appear. Now look at the formula itself. Notice that it says C2/B2; that is, when you copied the formula one cell below where it started, the spreadsheet automatically changed the cell addresses inside it by that amount. This automatic adjustment process is one of the most powerful features of the spreadsheet. Next we'll use it to get the price per ounce of *all* the other kinds of chips at once!

Repeated Copying

You can copy a cell entry over and over again, all at once, along as much of a row or column as you mark out. If the entry is a formula, the spreadsheet will automatically adjust the cell addresses in it at each step. In some spreadsheet programs (such as Excel), this is done by the Fill command. In others (such as Lotus 1-2-3), it is done as part of the Copy command, by highlighting the entire region of cells into which you want the formula copied.

Published by IT'S ABOUT TIME, Inc. © 2000 MATHconx, LLC

NOTES

Find out how this works for your spreadsheet. Then copy what's in D2 into cells D3 through D13 and watch all the per ounce prices appear immediately. At this point, your spreadsheet should look something like Display B.3.

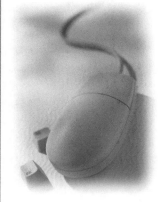

	A	B	C	D
1	Cape Cod	11	2.49	0.226
2	Eagle Thins	9.5	1.99	0.209
3	Humpty Dumpty	6	1.19	0.198
4	Humpty Dumpty	10	1.68	0.168
5	Lay's	6	0.95	0.158
6	Lay's	14	2.79	0.199
7	O'Boisies	14.5	2.79	0.192
8	Ruffles	6	1.39	0.232
9	Ruffles	14	2.79	0.199
10	Tom's	6	1.39	0.232
11	Tom's	11	1.69	0.154
12	Wise	6	1.39	0.232
13	Wise	10	1.48	0.148

Display B.3

 What *formula* is being used in cell D3? In D7? In D13?

B.4

Inserting Rows and Columns

Now let's put in column headings so that the spreadsheet is easier to understand. Move the cursor to the beginning of row 1 and use the Insert Row command of your spreadsheet to put in two rows at the very top. (Cape Cod should now be in cell A3.) We'll use the first row for headings and leave the second row blank. Enter Brand in A1, ounce in B1, price in C1 and enter $/oz. in D1. Change the width of column D to 7 spaces.

Because we've moved everything down, the row numbers no longer correspond to the number of brands listed. Make space to renumber the rows that list the brands, like this: Move the cursor to the top of the first column and use the Insert Column command to put two new columns at the far left. (Cape Cod should now be in cell C3.)

Appendix B: Using a Spreadsheet

Entering Numbers and Text

It is tempting not to repeat the brand names for different sizes of the same brand. However, it will cause serious problems when the rows are sorted because the brand name will not be carried along with the rest of the information for some rows. Discourage this particular economy of effort, or be prepared to have some students starting over again when they hit the sort step.

Repeated Copying

B.4

C3 /B3   C7 / D7   C13 /D13

NOTES
_____

Numbering Rows

Now let's try a little experiment. We'll number the brands in two different ways. Make the two new columns, A and B, only 4 spaces wide. Now put the numbers 1 through 13 down these two columns, starting at the third row, in these two ways.

B.5

- In column A, enter each number by hand—the number 1 in A3, the number 2 in A4, and so on, down to the number 13 in A15.

- In column B, enter the formula B2+1 in cell B3. The number 1 will appear because the spreadsheet treats the empty cell B2 as if it had 0 in it. Now copy this formula into all the cells from B3 through B15.

Do columns A and B match? (They should. If they don't, ask your teacher to help you find what went wrong.) At this point, your display should look like Display B.4.

	A	B	C	D	E	F
1			Brand	oz.	$/bag	$/oz.
2						
3	1	1	Cape Cod	11	2.49	0.226
4	2	2	Eagle Thins	9.5	1.99	0.209
5	3	3	Humpty Dumpty	6	1.19	0.198
6	4	4	Humpty Dumpty	10	1.68	0.168
7	5	5	Lay's	6	0.95	0.158
8	6	6	Lay's	14	2.79	0.199
9	7	7	O'Boisies	14.5	2.79	0.192
10	8	8	Ruffles	6	1.39	0.232
11	9	9	Ruffles	14	2.79	0.199
12	10	10	Tom's	6	1.39	0.232
13	11	11	Tom's	11	1.69	0.154
14	12	12	Wise	6	1.39	0.232
15	13	13	Wise	10	1.48	0.148

Display B.4

Numbering Rows

B.5

The experiment actually takes place in the next step, when the rows are placed in order with respect to the per ounce price of the various bags of chips. The numbers entered by hand will go along with their original rows; however, the numbers entered by formula will remain as they are, numbering the rows from best deal to worst.

NOTES

_____

_____

_____

_____

_____

_____

_____

_____

_____

_____

_____

_____

_____

_____

_____

Published by IT'S ABOUT TIME, Inc. © 2000 MATHconx, LLC

Appendix B: Using a Spreadsheet

Ordering Data

Another handy feature of an electronic spreadsheet is that it can put in order data that is listed in a column. It can put numbers in size order, either increasing or decreasing. Most spreadsheets can also put text entries in alphabetical order. To do this, you need to find the Sort command and tell it what list of data you want to rearrange. (In Excel, Sort is in the Data menu; in Lotus 1–2–3, it's in the Select menu.) The computer prompts you for a little more information, such as whether you want ascending or descending order, then does the sorting.

(Note: Some spreadsheets move entire rows when they sort; others can be told just to rearrange the data in a single column. Check your user's manual to see how your spreadsheet works. In this example, we assume that the spreadsheet moves entire rows when it sorts.)

Let's rearrange the potato chip list according to the price per ounce, from most expensive to least expensive. Follow your spreadsheet's instructions to sort the per ounce prices in column F in ascending order. Which kind of potato chip is the best buy? Which is the worst buy?

Look at columns A and B.

B.6

1. Do they still match? What has happened? Explain.

2. What would have happened if you had entered the number 1 in B3, then entered the formula $=B3+1$ in B4? Explain.

Now that we have all this information, how do we find out how much it will cost the club for the 2 pounds of potato chips? Here's one plan.

* Compute the number of ounces in 2 pounds.

* Multiply the cost of 1 ounce by the total number of ounces needed.

(*Warning*: There's something wrong with this approach; what is it?)

B.7

We'll do this on the spreadsheet because it provides an example of a different way to use cell addresses. To find the total number of ounces, we just multiply the number of pounds (2) by 16. Make these entries on the spreadsheet.

Ordering Data

1. See B.5.

2. Because the B3 entry is not a variable, the 1 would stay with the Cape Cod row, causing the numbering in column B to start over at that point.

   The problem with this approach is that the chip prices are not bulk prices. You can't buy the chips by the ounce; they're sold in bags of certain fixed sizes. We deal with this after discussing the use of constant cell addresses. Cell D18 will have the value 32.

NOTES

- In C17, enter number of lbs.; in D17, enter 2.

- In C18, enter number of oz.; in D18, put the formula that multiplies the entry in D17 by 16. (What is that value?)

Constant Cell Addresses

To find out how much 2 pounds of each kind of potato chip will cost, first set column G to display in currency format. Then move the cursor to cell G3. (This should be the first blank cell at the end of row 3.) We want this cell to show the number of ounces to be bought (in D18) multiplied by the price per ounce (in F3). Let's try it.

- Enter the formula D18*F3 in G3. The result should be $4.74.
  (Is your first kind the Wise 10 oz. bag?)

- So far, so good. Now copy this formula to the next line, in G4. What do you get? $0.00? How come?

- Look at the formula as it appears in G4. Does it say D19*F4? What happened?

Remember that when you shift a formula from one location to another, the spreadsheet automatically shifts every cell address in exactly the same way. We copied this formula to a location one row down from where it was, so the spreadsheet added the number 1 to the row number of each cell address in the formula. Now, we want that to happen to one of these addresses, but not to the other. That is, the cost of the kind of potato chip in row 4 should use the price per ounce in F4, but it should still use the total number of ounces from D18.

> To prevent the spreadsheet from automatically adjusting a cell address when a formula is moved, enter the cell address with a $ in front of its column letter and a $ in front of its row number.

B.8

This means that you should go back to cell G3 and enter the multiplication formula $D$18*F3. Now copy this to G4. Do you get $4.92? Good. (If not, what went wrong? Ask your teacher if you need help figuring it out.) Now copy this formula into cells G5 through G15. Column G now should show the cost for 2 pounds of each kind of potato chip in your list.

Published by IT'S ABOUT TIME, Inc. © 2000 MATHconx, LLC

Constant Cell Addresses

B.8

This is a fairly standard way of keeping a cell address fixed. However, some spreadsheets, particularly some older ones, use a different convention to distinguish between variable and constant cell addresses. Consult the user's manual; then advise your students accordingly.

NOTES

Go to G1, make this column 7 spaces wide and enter the word cost as the column heading.

B.9

1. According to column G, which kind of potato chip is the best buy?

2. Why is that *not* necessarily the best buy for your club?

3. What's wrong with letting this answer tell you what kind to buy? (*Hint*: How many *bags* would you have to buy?)

The INT Function

As the hint in the box above suggests, using the information in column G to guide your choice may not be a good idea because the supermarket sells potato chips by the bag. In order to know how much it will cost to get at least 2 pounds of chips, you first must know how many bags you'll need.

How do you do that? Easy, right? Just divide 32 oz. (2 lbs.) by the number of ounces in a single bag. If you get a mixed number, add 1 to the whole-number part.

For example, if you want at least 32 oz. in 10 oz. bags, divide 32 by 10. You get 3.2 as an answer, but, since you can't buy 0.2 of a bag of chips, you need 4 bags.

There's a spreadsheet function—called INT—that makes this very easy to compute automatically. The INT function gives you the greatest integer less than or equal to the number you put into it. For instance,

$$INT\left(3\frac{1}{3}\right) = 3$$

$$INT(2.98) = 2$$

$$INT(5) =$$

Let's use this function to carry out the computation we just did, finding how many 10 oz. bags of Wise potato chips we need in order to have at least 2 pounds. But instead of entering the numbers in separately, we'll get them from other cells on the spreadsheet. Move to cell H3 and enter the formula

$$INT(\$D\$18/D3) + 1$$

Published by IT'S ABOUT TIME, Inc. © 2000 MATHconx, LLC

B.9

1. Wise
2. and 3. See B.7 for explanations.

NOTES

Just to make sure you understand what we're doing, answer these questions before moving on.

1. What does $D$18 stand for?

2. Why are the $ symbols there?

3. What does D3 stand for?

4. What number is $D$18/D3?

5. What number is INT($D$18/D3)?

6. What number is INT($D$18/D3)+1?

7. If you copy this formula to cell H4, how will it read?

a

B.10

Now use the Fill command to copy this formula into cells H4 through H15. For each kind, the number you get says how many bags you need in order to have at least two pounds of chips. Put the heading bags at the top of this column H, and make the column 5 spaces wide.

Now we can finish the problem. To find the cost of at least 2 lbs. of each kind of chip, multiply the number of bags you need by the cost of a single bag. Enter a formula in I3 that does this; then copy it into I3 through I15. Finish your spreadsheet display by renaming column G cost 1 and naming column I cost 2 and changing the width of column I to 6 spaces.

1. If you *must* get at least 2 lbs. of chips and you want to spend as little as possible, which kind do you buy?

2. How many bags do you buy?

3. What does it cost you?

b

B.11

The next questions show the power of spreadsheets for testing out different variations of a situation. Each part is exactly the same as above, except that the total number of pounds of chips is different. Answer each one by changing as little as possible on your spreadsheet.

1. If you *must* get at least 3 lbs. of chips and you want to spend as little as possible, which kind do you buy? How many bags do you buy? What does it cost you?

2. If you *must* get at least 4 lbs. of chips and you want to spend as little as possible, which kind do you buy? How many bags do you buy? What does it cost you?

c

B.12

The INT Function

B.10

1. 32 (the number of ounces in 2 pounds).
2. The $ symbols keep this cell address from changing when the formula is copied elsewhere.
3. 10 (the number of ounces in a single bag)
4. 3.2
5. 3
6. 4
7. INT($D$18/D4)+1.

B.11

1. Tom's 11 oz. size
2. 3
3. $5.07. The finished spreadsheet should look like Display B.1T.

	A	B	C	D	E	F	G	H	I
1			Brand	oz.	$/bag	$/oz.	cost 1	bags	cost 2
2									
3	13	1	Wise	10	1.48	0.148	4.74	4	5.92
4	11	2	Tom's	11	1.69	0.154	4.92	3	5.07
5	5	3	Lay's	6	.95	0.158	5.07	6	5.70
6	4	4	Humpty Dumpty	10	1.68	0.168	5.38	4	6.72
7	7	5	O'Boisies	14.5	2.79	0.192	6.16	3	8.37
8	3	6	Humpty Dumpty	6	1.19	0.198	6.35	6	7.14
9	6	7	Lay's	14	2.79	0.199	6.38	3	8.37
10	9	8	Ruffles	14	2.79	0.199	6.38	3	8.37
11	2	9	Eagle Thins	9.5	1.99	0.209	6.70	4	7.96
12	1	10	Cape Cod	11	2.49	0.226	7.24	3	7.47
13	8	11	Ruffles	6	1.39	0.232	7.41	6	8.34
14	10	12	Tom's	6	1.39	0.232	7.41	6	8.34
15	12	13	Wise	6	1.39	0.232	7.41	6	8.34
16									
17			number of lbs.:	2					
18			number of oz.:	32					

Display B.1T

B.12

If the spreadsheet has been set up as described, each group of questions can be answered by changing only one entry on the spreadsheet, the total number of pounds shown in D17. Here are the answers.

1. 3 lbs.: 5 bags of Wise 10 oz. size, at a total cost of $7.40

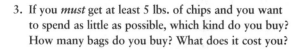

3. If you *must* get at least 5 lbs. of chips and you want to spend as little as possible, which kind do you buy? How many bags do you buy? What does it cost you?

Problem Set: Appendix B

1. These two questions refer to the potato chip spreadsheet that you just made.

   (a) Add a column J that shows the total number of ounces of potato chips of each kind that you get when you buy enough bags to get at least two pounds of them. What formula will compute these numbers?

   (b) Add a column K that shows the number of *extra* ounces (more than 2 pounds) that you get when you buy enough bags to get at least two pounds of potato chips. What formula will compute these numbers?

2. Make a spreadsheet like the one for the potato chips to deal with this problem.

   (a) Your favorite aunt runs a shelter for homeless cats. As a present for her birthday, you decide to give her 5 pounds of canned cat food. You want to spend as little money as possible. The brands, sizes, and prices for the canned cat food at the supermarket are shown in Display B.5. What brand and size is the best buy, and how many cans of it should you get? What will it cost?

   (b) Your best friend thinks you have a great idea. She decides to buy your aunt 5 pounds of canned cat food, too. If you both chip in and buy a combined present of 10 pounds of canned cat food, what is the best buy of canned cat food for your combined present? Explain your answer?

Published by IT'S ABOUT TIME, Inc. © 2000 MATHconx, LLC

2. 4 lbs.: 6 bags of Tom's 11 oz. size, at the total cost of $10.14
3. 5 lbs.: 14 bags of Lay's 6 oz. size, at the total cost of $13.30

Problem Set: Appendix B

1. (a) Formula for column J (as it appears for J3): H3*D3
   (b) Formula for column K (as it appears for K3): J3—$D$18
      Note that the formula J3—32 also works for the original 2 lb. problem, but it doesn't adjust automatically for the 3, 4, and 5 lb. problems.

2. A spreadsheet solution for this problem is shown in Display B.2T. Much of it can be constructed by mimicking the potato chip spreadsheet setup. The only new wrinkle is the "*n* for *x*" pricing. This invites the insertion of two new columns, F for the number of cans that are bought for the price shown in column E; column G is for the price per can. The row order shown in Display B.2T results from sorting with respect to price per oz. (column H), from least to most.

Here is how each column is constructed. Except for "number of oz.," each formula is shown as it would be in row 3; the data in its column is obtained by using Fill to copy the formula to the other cells in the column.

A Number entered consecutively from 1 to 12.

B Formula: B2+1

C Brand entered individually, in alphabetical order (smaller size first within brands).

D Ounces per can entered individually. Number of lbs. entered by hand in D16. Formula for number of oz.: D16*16

E Prices entered individually; total price for group of cans, where appllicable.

F Number of cans bought for price in column E; entered individually.

G Display set to 3 decimal places. Formula: E3/F3

H Display set to 3 decimal places. Formula: G3/D3

I Cost (if sold in bulk) for D16 pounds. Formula: $D$17*H3

J Number of cans required for at least D16 pounds. Formula: INT($D$17/D3) +1

K Cost for number of cans in column J. Formula: J3*G3
   (a) The best deal for 5 lbs. is the 14 oz. size of Puss'n Boots; 6 cans are required
   (b) However, the best deal for 10 lbs. is the 13 oz. size of 9 Lives; 13 cans are required, for a total cost of $6.37

Brand of Cat Food	No. of Ounces	Cost
Alpo	6 oz.	3 for $1.00
Alpo	13.75 oz.	$ .65
Figaro	5.5 oz.	$ .37
Figaro	12 oz.	$ .66
Friskies	6 oz.	$ .35
Friskies	13 oz.	$ .58
Kal Kan	5.5 oz.	4 for $1.00
Puss 'n Boots	14 oz.	$ .55
9 Lives	5.5 oz.	3 for $ .88
9 Lives	13 oz.	$ .48
Whiskas	5.5 oz.	3 for $1.00
Whiskas	12.3 oz.	$ .55

Display B.5

3. Here's a bonus question.

(a) Invent a problem about breakfast foods that is like the potato chip and cat food problems.

(b) Go to your local supermarket and gather the brand, size, and price information that you will need to solve your problem.

(c) Using the data you gather for part (b), set up a spreadsheet that solves the problem you invented in part (a).

Published by IT'S ABOUT TIME, Inc. © 2000 MATHconx, LLC

	A	B	C	D	E	F	G	H	I	J	K
1			Brand	oz.	price	#	$/can	$/oz.	cost 1	bags	cost 2
2											
3	10	1	9 Lives	13	0.49	1	0.490	0.038	3.02	7	3.43
4	8	2	Puss'n Boots	14	0.55	1	0.550	0.039	3.14	6	3.30
5	6	3	Friskies	13	0.58	1	0.580	0.045	3.57	7	4.06
6	12	4	Whiskas	12.30	0.55	1	0.550	0.045	3.58	7	3.85
7	7	5	Kal Kan	5.50	1.00	4	0.250	0.045	3.64	15	3.75
8	2	6	Alpo	13.75	0.65	1	0.650	0.047	3.78	6	3.90
9	9	7	9 Lives	5.50	0.88	3	0.293	0.053	4.27	15	4.40
10	4	8	Figaro	12	0.66	1	0.660	0.055	4.40	7	4.62
11	1	9	Alpo	6	1.00	3	0.333	0.056	4.44	14	4.67
12	5	10	Friskies	6	0.35	1	0.350	0.058	4.67	14	4.90
13	11	11	Whiskas	5.50	1.00	3	0.333	0.061	4.85	15	5.00
14	3	12	Figaro	5.50	0.37	1	0.370	0.067	5.38	15	5.55
15											
16			number of lbs.:	5							
17			number of oz.:	80							

Display B.2T

Note that there is an interesting detail here, if you want to take the time to pursue it. Notice that the 5th, 6th and 7th kinds of cat food show the same price per oz. in column H. But the bulk costs of D17 ounces in column I is different! How can that be? A brief discussion can lead to the fact that the computer stores the answers to its computations in a much more exact form than it usually shows.

3. Students' answers to the bonus question will vary.

Appendix C: Programming the TI-82 (TI-83)

After you have been using the TI-82 (TI-83) for a while, you may notice that you are repeating certain tasks on your calculator over and over. Often you are repeating the same sequence of keystrokes, which can become very tiresome. Programs give you a way to carry out long sequences of keystrokes all at once, saving you a great deal of time and energy.

In this appendix we will show you some simple TI-82 (TI-83) programs and how to enter and use them. In the textbook, there are some other programs which you will find useful in solving problems.

Correcting Mistakes

When you enter a program you will almost surely make some keying mistakes. You can use the arrow keys to back up and key over any mistakes. To insert something new, rather than keying over what is already there, give the insert command (INS above the ⌜ DEL ⌝ key) by keying

⌜ 2nd ⌝   ⌜ DEL ⌝

Use the ⌜ DEL ⌝ key to delete the current character.

Entering Programs

To enter a program, give the ⌜ NEW ⌝ command under the ⌜ PRGM ⌝ menu by keying

⌜ PRGM ⌝   ⌜ ◁ ⌝   ⌜ ENTER ⌝

Your calculator should look like Display C.1. You are now in program writing mode. Whatever you key in will be stored in the program you are creating, rather than being executed directly. To get out of program writing mode, give the QUIT command.

Published by IT'S ABOUT TIME, Inc. © 2000 MATHconx, LLC

APPENDIX C

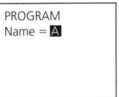

Display C.1

Next you need to name your program so that you can use it. We will start with a very short and not very useful program just to test your ability to enter a program and run it. Give your program the name ADD. Normally, to enter the capital letters that are above and to the right of some of the keys you must press the [ALPHA] key first. When naming a program, however, the calculator goes into **ALPHA** mode automatically. This means you *don't* have to press the [ALPHA] key when entering the letters in the program name. Key in

Your calculator screen should now look like Display C.2

Display C.2

We now need to enter the actual program commands. Our test program will ask for two numbers and then add them. Each line of the program begins with a colon. At the end of each line of the program press [ENTER]. The first line of our program asks for the first of the two numbers it will add. The two numbers that we tell the calculator to add are called the *input* for the program. The Input command is the first item under the I/O section of the **PRGM** menu. We will store the input in memory A. Key in

Published by IT'S ABOUT TIME, Inc. © 2000 MATHconx, LLC

MATH *Connections*: A Secondary Mathematics Core Curriculum

The TI screen should now look like Display C.3.

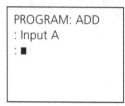

```
PROGRAM: ADD
: Input A
: ■
```

Display C.3

The next line in the program asks for another number and stores
it in memory location B. Enter the second line now, based on the
way you entered the first line. Your screen should now look like
Display C.4.

```
PROGRAM: ADD
: Input A
: Input B
```

Display C.4

The third line adds the numbers stored in memories
A and B and stores the result in memory C. Key in

ALPHA   A   +   ALPHA   B   STO ▷   ALPHA   C   ENTER

The new screen is in Display C.5.

```
PROGRAM: ADD
: Input A
: Input B
: A + B→C
: ■
```

Display C.5

We finish our program with a statement which displays the
result of adding the two numbers (now stored in memory C).
The display command Disp is the third item under the I/O
section of the PRGM menu.

C-3

APPENDIX C

Appendix C: Programming the TI-82 (TI-83)

Key in

PRGM   ▷   3   ALPHA   C

The resulting screen is in Display C.6

```
PROGRAM: ADD
: Input A
: Input B
:A + B──▶C
:Disp C
:■
```

Display C.6

We are done writing the program! To quit programming mode
use the [QUIT] key (above the [MODE] key). The program is
automatically saved. Key in

2nd   QUIT

You are now back to the Home Screen, where you started.

Running Programs
To run the program we just keyed in, we go to the EXEC
section of the PRGM menu, key in the number of the program we
want to run, and then press [ENTER]. We will assume that the
program named ADD that we just entered is program number 1.
Key in

PRGM   1   ENTER

If you entered the program correctly, a question mark appears
asking for input. (If there is an error, look at the next section
on editing programs.) This question mark is produced by the
first line of your program. You are being requested to type
in the first of two numbers, which will then be added by
the program. Let's suppose that we want to add the numbers
4 and 5. Press [4] then press [ENTER]. A second question
mark appears asking for the second number. Press [5]
and press [ENTER] again. The result, 9, should appear.
The screen now looks like Display C.7.

C-4

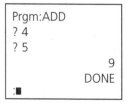

Display C.7

To run the program again, just press the [ENTER] key. You don't have to go through the PRGM menu to run the program the second time, as long as no other calculations have been performed in between. Try adding two other numbers to see how this works.

Quitting Programs

If you are in the middle of running a program and you want to stop the program, press [ON] key. To try this out, run the ADD program again, but this time when the first question mark appears, press [ON] . The screen should look like Display C.8.

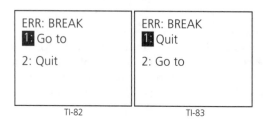

Display C.8

Press [2] to quit the program and return to the Home Screen. Pressing [1] puts you back in program writing mode at the point in the program where you stopped the program.

Editing Programs

If your program doesn't work, or if you just want to make changes to a program, you use the EDIT section of the PRGM menu. Key in

Published by IT'S ABOUT TIME, Inc. © 2000 MATHconx, LLC

C-5

APPENDIX C

This should put you back in the ADD program (assuming it is program 1). Your screen should look just as it did when you left the program writing mode (see Display C.6). Use the arrow keys and the insert ⌐INS⌐ and delete ⌐DEL⌐ keys as explained in the Correcting Mistakes section.

To open space for a new line, put the cursor at the beginning of a line, give the insert command ⌐INS⌐ and then press ⌐ENTER⌐. To try this on your ADD program, use the arrow keys to put the cursor at the beginning of the second line of the program and key in

⌐2nd⌐ ⌐DEL⌐ ⌐ENTER⌐

Your screen should look like Display C.9.

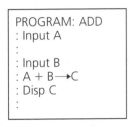

PROGRAM: ADD
: Input A
:
: Input B
: A + B→C
: Disp C
:

Display C.9

The blank line we just created will not affect the program, so we can just give the QUIT command to leave the program writing mode.

A Useful Program

Now that you have some practice with writing, editing and running programs, let's take a look at a program that you might really find useful.

Graphing With Parameters

Suppose that we want to graph the equation of a straight line, say $y = ax + 5$, for several values of a. The constant a is called a *parameter*. First we can enter the expression AX + 5 as expression Y1 under the Y= menu. We can then store numbers in memory A and press ⌐GRAPH⌐. The problem is that we only see the graph for one value of A at a time. The following program allows you to easily produce graphs for many values of A and keep all of the graphs on the screen together.

Published by IT'S ABOUT TIME, Inc. © 2000 MATHconx, LLC

C-6

Enter the program shown in Display C.10, using what you learned from the **Entering Programs** section. Name the program PARAMS. Note: DrawF is item 6 under the DRAW menu (above the [PRGM] key) for the TI-82. Y_1 is item number 1 of the Function sub menu under the **Y-vars** menu (above the [DRAW] [VARS] key). For the TI-83, Y_1 is found by keys

[VARS] [▷] [1] [1] [ENTER]

```
PROGRAM: PARAMS
: Input  A
: DrawF  Y1
: ■
```

Display C.10

The program is simple, but saves quite a few keystrokes. You put in a value for A, and then the function is graphed using the DrawF command.

To use this program you must store your function in function memory Y_1 and then *turn off* Y_1 (put the cursor on the = and press ENTER). Your Y= WINDOW should look like Display C.11. Notice that the = is *not* highlighted, indicating the function is off.

```
Y1  =  AX + 5
Y2  =
Y3  =
Y4  =
Y5  =
Y6  =
Y7  =
Y8  =
```

Display C.11

To set the graph WINDOW to the Standard setting, press the 6 under the ZOOM menu. Your WINDOW settings should appear as in Display C.12.

APPENDIX C

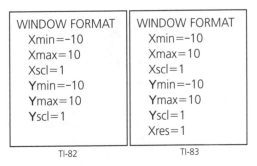

Display C.12

Now run the program. Try starting with an A value of 1. Press
[1] then [ENTER] in response to the question mark. You
should see a graph of the function $y = 1x + 5$. To run the
program again, first press the [CLEAR] key (to get back to the
Home Screen) then [ENTER]. Try an A with a value of -2 this
time. Now both the graphs of Y = AX + B for A = 1 and for
A = -2 should be on the screen, as shown in Display C.13.

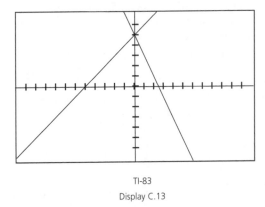

TI-83
Display C.13

If you want to clear the graph screen, use ClrDraw, which is
item 1 under the DRAW menu.

You only need to change the function stored in Y1 to graph any
other function with one parameter. For instance, try graphing
the function $y = a^x$ for various a values.

C-8

Appendix D: Linear Programming With Excel

The graphical method of solving Linear Programming problems is explained in the text. The limitation of this method is that it is applicable only to problems with two variables.

Linear Programming is one of the most used mathematical methods in real world problem solving. Only rarely are two variables present in a realistic problem. Mathematicians have developed methods to solve Linear Programming problems with any number of variables; a computer program is then needed to carry out the computations. One commonly used computer program which solves Linear Programming problems is called LINDO™.

Several spreadsheet programs can solve Linear Programming problems. Microsoft Excel, Lotus 1-2-3, and Quattro Pro are such spreadsheet programs. We will look at how Excel 5.0 solves linear programming problems, but we will not explain the method Excel uses. This is known as treating the program as something mysterious—we can follow the instructions for using the program, but we don't know how the program gives a solution to the problem. Knowing how to solve simple problems graphically helps us to understand the output from the program, even if we don't understand the exact method the computer is using. Just imagine that Excel is using a method similar to the one you learned for two variable problems.

Our first example will be a very simple problem with only two variables. Then we can solve the problem graphically as in the textbook and check our answer against the answer from the computer. We will then extend the problem to three variables; here a graphical solution is not possible.

Problem

A new housing development is being built near Bart's house. Bart has noticed that the construction workers often leave the site to get lunch. Always on the lookout to earn money, Bart figures that he can make lunches for the workers and sell them at a

Published by IT'S ABOUT TIME, Inc. © 2000 MATHconx, LLC

APPENDIX D _____

profit. We will assume there are enough workers and Bart can sell all of the sandwiches that he can make.

Bart decides to make two types of lunches. The first lunch will have two sodas and one sandwich (the Thirsty Worker Lunch) and the second type of lunch will have two sandwiches and one soda (the Hungry Worker Lunch). Bart plans to buy the sandwiches and sodas from a local deli for $3.00 per sandwich and $0.50 per can of soda. He will sell the Thirsty Worker Lunches for $5.00 and the Hungry Worker Lunches for $8.00. This means that Bart's profit for each Thirsty Worker Lunch is $1.00 and his profit on each Hungry Worker Lunch is $1.50.

Bart has one problem. The deli doesn't open until 11:30 a.m., which is too late to make lunches and have them ready for the workers. He figures he needs to buy his supplies the night before, but that means he needs to keep the (24 cans) of soda in the family refrigerator. Bart finds a cooler in the basement that will hold 20 sandwiches.

How many Hungry Worker Lunches and how many Thirsty Worker Lunches should Bart prepare in order to make the most money? His constraints are that he can use only 20 sandwiches and 24 sodas as explained above.

Solution

SETUP

First we need to formulate the problem as a linear programming problem. Let x represent the number of Thirsty Worker Lunches and y the number of Hungry Worker Lunches that Bart prepares. Then Bart's profit P is

$$P = 1.00x + 1.50y$$

The soda constraint would be

$$2x + y \leq 24$$

and the sandwich constraint would be

$$x + 2y \leq 20$$

The other two constraints which we don't want to forget are

$$x \geq 0 \text{ and } y \geq 0$$

Published by IT'S ABOUT TIME, Inc. © 2000 MATHconx, LLC

Computer Solution

Our goal is to set up the spreadsheet as shown in Displays D.1 and D.2 and to show the formulas that you actually key into each cell. Display D.2 shows what the result should look like on your spreadsheet. To get this result, follow the steps below.

1. Key in a title for the spreadsheet in cell A1; we used "Bart's Lunch Business." Don't key in any quotation marks. We dressed up the title a bit by making the font larger (12 point) and using the Outline style; just select cell A1 and choose Font from the Format menu.

2. Key in x in cell A3, y in cell B3 and P = 1.00x + 1.50y in cell C3.

3. Cells A4 and B4 will represent the initial guess for x and y. These guesses don't have to be good; they just have to make sense. The easiest guesses are 0 for both x and y, since they satisfy all the constraints. Therefore, key in 0 in A4 and 0 in B4.

4. Key in = 1.00 * A4 + 150 * B4 in cell C4. This represents the profit for the x and y you chose. You should see 0 appear since x and y are now 0.

5. Key in the word constraints in cell A6. We will use row 7 to label the constraints, and row 8 to put in the constraints as formulas.

6. Key in $x + 2y < = 20$ into cell A7 and $2x + y < = 24$ into cell B7 as labels for the sandwich and soda constraints.

7. Key in the **left side only** of the sandwich constraint into cell A8. Therefore you key in = A4 + 2 * B4 into cell B7. Similarly key in = 2 * A4 + B4 into cell B8. At this point zeros should appear in these cells since x and y are both 0.

	A	B	C
1	Bart's Lunch Business		
2			
3	x	y	P=1.00x+1.50y
4	0	0	=1*A4+1.5*B4
5			
6	constraints		
7	x+2y<=20	2x+y<=24	
8	=A4+2*B4	=2*A4+B4	

Display D.1

APPENDIX D

8. Select cells A3 through C8 and choose CELLS: Alignment from the Format menu. Choose center for horizontal alignment. Then choose CELLS: Number from the Format menu and select 0.00. Your spreadsheet should now look like Display D.2

	A	B	C
1	Bart's Lunch Business		
2			
3	x	y	P=1.00x+1.50y
4	0.00	0.00	0.00
5			
6	constraints		
7	x+2y<=20	2x+y<=24	
8	0.00	0.00	

Display D.2

We are now ready to tell the computer to find the solution to our linear programming problem. We use a capability of Excel called the Solver to get the solution. You may have noticed that we have not actually specified all of the information for the constraints. In particular, we need to tell Excel that cells A4 and B4 need to be positive (x and y should be positive), that cell A8 should be less than 20 and cell B8 should be less than 24 (the sandwich and soda constraints). We do this as shown below.

9. Choose Solver from the Tools menu. The goal is to get the Solver dialogue box to look like Display D.3.

10. Type C4 in the Set Target Cell box. You don't need to type the $s; they will be added by Excel automatically. This box tells the computer which cell to maximize; in our problem C4 is the cell that represents Bart's profit.

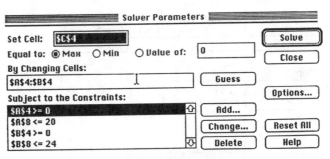

Display D.3

11. Type A4:B4 in the By Changing Cells box. This instructs the computer to adjust the cells in this range (the number of

Published by IT'S ABOUT TIME, Inc. © 2000 MATHconx, LLC

Thirsty Worker and Hungry Worker Lunches in our problem) until cell C4 is at a maximum.

12. Put the constraints into the Subject to the Constraints box. For each you must click the Add button, which brings up another dialogue box as shown in Display D.4. Key in the Cell Reference and the Constraint value (the right-hand side of the constraint inequality). Click on the < = drop-down menu to choose < = or > = (you can also choose = or int for integer).

Cell Reference:		Constraint:
$A$8	<= ⬇	=20

OK	Cancel	Help

Display D.4

13. Click the Max button in the Equal To line to indicate that this is a maximization problem (it is probably already chosen).

14. Click the Solve button, and Excel will attempt to solve the problem. Excel will then report whether it found a solution or not. If a solution was found, just click OK in the dialogue box that pops up, and the values of the variables which provide the solution will have replaced the initial guesses that you gave in step 3.

Bart's Lunch Business

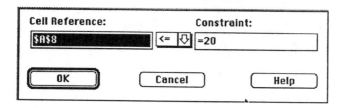

x	y	P=1.00x+1.50y
9.33	5.33	17.33

Constraints:

x+2y<=20	2x+y<=24
20.00	24.00

Display D.5

15. Summarize your findings and explain the spreadsheet for someone who might not be familiar with the problem. Display D.5 shows the solution to our problem, as well as a short verbal description of the results. We also dressed up

APPENDIX D _____

Appendix D: Linear Programming With Excel

our report with a little formatting (using Borders from the Format menu and turning off the gridlines with Display from the Option menu). Note that, in this problem, we needed to round our results since fractional parts of a lunch do not make sense.

Summary of Findings

Bart is going to sell two types of lunches. We let x represent the number of Thirsty Worker Lunches, which contain two sodas and one sandwich. We let y represent the number of Hungry Worker Lunches, which contain one soda and two sandwiches. Bart's constraints are that he can use at most twenty sandwiches and twenty-four sodas. His profit P is $1.00 on each Thirsty Worker Lunch and $1.50 on each Hungry Worker Lunch.

The tables in Display D.5 resulted from performing a linear programming analysis of the problem. What they show is that Bart should make 9 Thirsty Worker Lunches and 5 Hungry Worker Lunches. The values in the table need to be rounded since Bart can't make a fraction of a lunch. His profit will be less than the $17.33 shown in the table; his actual profit will be

$$1(9) + 1.5(5) = \$16.50$$

Extension to Three Variables

Suppose that Bart decided to make a third type of lunch with 2 sandwiches and 2 sodas (Super Lunches). Let z represent the number of Super Lunches. Bart will sell Super Lunches for $9.00; his profit on these lunches would be $2.00. Then the new profit formula would be

$$P = 1.00x + 1.50y + 2.00z$$

The new constraints would be

$$2x + y + 2z \leq 24 \text{ and } x + 2y + 2z \leq 20$$

for soda and sandwiches. Display D.6 shows what formulas you would type into the cells. Display D.7 shows how the spreadsheet would look before solving.

Published by IT'S ABOUT TIME, Inc. © 2000 MATHconx, LLC

MATH Connections: A Secondary Mathematics Core Curriculum

	A	B	C	D
1	Bart's Lunch Business			
2				
3	x	y	z	P=1.00x+1.50y+2.00z
4	0	0	0	=1*A4+1.5*B4+2*C4
5				
6	constraints			
7	x+2y+2z<=20	2x+y+2z<=24		
8	=A4+2*B4+2*C4	=2*A4+B4+2*C4		

Display D.6

	A	B	C	D	E
1	Bart's Lunch Business				
2					
3	x	y	z	P=1.00x+1.50y+2.00z	
4	0.00	0.00	0.00	0.00	
5					
6	constraints				
7	x+2y+2z<=20	2x+y+2z<=24			
8	0.00	0.00			

Display D.7

When you now invoke the Solver, there are several changes you
must make in the Solver dialogue box. You need to add the
constraint C4 > = 0 ($z \geq 0$ in the problem) into the Subject to
Constraints box, change the By Changing Cells box to A4:C4,
and change the Set Target Cell box to D4. After these changes,
click Solve; the spreadsheet displays the solution to the new
problem as shown in Display D.8.

	A	B	C	D	E
1	Bart's Lunch Business				
2					
3	x	y	z	P=1.00x+1.50y+2.00z	
4	4.00	0.00	8.00	20.00	
5					
6	constraints				
7	x+2y+2z<=20	2x+y+2z<=24			
8	20.00	24.00			

Display D.8

We see by looking at Display D.8 that the addition of a new type
of lunch has a significant effect on the solution to Bart's problem.
Now Bart should make 4 Thirsty Worker Lunches and
8 Super Lunches and no Hungry Worker Lunches. His profit goes
up from $16.50 to $20. We don't need to round this time since
the values of the variables come out as whole numbers.

Published by IT'S ABOUT TIME, Inc. © 2000 MATHconx, LLC

APPENDIX D

A big advantage to using the computer solution for this type of problem (besides being able to solve problems with more than two variables) is that we can easily change any part of the problem and get instant feedback on how the answer changes. For example, if Bart increases the amount he charges for the Hungry Worker Lunches, common sense tells us that at some number it would become profitable to make some of these lunches. What is that number?

To answer the previous question, you could try increasing the selling price of Hungry Worker Lunches to $8.60 to see if this changes the solution. This changes the profit on these lunches to $2.10. The only change in the problem is that the new profit formula is

$$P = 1.00x + 2.10y + 2.00z$$

Just change cells D3 and D4 to reflect the new information and run the Solver again. We find that, indeed, it is now profitable to provide some Hungry Worker Lunches (the result is interesting—check it out). With more guess and check we could pin down exactly the point where the y variable enters the problem.

Follow up Exercise

Carry out the process that was started. Keeping all other quantities in the problem constant, try different values for the price for the Hungry Worker Lunches until you find the number at which it just becomes profitable to make some of these lunches. Thus, it should be the case that if you lower the price of these lunches by any small amount, the solution to the Linear Programming problem has $y = 0$ in it, but if you increase the price by any small amount the solution for y is greater than 0. Discuss what happens to all variables and the total profit as you make these changes.

D-8

Published by IT'S ABOUT TIME, Inc. © 2000 MATHconx, LLC

GLOSSARY   2.a. and 2.b.

acute angle An angle of measure less than 90° (but more than 0°).

alternate interior angles Two nonadjacent interior angles on opposite sides of a transversal that intersects two lines.

altitude (of a triangle) A line segment drawn from one vertex to the opposite side.

angle A figure formed by two line segments or rays with a common endpoint.

angle bisector A line or ray that divides an angle in half.

arc A portion of a circle.

axis of revolution The line about which a planar figure is revolved to form a solid of revolution.

axis of symmetry A line that divides a figure in half so that each half is a mirror image of the other.

base unit In a system of measurement, a unit which is chosen (for convenience) and by which other units of measure are defined.

center of gravity (of a planar region) The point (centroid) at which the region would balance if it were cut out of a sheet of uniform density material.

central angle An angle formed by two radii of a circle.

centroid The center of gravity of a planar region (its balance point).

chord A line segment with both its endpoints on the same circle.

circle The set of all points in a plane that are a fixed distance from a particular point.

circumference The distance around a circle.

collinear (points) Points on the same straight line.

column matrix A matrix with only one column. That is, an $m \times 1$ matrix.

Published by IT'S ABOUT TIME, Inc. © 2000 MATHconx, LLC

Glossary 2.a. and 2.b.

complementary angles Two angles whose measures have a sum of 90°.

concentric circles Two or more circles in the same plane that have the same center.

conclusion The statement q in the conditional statement "If p, then q."

conditional (statement) A statement that can be put in the form "If p, then q," where p and q are themselves statements.

congruent angles Two angles that have the same measure.

congruent figures Figures that have the same shape and size.

consistent system (of linear equations) A system of equations that has only one solution.

constant A number or symbol representing a value that doesn't change.

constant of proportionality The number that one variable is multiplied by in a direct proportion, or the number that is divided by one variable in an inverse proportion.

contour An imaginary line/curve every part of which is at the same altitude at above or below sea level.

contour interval The vertical distance between one contour line and the next.

contour map A map that shows contour lines and intervals; a topographical map.

converse (of a universal statement) A statement that interchanges the subject and predicate of the given universal statement.

convex polygon A polygon in which the measure of each angle is less than 180°.

coordinate plane A plane that contains a pair of coordinate axes.

corresponding angles The angles, one interior and one exterior, on the same side of the transversal that intersects two lines.

cosine (of an acute angle) In a right triangle, the cosine of an acute angle is the ratio of the length of the adjacent leg to the length of the hypotenuse.

Published by IT'S ABOUT TIME, Inc. © 2000 MATHconx, LLC

counterexample An example that proves a universal statement to be false.

cross section The intersection of a plane and a three dimensional object.

cubic inch A cube with each edge one inch long.

cubic meter A cube with each edge one meter long.

dependent system (of linear equations) A system of equations with finitely many solutions.

diameter A line segment that passes through the center of a circle and has both endpoints on the circle; also, the length of such a segment.

direct variation A relationship between two variables such that one is a positive constant times the other.

directly proportional (variables) Two variables such that one is a positive constant times the other.

discrete mathematics The branch of mathematics that deals with processes related to counting separate, distinct objects.

disk A circle together with all the points inside it.

domain (of an equation) The set of all numbers for which an equation makes sense (is either true or false).

elementary row operation Any one of the following operations performed on a matrix:

I) Interchange two rows.

II) Multiply a row by a nonzero constant.

III) Multiply one row by a nonzero constant and add the new equation to another row.

equilateral polygon A polygon in which all sides have the same length.

exterior angle (of a polygon) An angle formed by one side of a polygon and the extension of an adjacent side.

geometric construction A procedure for creating a geometric object with a compass and an unmarked straightedge.

gnomon The upright, triangular component of a sundial.

Published by IT'S ABOUT TIME, Inc. © 2000 MATHconx, LLC

hypotenuse The longest side of a right triangle; the side opposite the right angle.

hypothesis The statement p in the conditional statement, "If p, then q."

identity (equation) An equation that is true for all numbers in its domain.

identity matrix A square matrix in which each entry of the main diagonal is 1 and all other entries are 0.

if and only if A logical connective which says that each one of the two statements it connects implies the other one.

inconsistent system (of linear equations) A system of equations which has no value that will satisfy all of the equations.

inscribed angle (in a circle) An angle formed by two chords with an endpoint in common.

inscribed polygon (in a circle) A polygon with all its vertices on a circle.

inscribed polyhedron (in a sphere) A polyhedron placed inside a spherical shell in such a way that all its vertices "touch" (are points of) the sphere.

interior angles (of a polygon) The angles that open toward the inside of the polygon.

inverse cosine An acute angle whose cosine is a specified number between 0 and 1.

inverse sine An acute angle whose sine is a specified number between 0 and 1.

inverse tangent An acute angle whose tangent is a specified positive number.

inverse variation A relationship between two variables such that one is a positive constant divided by the other.

inversely proportional (variables) Two variables such that one is a positive constant divided by the other.

isosceles triangle A triangle with two sides of equal length.

leg (of a right triangle) One of the sides of a right triangle that is not the hypotenuse.

Published by IT'S ABOUT TIME, Inc. © 2000 MATHconx, LLC

line segment A set of points on a line consisting of two endpoints and all the points between them.

linear equation An equation in which the exponent of each variable is one and whose graph is a line.

main diagonal The entries of a matrix in the first row-first column, second row-second column, third row-third column, etc.

major arc (of a circle) An arc that is larger than a semicircle.

matrix (matrices) Rectangular array(s) of numbers.

median (of a triangle) A line segment with one end point, a vertex of a triangle, and the other endpoint the midpoint of the side opposite the vertex.

Mercator map A flat map of the world that represents directions accurately.

method of elimination A method of solving systems of linear equations based on eliminating one or more of the same variables from each of the equations.

minor arc (of a circle) A portion of a circle that is smaller than a semicircle.

model A representation of something physical or mathematical considered important for a particular purpose.

obtuse angle An angle of measure greater than 90° and less than 180°.

ordered triples A set of three numbers or letters which represents the position of a point in three dimensional space i.e., (x, y, z).

parallel lines Straight lines in a plane that never intersect, no matter how far they are extended.

parallelogram A quadrilateral with both pairs of opposite sides congruent; equivalently, a quadrilateral with both pairs of opposite sides parallel.

parameter An independent variable used to express the coordinates of variable points and the function of them.

parametric equations Any equations that involve parameters.

Published by IT'S ABOUT TIME, Inc. © 2000 MATHconx, LLC

G-5

perimeter The distance around (the boundary of) a figure; its length as a path.

perpendicular bisector A line that intersects a given segment at right angles and divides it into two equal parts.

perpendicular lines Two lines that intersect at right angles; i.e., two lines that form congruent adjacent angles.

plane A set of points in space that satisfy an equation of the form $ax + by + cz + d = 0$, where a, b, c cannot all be zero.

polygon A polygonal path that starts and ends at the same place and doesn't intersect itself anywhere in between.

polygonal path A sequence of line segments each connected to the next by a common endpoint.

polyhedron A three dimensional shape made up of polygonal pieces joined at their edges.

prism A three dimensional shape formed by two congruent polygons (bases) in parallel planes and parallelograms on all the other faces.

proportion An equality between two ratios.

Pythagorean triple Three positive numbers a, b, and c such that $a^2 + b^2 = c^2$.

radius The distance or segment from the center to any point of a circle.

ratio One number divided by another; a common fraction.

ray Part of a line that starts at a particular point on the line and extends infinitely far in one direction.

reflex angle An angle of measure greater than 180° and less than 360°.

regular polygon A polygon with all sides the same length and all angles equal in measure.

rhombus A quadrilateral with all its sides congruent (the same length).

right angle An angle of measure 90°.

right triangle A triangle that has a right angle as one of its three angles.

Published by IT'S ABOUT TIME, Inc. © 2000 MATHconx, LLC

rotation symmetry The property of figures by which a copy can be made to coincide with the original figure by a rotation (around a point) of less than 360°.

row matrix A matrix with only one row. That is a $1 \times n$ matrix.

scalar A number. This word represents a number frequently used to distinguish numbers from matrices or other mathematical objects.

scaling factor The constant that describes the size relationship between two similar objects.

sector (of a circle) The region enclosed by two radii and the arc between them.

segment (of a circle) The region enclosed between an arc and a chord.

side (of a polygon) Any one of the line segments that determines the polygon.

similar objects Two objects such that the distance between any two points of one object is a particular constant times the distance between the corresponding points of the other object. Objects that have the same shape.

sine (of an acute angle) In a right triangle, the sine of an acute angle is the ratio of the length of the side opposite the angle to the length of the hypotenuse.

slope measure (of an angle) The perpendicular distance from any chosen point on one ray of the angle to the other ray divided by the distance from the vertex to the foot of the perpendicular of the other ray.

solid of revolution A three dimensional shape formed by revolving a two dimensional figure around a straight line (axis of revolution).

sphere The set of all points in space that are a fixed distance from a particular point.

straight angle An angle of measure 180°.

substitution A method of solving systems of linear equations by equating two expressions equal to the same variable.

supplementary angles Two angles whose measures have a sum of 180°.

Published by IT'S ABOUT TIME, Inc. © 2000 MATHconx, LLC

system of equations A set of at least two equations having the same variables, the solution of which may make all of the equations true.

tangent (of an acute angle) In a right triangle, the tangent of an acute angle is the ratio of the length of the side opposite the angle to the length of the adjacent leg.

three dimensional space (3-space) A space in which there are three independent directions such as length, width, height.

topographical map A map that shows contour lines and intervals which indicate elevations in reference to sea level; a contour map.

torus A three dimensional figure shaped like a round doughnut.

transversal A straight line that intersects two or more other coplanar straight lines at distinct points.

triangulation The process of dividing a polygon into nonoverlapping triangles.

unique solution The only possible solution (one and only one point) to a system of linear equations.

unit circle A circle of radius 1 (of whatever unit length is being used). In a coordinate plane, the unit circle has center (0,0).

unit cube A cube that measures one unit of length along each edge.

universal statement A statement of the form: "All [SOMETHING] are [SOMETHING ELSE]."

vertex A common endpoint of two sides of a polygon or an angle.

vertical angles Two angles such that the angles formed by two intersecting lines do not have a common side.

volume (of a three dimensional object) The number of unit cubes (of some unit length) needed to fill up the space it occupies.

wire frame pictures A partial picture of a three dimensional object drawn on a two dimensional surface by using a grid system of lines.

z-axis In 3-space, the axis perpendicular to the xy-plane.

Published by IT'S ABOUT TIME, Inc. © 2000 MATHconx, LLC

Index 2.b.

A

Abbott, Edwin A., 478
About Words
  annulus, 347
  axes, 301
  axis, 301
  circumference, 327
  circumvent, 327
  contorno, 383
  contour, 383
  disc, 287
  diskettes, 287
  dodeca-, 379
  equi-, 300
  equidistant, 300
  -hedron, 378
  hexa-, 379
  hypotheses, 301
  hypothesis, 301
  icosa-, 379
  inscribed angle, 352
  inscription, 352
  intercept, 352
  octa-, 379
  perpendicular, 395
  poly-, 378
  polyhedra, 378
  polyhedron, 378
  radii, 288
  radius, 288
  tetra-, 379
  upright, 395
Alpha (α), 351
Angle
  central, 339, 357
  face, 378
  inscribed, 352, 357
Annulus, 347
  area of, 347
Arc, 339
  length of, 339, 343
  major, 339
  minor, 339
Archimedes, 333, 409
Area, 323-325
  of annulus, 347
  of circle, 325, 334, 335
  of decagon, 331
  of n-gon, 331
  of octagon, 331
  of triangle, 331
Axis, 301
  of circle, 287
  of revolution, 422
  of symmetry, 287, 289

B

Balance point, 429
Bisector
  perpendicular, 300
Black Elk, 285
Blackwell, William, 284

C

CAD, 489
  wire frame pictures, 489
Calculus, 448
Cartesian
  coordinate system, 438, 439
  coordinate plane, 439
  Rene Descartes, 448
Cavalieri's Principle, 402, 403, 438
Cavalieri, Bonaventura, 402, 438
Center, 286, 409
  of gravity, 429
  centroid, 429
Center-radius, 286
Central angle, 339, 357
Centroid, 429
  center of gravity, 429

Chord, 288, 292
Circle, 283
  arc, 339
  area, 323-325
  axis of symmetry, 287, 289
  center, 286
  central angle, 339
  chord, 288, 292
  circumference, 286, 287, 323, 327
  concentric, 328, 396
  constant width, 292, 360, 362
  diameter, 287
  disk, 287, 292
  equation of, 461, 462
  function, 287
  half, 357
  radius, 286, 461
  sector, 339
  segment, 339
  similar, 324
  tangent line, 292
  unit, 312, 325
Circumference, 286, 327, 328, 329, 334, 335
Collinear, 302
Column matrix, 532-533
Concentric circle, 396
Conclusion, 301
Conditional statement, 301
  conclusion, 301
  conditional, 301
  converse, 301
  equivalent, 301
  hypothesis, 301
  if and only if, 301
  if, then, 301
Cone, 375
  steepness, 375

Published by IT'S ABOUT TIME, Inc. © 2000 MATHconx, LLC

I-1

Index 2.b.

volume, 415
Consistent system(s), 493
Constant width, 286, 291, 292, 360, 362
    Reuleaux triangle, 362
Contour, 383
    contour interval, 383
Converse, 301
Coordinate plane(s), 438-439
    2-space, 373, 438, 468-476
    3-space, 373, 439-444
    rectangular, 438-439
Coordinate system
    Cartesian, 439
    coordinate pair, 311-313
    oblique, 455
    rectangular, 441
Cosine, 314
Cross section, 387, 388, 395
CAT scan, 395
Cube, 370
Curve of constant width, 292, 360, 362, 365
Cylinder, 393
    oblique, 394
    right, 394
    volume, 406

D
Davis, Professor Philip J., 285-287, 289-291
Decagon, 330
Dependent systems, 508
Descartes, Rene, 448
Diameter, 287
Disk, 287, 292, 474
Diskettes, 287
Distance formula,
    2-space, 442
    3-space, 444
Dodecagon, 379
Doppler weather, 336-337
Dow Jones Industrial Average, 468-469

E
Einstein, Albert, 476
Elementary row operation, 515
Equation
    circle, 461-462
    linear, 450-451
    parametric, 313
    plane, 439, 490
Equi-, 300
Equidistant, 300, 301
Equilateral triangle, 308, 382
Equivalent statement, 301
Euclid's *Elements*, 378, 379
Explorations
    circle translations, 319
    circles through two points, 299
    circles with same radius, 298
    cone, 375
    contour, 390
    fold line, 299
    parametric equations, 314, 315, 316, 318
    properties of balance line, 431
    pyramid, 373

F
Face angles, 378
Facts to Know
    circles are similar, 324
    distance formula (2-space), 442
    distance formula (3-space), 444
    equation of a circle, 461, 462
    equation of a plane, 490
    equation of a sphere, 464
    face angles of polyhedra, 378
    median of a triangle, 432

radius, 461-462
    volume of a prism and a cylinder, 406
    volume of a solid of revolution, 430
Fold line, 299
Four dimensions, 475
Function, 313, 377, 484
    angle size, 377
    circle, 287
    cosine, 313
    sine, 312

G
Galilei, Galileo, 402
Gaussian elimination
    back solving, 517
Go-Kart track, 340-341, 344-348
Gravity,
    center of, 429
    centroid, 429
Guldin, Paul, 430

H
Half-squares, 326
Hexa-, 379
Hexagon, 330, 374
Hexahedron, 379
Hypothesis, 301

I-J
Icosa-, 379
Icosahedron, 379
If and only if, 301
Inconsistent system(s), 493
Independent systems, 508
Indivisibles, 402
Inscribed angle, 352, 357
    decagon, 330
    hexagon, 330
    octagon, 330
    polygons, 330
    semicircle, 357
Instant of time, 471
Intercepts, 352
Irrational number, 334

Published by IT'S ABOUT TIME, Inc. © 2000 MATHconx, LLC

MATH *Connections*: A Secondary Mathematics Core Curriculum

K

Kepler, Johannes, 402

L

Length of circular arc, 339,
    343
Limit, 414
Linear equation
    2-space, 450-451
Linear equations,
    systems of, 438, 511-513
    2-space, 450, 511
    3-space, 438, 450,
        511-513
    Gaussian elimination, 517

M

Main diagonal, 538
Major arc, 339
Maps,
    contour, 384
    topographical, 384
Matrices, 508, 513
    column, 532, 533
    identity, 538
    main diagonal, 538
    row, 532, 533
    zero matrix, 527
Matrix operations, 514-521,
        525-527, 529-538
    elementary row
        operation, 515
    Gaussian elimination, 517
    multiplication, 536
    negative, 527
    opposite, 527
    product, 536
    scalar, 531
    subtraction, 529
    sum, 525
    TI-82 (TI-83), 518, 519
Median,
    of a triangle, 432
Method of elimination, 501
Michelson, 476
Minor arc, 339

Money, 468
Morley, 476

N-O

N-gon, 296
Newton, Isaac, 475-476
Number
    irrational, 334
Oblique coordinate system,
    455
Octa-, 379
Octagon, 289, 295, 330
Octahedron, 379
Operations,
    elementary row, 515
Ordered pairs, 438

P-Q

Pappus, 430
Pappus-Guldin Theorem,
    430, 432
Parameter, 313
Parametric equations, 313
Pentagon, 374
Perimeter, 330, 331
Perpendicular, 394
    bisector, 300, 301,
        303-304
    radii, 292
    tangents, 292
Pi (π), 323, 334
    history of, 333
Pixel, 311
Planar region, 324
Plane
    equation of, 439, 490
Plato, 379-380
Platonic Solids, 380
    dodecahedron, 380
    hexahedron, 380
    icosahedron, 380
    octahedron, 380
    tetrahedron, 380
Poly-, 378
Polygon
    decagon, 330

hexagon, 330
    octagon, 289, 295, 330
    pentagon, 374
    regular, 330
Polygonal region, 393
Polygons, inscribed, 330
    decagon, 330
    hexagon, 330, 374
    octagon, 330
    pentagon, 374
    square, 374
    triangle, 374
Polyhedron, 378
    cube, 370
    dodecahedron, 379
    face of, 378
    hexahedron, 379
    icosahedron, 379
    octahedron, 379
    regular, 378
    tetrahedron, 378-380
    vertex of, 374, 378
Prism, 393, 403
    oblique, 394, 403
    right, 394
    slant height, 405
    volume, 406
Pyramid, 373, 413
    volume, 413
Pythagorean Theorem,
    379, 441
Quadrilateral, 302

R

Radius, 286, 288, 292, 328,
    461
Rectangular coordinate
    system, 439, 441
Reflection symmetry, 289
Regular polygons, 295
Reuleaux, Franz, 362
Reuleaux triangle, 362
Revolution,
    axis of, 422
    solid of, 422
Ring, 347

Index 2.b.

Rotation symmetry, 289-290
  angle of, 290
  center of, 290
  point of, 290
Row matrix, 532-533

S
Scalar product, 531
Scaling factor, 409
Sector of a circle, 339
Segment of a circle, 339
Semicircle, 297, 354, 357
Set-builder notation,
  449-459
  3-space, 463
Similar
  circles, 324
Sine, 312
Slant height, 405
  cylinder, 405
  prism, 405
Slope, 375, 376, 451-453
Solid figure, 402
  solids, 402
Solid of revolution, 422
  axis of revolution, 422
  volume of, 430
Space, 373
  2-space, 373, 468-470
  3-space, 369
  three dimensional, 369
  two dimensional, 373
Sphere, 409, 416, 463
  center, 409, 463
  equation of, 463
  radius, 462, 463
  volume, 419
Square, 374
Square base pyramids, 415
Stacking lines, 456-458
Statement, equivalent, 301
  conditional, 301
Statistical forecasting, 309
Steepness, 375
Stonehenge, 284
Substitution, 485

Symbols
  α (alpha), 351
  θ (theta), 313
  π (pi), 323
Symmetry
  axis of, 287, 289
  center of, 290
  point of, 290
  reflection, 289
  rotation, 289
Systems of equations, 483,
  511-513
  consistent, 493, 498
  dependent, 508
  inconsistent, 493, 498
  independent, 508
  linear, 438, 511-513
  method of elimination,
    501, 506
  2-space, 511
  3-space, 511-513
  two equations in two
    unknowns, 483

T
Tangents
  perpendicular, 292
Temperature, 468
Tetra-, 379
Tetrahedron, 378, 379
Theta (θ), 313
Three dimensional space,
  (3-space), 369
  graphing, 485-490,
    496-497
  plane, equation of, 490
  solving system of,
    511-513
  wire frame pictures, 489
Tic-tac-toe, 445
Time, 468
  instant of, 471
Time-space figures, 471, 473
  cone, 475
Topographical maps, 384
  configuration, 398

contour, 383
  contour maps, 384
  relief, 398
Topography, 384, 398
Torus, 425
Triangle, 374
  equilateral, 308, 382
  median, 432
  Reuleaux, 362
Two dimensional space, 373
  (2-space), 373, 468-470
  distance formula, 442

U-V
Unit circle, 312-313, 325,
  332, 334
Units of measure, 468
Volume, 406
  cone, 415, 418
  cylinder, 406
  hollow core cylinder, 429
  prisms, 406
  pyramid, 412, 413, 418
  solid of revolution, 430
  spheres, 410, 416, 419

W-X-Y-Z
Weight, 468
Width, constant, 292, 360,
  362
Wire frame pictures, 489
Wobbit, 399
Words to Know
  center of gravity, 429
  centroid, 429
  median, 432
X rays, 395
y-intercepts, 451-453
z-axis, 439
Zero matrix, 527

Published by IT'S ABOUT TIME, Inc. © 2000 MATHconx, LLC

NOTES

MATH *Connections*: A Secondary Mathematics Core Curriculum